Lecture Notes in Networks and Systems

Volume 61

The series "Lecture Notes in Networks and Systems" publishes the latest developments in Networks and Systems—quickly, informally and with high quality. Original research reported in proceedings and post-proceedings represents the core of LNNS.

Volumes published in LNNS embrace all aspects and subfields of, as well as new challenges in, Networks and Systems.

The series contains proceedings and edited volumes in systems and networks, spanning the areas of Cyber-Physical Systems, Autonomous Systems, Sensor Networks, Control Systems, Energy Systems, Automotive Systems, Biological Systems, Vehicular Networking and Connected Vehicles, Aerospace Systems, Automation, Manufacturing, Smart Grids, Nonlinear Systems, Power Systems, Robotics, Social Systems, Economic Systems and other. Of particular value to both the contributors and the readership are the short publication timeframe and the world-wide distribution and exposure which enable both a wide and rapid dissemination of research output.

The series covers the theory, applications, and perspectives on the state of the art and future developments relevant to systems and networks, decision making, control, complex processes and related areas, as embedded in the fields of interdisciplinary and applied sciences, engineering, computer science, physics, economics, social, and life sciences, as well as the paradigms and methodologies behind them.

** Indexing: The books of this series are submitted to ISI Proceedings, SCOPUS, Google Scholar and Springerlink **

More information about this series at http://www.springer.com/series/15179

Paweł Gburzyński

Modeling Communication Networks and Protocols

Implementation via the SMURPH System

 Springer

Paweł Gburzyński
Computer Engineering Department
Vistula University
Warsaw, Poland

ISSN 2367-3370   ISSN 2367-3389   (electronic)
Lecture Notes in Networks and Systems
ISBN 978-3-030-15390-8   ISBN 978-3-030-15391-5   (eBook)
https://doi.org/10.1007/978-3-030-15391-5

Library of Congress Control Number: 2019933844

This Springer imprint is published by the registered company Springer Nature Switzerland AG
The registered company address is: Gewerbestrasse 11, 6330 Cham, Switzerland

Contents

Chapter 1
Introduction

This book introduces a software system, dubbed SMURPH,[1] comprising a programming language, its compiler, and an execution environment, for specifying communication networks and protocols and executing those specification in virtual worlds mimicking the behavior of (possibly hypothetical) real-life implementations. The level of specification is both high and low at the same time. High in that the specification, including the network configuration and its protocol (program) is expressed in a self-documenting, structural fashion; low in that the level of detail of the specification corresponds to a friendly view of the implementation, where all the relevant algorithmic issues can be seen without tying the description to specific hardware. The essential features of SMURPH are illustrated with examples presented as an introduction to event-driven simulation with a focus on telecommunication systems.

SMURPH was originally designed more than twenty years ago to assist my (then purely academic) research in the relatively narrow domain of shared-media, wired, local area networks (LANs). Over the years of its application and evolution it has undergone many enhancements and extensions adapting it to other domains of telecommunication, including wireless networking. It has become considerably more than a simulator. For quite some time now, I have been using it as an invaluable tool in my quite practical (shall we say "industrial") endeavors in wireless sensor networks (WSNs), where it has acted as a friendly replacement for the mundane, real, physical world needed for the actual, real-life deployments. Normally, to carry out an authoritative experiment involving a non-trivial WSN, one would have to place hundreds of boxes, possibly in several buildings, then run the network for some time (perhaps a few days), then (most likely) reload the boxes with new firmware (after detecting problems), then run more tests, and so on. The realism and completeness of the virtual execution in SMURPH mostly obviates the need for real-life experiments. This is because the network model can be an authoritative replica of the real network, in the sense that its actions (responses) perceivable by the user are basically indistinguishable in the two worlds. A truly revolutionary add-on to the system was a

---

[1] The software can be downloaded from https://github.com/pkgn/SMURPH.

© Springer Nature Switzerland AG 2019
P. Gburzyński, *Modeling Communication Networks and Protocols*,
Lecture Notes in Networks and Systems 61,
https://doi.org/10.1007/978-3-030-15391-5_1

1

layer of glue between SMURPH and the software development kit for real-life WSN nodes allowing the programmer developing applications for real networks to run the very same programs in the virtual world created by SMURPH. That mostly eliminated the need for tedious and troublesome tests involving potentially large networks consisting of real devices by replacing those tests with pressing the touch-pad on a laptop from a comfortable armchair.

This book focuses on SMURPH as such, without completely describing its present most practically useful context (which is introduced elsewhere). Its objective is twofold. First, I would like it to constitute a reasonably friendly manual of SMURPH. This means that the book should be a complete functional description of the package. Second, partially by fulfilling the "friendliness" postulate of the first objective, I would like to make this book educational, e.g., serve as a student text in an advanced course on network simulation, with SMURPH acting as the protocol/network modeling vehicle. Instead of trying to make this book as complete as possible, and, e.g., inflating it with a discussion of the popular WSN techniques (which can be found in many other books easily available on the market), I want instead to confine it to an introduction of "pure" SMURPH as a virtual lab tool. That tool will greatly increase the range of experiments that a student can comfortably conduct, compared to those doable with any base of physical devices that can be made available to the student in a real-life academic lab. The issue is not just the availability of hardware and firmware, but also its malleability (new virtual hardware can be built for free and with little effort) and the scope/diversity of the experimental environment.

In my attempts to meet the above-stated objectives, I will try to organize the material of this book around a complete description of SMURPH functionality, basically in a manual-like fashion. As presentations of this sort tend to be dry and unentertaining, I will attempt to lighten it up with illustrations and case studies. More attention will be given to the wireless modeling capabilities of SMURPH (than to the tools supporting wired network models); nonetheless, we shall see to it that nothing important is missing.

1.1 A Historical Note

The present package is the outcome of an old and long project which started in 1986 as an academic study of collision-based (CSMA/CD) medium access control (MAC) protocols for wired (local-area) networks. About that time, many publicized performance aspects of CSMA/CD-based networks, like the Ethernet of those days [1], had been subjected to heavy criticism from the more practically inclined members of the community [2], for their apparent lack of practical relevance and overly pessimistic, confusing conclusions. The culprit, or rather culprits, were identified among the popular collection of analytical models, whose nonchalant application to describing poorly understood and crudely approximated phenomena had resulted in worthless numbers and exaggerated blanket claims.

My own studies of access-control protocols for local-area networks, on which I embarked at that time with my collaborators, e.g., [3–14], were aimed at devising novel solutions, as well as dispelling controversies surrounding the old ones. Because exact analytical models of the interesting systems were nowhere in sight, the performance evaluation component of our work relied heavily on simulation. To that end, we developed a detailed network simulator, called LANSF,[2] which, in addition to facilitating performance studies, was equipped with tools for asserting the conformance of protocols to their specifications [15–21]. To be able to make quantitative claims about the low-level medium access schemes, whose behavior (and correctness, i.e., conformance to specification) strongly depended on the precise timing of signals perceived in the shared medium (cable or optical fiber), we made sure from the very beginning that all the relevant low-level phenomena, like the finite propagation speed of signals, race conditions, imperfect synchronization of independent clocks, and so on, were carefully reflected in the models. In consequence, those models received the appearance of programs implementing the protocols on some abstract hardware. Thus, our system was often referred to as an *emulator* rather than a simulator.

Around 1989 I became involved in a project with Lockheed Missile and Space (now Lockheed-Martin) aimed at the design of a high-performance reliable LAN for space applications [22]. To that end, between 1989 and 2003, LANSF was heavily reorganized, generalized, documented, and re-implemented in C++ (it was originally programmed in plain C) receiving the look of a comprehensive programming environment with its own language [23–26]. Under the new name SMURPH,[3] in addition to assisting our R&D work on new networking concepts, the package was used for education, e.g., in a graduate course on protocol design which I taught between 1992 and 2008 at the University of Alberta. The protocols discussed in that course were put in a book [27] which also included a complete description of the SMURPH package as of 1995. That book must be considered aged today, because SMURPH has evolved, primarily in response to the shifting focus of my research. For one thing, the present version of SMURPH offers elaborate hooks for modeling wireless channels. That alone provides enough excuse for a new presentation.

Chronologically, the path toward accommodating wireless channels in SMURPH took an important detour. In mid-nineties, while still loosely collaborating with the group at Lockheed and thinking about software for automatically controlling space equipment, we noticed that the close-to-implementation look of SMURPH models offered a bit more than a specification environment for low-level network protocols. It appeared to us that the programming paradigm of SMURPH models might be useful for implementing certain types of real-life applications, namely those that are reactive in nature, i.e., their operation consists in responding to events, possibly arriving from many places at the same time. The first outcome of this observation was an extension of SMURPH into a programming platform for building distributed controllers of physical equipment represented by collections of sensors and actuators

---

[2]Local Area Network Simulation Facility.
[3]A System for Modeling Unslotted Real-time PHenomena.

[23–30]. Once again, the package was renamed, to SIDE,[4] and extended by means of interfacing its programs to real-life devices.

In 2002, I became interested in wireless networks from an angle that was drastically more practical than all my previous involvements in telecommunications. With a few friends we started Olsonet Communications Corporation,[5] a Canadian company specializing in custom solutions in wireless sensor networking. Owing to the very specific scope of our initial projects, we needed efficient programming tools to let us put as much functionality in tiny microcontrollers (devices with a few kilobytes of RAM) as possible without sacrificing the high-level organization of programs, modularity, flexibility, etc. Having immediately noticed the reactive character of the software (firmware) to be developed, we promptly revisited the last Lockheed project and created PicOS, a small-footprint operating system for tiny microcontrollers with the thread model based on the idea of SMURPH/SIDE processes [31, 32]. Note that that wasn't yet another extension of SMURPH, but rather an extrapolation of some of its mechanisms into a (seemingly) drastically different kind of application. Of course, working on novel communication schemes in WSNs, we also needed tools to verify our concepts without having to build and deploy (often quite massive)[6] networks of real devices. So SMURPH was used alongside PicOS as a separate tool. The protocols and networks being developed (under PicOS) for real-life deployments were manually reprogrammed into SMURPH models to be evaluated and verified in a convenient virtual setting. That inspired the next substantial extension of SMURPH, by tools for modeling wireless channels and node mobility [33].

Then, the closeness of PicOS to SMURPH allowed us to close the gap between the two with a layer of glue to mechanically translate programs for wireless nodes (written for PicOS) into SMURPH models. The intention was to be able to virtually run complete network applications (called praxes in PicOS parlance) under easily parameterizable, artificial deployments. That glue, dubbed VUE^2,[7] was more complicated than a (rather simple in this case) compiler from one language to another, because it had to include a complete virtual environment of the node: peripheral devices, sensors, UARTs, and so on [34, 35]. A related feature added to SMURPH at that time was the so-called visualization mode for simulation. When run in this mode, the simulator attempts to synchronize the virtual time of the model to the prescribed intervals of real time. The purpose of the visualization mode is to demonstrate (visualize) the behavior of the modeled system by making intervals of virtual time separating the modeled events appear as actual time intervals of the same (albeit real) duration. This is only possible if the simulator can "catch up" with real time. If this is not the case, the program can still render a scaled, slow-motion visualization of the model.

---

[4]Sensors In a Distributed Environment.

[5]http://www.olsonet.com.

[6]The first commercial deployment involved a network consisting of 1000+ nodes.

[7]Note that the superscript (^{2}) at VUE is part of the name. The acronym reads: Virtual Underlay Execution Engine, which is sometimes shortened to VUEE.

Even though the package has been known under at least two names (SMURPH and SIDE), and closely intertwined with some other tools (VUE[2]), we shall stick in this book to its single most prominent name SMURPH. Whenever you see the name SIDE (some files and directories may include that string) please remember that it refers to SMURPH. More generally: SMURPH, SIDE, as well as SMURPH/SIDE all denote the same package.

1.2   Event-Driven Simulation: A Crash Course

The objective of computer simulation is to formally mimic the behavior of a physical system without having to build that system to see what it *really* does. Depending of the system type, as well as the goal of our study, the aspects of the system's behavior that we would like to model may vary. Suppose that somebody builds a simulation model of a computer network. The most natural motivation behind such a model would be to study the performance of the network, e.g., to estimate how much data it can pass per time unit, how many users it can handle simultaneously without saturating itself, how well it will meet the expectations of those users regarding some measures of the quality of service. This kind of objective, i.e., to obtain a quantified measure of service useful from the viewpoint of the system's user, is typical for many simulation models, including models of communication networks. A simulation study within such a model usually assumes some hypothetical load of the system, submits the model to that load, and amasses performance data consisting of counters and time intervals describing numerically a collection of the model's behaviors observed during its execution. Those performance data are then analyzed statistically to produce simple (easily comprehended) measures, like maximum throughput, average delay, delay jitter, packet delivery fraction, and so on. The common characteristics of all such models is that they strive to capture a sequence of well-defined, individual (discrete) events in the system and are mostly interested in the timing of those events. For a network model, those events may represent packet transmission and reception, interference (bit error), new user arrival, and so on.

An important subset of performance measures that can be studied in such models are compliance assertions, i.e., statements expressing the desired properties of the system. For example, for a communication network catering to a real-time application, such a statement may say that the amount of time needed to pass a message between a given pair of nodes never exceeds some critical value. One can also think of assertions that do not directly relate to numerical measures (like explicit time intervals), but express formal correctness criteria of the communication protocol assumed by the designer. For example, the designer may want to verify that a certain succession of states at one node (e.g., indicating a dynamically arising exceptional situation) is followed by another specific succession of states, possibly involving other nodes (characteristic of the expected reaction to the emergency). Models that well capture the discrete sequences of events (states) that the underlying system experiences during its operation are nicely amenable to this kind of studies. They can be viewed as virtual tests of the system. Note that some of those tests might be difficult (or even impossible) to carry out in the real system, because of their overly intrusive or even unphysical nature. It is quite easy in the model to have an instantaneous view of the entire system (by momentarily holding its evolution in time), which may not be possible in real life, e.g., because of the relativistic limitations.

Another possible application of a simulation model is to substitute for the physical system in a situation where the physical system is too expensive or cannot be easily subjected to the interesting kinds of load. Flight simulators (and generally games) fall into this category. The model's objective is then to create an illusion of interacting with the real system, be it for education or entertainment. A video game may create a model of a system that does not exist in real life, perhaps one that cannot exist even in principle, e.g., because it operates according to its own laws of physics. A flight simulator may produce a very realistic impression of flying a real aircraft. In addition to providing a training tool much cheaper than an actual aircraft, the simulator can be harmlessly subjected to stresses (we should call them loads) that would be dangerous in a real flight. Models of this kind are called real-time models. We do not expect from them simple numerical performance data as their final output. Also, there are no obvious discrete events that would clearly mark some isolated, interesting moments in the model's evolution: what matters is basically the entire (continuous) process unfolding before the eyes of the (usually interacting) beholder.

The third category of models, somewhat related to the ones mentioned in the previous paragraph, are non-real-time models of complex processes, e.g., weather or climate. One difference is that real-time execution of such a model would be irrelevant or absurd. In fact, a useful climate model must execute many orders of magnitude faster than the real system. This usually implies that it must be considerably simplified. The simplification is accomplished by analytical formulas intended to capture the evolution of large parts of the system, typically into differential equations, that can be approximately solved by advancing the system in time by discrete steps. The same general approach is used in flight simulators, but the objectives are different. For a flight simulator, the solution must be delivered in real time, so the accuracy of the representation of the physical system is constrained by what can be achieved without saturating the computer. The equations describing the system evolution can be, at least in principle, as accurate as we want them to be. For a climate model, the solution must be produced in reasonable time, while the equations cannot possibly be accurate: there are too many factors which, to make matters worse, may affect the system in a chaotic (essentially non-computable) way.

1.2.1 Events

In this book, we deal with models of the first kind. We basically say that we only care about discrete events in the system, and the primary purpose of modeling those events is to see how they are spaced in time. Thus, the simulation looks like traversing a sequence of pairs: (t_i, e_i), where t_i is time and e_i is some event description. The sequence represents a realistic evolution of some actual system in time, so we should postulate $t_i < t_{i+1}$, which is a way of saying that in a sane system time always flows forwards. In fact, we shall opt for a weaker condition, namely $t_i \leq t_{i+1}$, admitting "time standing still." This weakening will let us conveniently model different things happening in parallel within a single execution thread.

The history of past events in the modeled system determines its present state and thus affects the timing of future events. This is the essence of how we build the system's model whose role is to walk through the sequence of pairs (t_i, e_i). Needless to say, the complete path is never given in advance (if it were, there would be no need to run the model). The way it works is that the future steps of the "walk" are dynamically derived from the ones already traversed. When processing an event (e_i, t_i) the model constructs its idea of the system's state at time t_i. That state often leads to other (subsequent) states, so the model does something *now* to make sure that the event sequence includes at least one continuation event (t_{i+1}, e_{i+1}) in the future. If any action is needed to arrive at that event, the model doesn't really have to go through the exact motions of that action, as long as it can tell how much time δ the action would take in real life. Then it knows that the new event should be timed at $t_{i+1} = t_i + \delta$. In the program, the unveiling sequence of future events is represented as a list (commonly called the event queue) sorted in the non-decreasing order of t_i. The head of this list points to the nearest event, i.e., the one to be handled next. This is in fact the most succinct expression of the methodology of discrete-time, event-driven simulation: the model operates by computing "deltas" amounting to increments from t_i to t_{i+k}, where k can be anything greater than 0.

1.2.2   The Car Wash Model

This general idea is best illustrated with a simple example. Suppose that we want to model a car wash. First, we must ask ourselves what processes in the underlying physical system are relevant and should be captured by the model. This depends on the insight we expect to acquire from the model, i.e., what performance measures we would like to collect, and which aspects of the physical system can be safely ignored as having no effect on those measures. For example, if we want to study the impact of the car wash on the environment, we may want to keep track of the amount of water and detergents needed to service a given car and, possibly, the amount of pollutants released by the cars while idling in the queue (or inside the wash). We might also want to account for the energy spent on powering the machinery of the wash. If, on the other hand, our objective is merely to see how many cars the wash can handle per time unit (the throughput) and investigate the statistics of service delay (as a measure of customer satisfaction), then we can ignore the detailed physics and chemistry of the processes involved and focus exclusively on their timing. The model can be made

Fig. 1.1 The car wash model

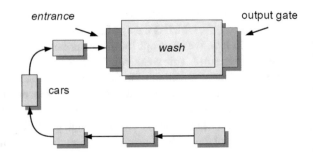

simpler, but, as long as it accurately captures the timing of the relevant processes, it will be no less accurate (for our purpose) than any more complicated model.

To make this example simple, and to clearly see the essence of the discrete-time, event-driven approach without diluting it in unnecessarily arcane details, let us assume that we have some idea regarding the timing of the wash service (say as a function of the options offered to the customers) and want to investigate the maximum throughput of our system, i.e., its maximum sustainable service rate. We view the car wash as shown in Fig. 1.1. There is an *entrance* process whereby cars arrive and are queued for service, and a *service* (proper wash) process where they are serviced (washed) in the order of their arrival. The two processes represent two places in our model where time is advanced, i.e., future events are generated based on some parameters. The entrance process is responsible for modeling the action of customer (car) arrival to the wash. We want it to operate according to this pseudocode:

```
Loop:
    car = generate_a_car ( );
    append (car_queue, car);
    delta = time_until_next_car_arrival ( );
    delay (delta);
    goto Loop;
```

where *generate_a_car* is a function that creates a car showing up in the input queue, and *time_until_next_car_arrival* is another function determining the time interval until the next car arrival. The scheme describes a stochastic process driven by two random distributions (hidden inside the two functions). Those distributions (and their parameters) will reflect our idea of the car arrival process to the wash. The wash process will then run like this[8]:

---

[8]People allergic to *goto* statements and labels will easily rewrite the pseudocode into two nested *while* loops. We prefer the politically incorrect version, with the label and *goto* statements emphasizing the reactive stature of the code, where the label represents the "top" and the *goto* statements mark "taking it from the top" after an event.

Loop:

```
if (car_queue == NULL) {
        wait (car_arrival);
        goto Loop;
}
car = remove_top (car queue);
delta = time_of_service (car);
delay (delta);
goto Loop;
```

It calls another timing function, *time_of_service*, which determines the amount of time needed to handle the given car. Again, the value returned by that function will be derived from some distribution, depending on the type of service requested by the car. Let's leave those functions until later. They are not completely uninteresting, far from that; however, being straightforward functions, they *are* in fact uninteresting from the viewpoint of the program's structure, which we want to see first.

As most physical systems that one would like to model, our car wash consists of multiple (two in this case) processes operating independently. From the viewpoint of the model, the most interesting aspects of those processes are the events that make them tick. Note that our processes in fact do *tick*, i.e., carry out discrete actions separated by idle intervals of *waiting* for something to happen. That something is mostly the time reaching a prescribed value, which looks like responding to an alarm clock going off. One exception is the car arrival event in the wash process. Note, however, that that event also coincides with an event of the first kind (an alarm clock going off) occurring inside the first process.

When we say "waiting," we mean it conceptually, i.e., the time spent on that waiting is meant to flow in the real system rather than in the model. The model never waits for the modeled time to reach a specific target value: it just grabs events from the event queue skipping over the idle periods. The execution time of the simulator depends on the complexity of the computations needed to determine the timing of the consecutive events, not on the amount of modeled time separating them.

The expressing power of event-driven simulation is in the way to program the multiple processes of the model such that their inherent parallelism is captured by a rather straightforward, sequential program without missing anything of merit. This is possible precisely because of the discrete nature of the actions carried out by those processes, which actions are triggered by events. The model will be accurate. if the time stamps formally assigned to those events well match the actual timing of their counterparts in the corresponding physical system.

To transform the above processes of our car wash model into a simulator constructed around the concept of event queue (Sect. 1.2.1), we shall assume that an event (an item kept in the event queue) has the following layout:

```
struct event_struct {
        TIME t;
        state where;
        process what;
};
```

where the last two attributes describe the event (formally corresponding to e_i from Sect. 1.2.1). The description consists of two parts, because it identifies the process to which the event pertains, as well as the action that the process should carry out in response to the event. We shall refer to that action as the process's *state*.

Still informally (for as long as possible we want to abstract from an implementation of these concepts) we will thus reformulate the processes as follows:

```
process entrance (state ignored) {
        car = generate_car ( );
        append (car_queue, car);
        delta = time_until_next_car_arrival ( );
        schedule (NEXT_CAR, wash, now);
        schedule (NONE, entrance, now + delta);
};


process wash (state where) {
        switch (where) {
                case NEXT_CAR:
                        if (wash_busy) return;
        Wash_it:
                        car = remove_top (car_queue);
                        wash_busy = YES;
                        delta = time_of_service (car);
                        schedule (END_WASH, wash, now + delta);
                        return;
                case END_WASH:
                        wash_busy = NO;
                        if (car_queue != NULL) goto Wash_it;
        }
};
```

The *schedule* function creates an event (with the specified attributes) and inserts it into the event queue according to the time (the last argument). To make the simulator run, we must insert a starting event into the queue and run the main loop (Sect. 1.2.1), just like this:

```
wash_busy = NO;
schedule (NONE, entrance, 0);
while (1) {
        event = remove_top (event_queue);
        now = event->t;
        event->what (event->where);
}
```

The "processes" are nothing more than functions with some standard layout. A process function accepts an argument indicating the state (the action) to be carried out by the process in response to the event. The entrance process consists of a single state (it executes a single type of action), so it ignores the argument. Generally, this kind of (forced) layout is not necessary. We can think of an implementation where the sole attribute of an event is the function to be called, and different functions carry out the different actions. The multi-state view of a process function is nonetheless quite useful and closely related to how SMURPH models are organized. A process naturally corresponds to a part of the modeled system which performs only one action at a time, like a localized, sequential component of a distributed system. It represents a single thread of execution which cycles through a number of stages separated by intervals of modeled time.

From the viewpoint of that modeled time, different processes of the model can execute in parallel. How is that possible, if the simulator itself executes a single sequential loop? Note that what matters for the accuracy of the model is the accuracy of the time stamps in the series of events traversed by the simulator in that single loop. It doesn't matter how fast (or slow) that loop is run and when the events are processed in the program's real time. In the above pseudocode, the current time of the modeled system is denoted by variable *now*. Its value only changes when a new event is picked from the queue. It is possible to schedule multiple processes for the same *now* which means that the model can account for the distributed nature of the modeled system.

For example, in our car wash model, a car arrival event handled by the entrance process results in another event being scheduled by that process for the same current time *now*. The purpose of that event is to notify the service process that a car has arrived, so, unless the service process is busy, it should take care of the car right away. Of course, there is more than one way to program this case to the same end. For example, the entrance process itself could find out that the wash is idle and formally start the service, without going through the extra event, which doesn't advance the modeled time and is there solely to "pass a message" to the other process. Some "methodologists" of event-driven simulation would insist that the time markers in the event queue should always be strictly increasing, and that such trivial actions should be carried out without pretending to involve the modeled time. Our philosophy is different and favors the kind of encapsulation of responsibility in the model where different components of the distributed system are represented by separate processes clearly isolated from other processes, according to their separation in the underlying real system. Calling those functions "processes" may seem extravagant (after all, they are just regular functions), but as representatives of the functionality of the real system they are exactly that.

What is still missing in the (pseudo) code of our car wash simulator? The superficial blanks that remain to be filled include the declarations of types and variables, the implementation of the event queue, and the definitions of two functions generating time intervals: *time_until_next_car_arrival* and *time_of_service*. We can imagine that they generate random numbers based on some distributions driven by some parameters, e.g., provided in the input data to the model. For example, the car wash may offer different types of service (e.g., undercarriage, waxing) resulting in different expected (mean) time of service. In all cases, it makes sense to randomize the actual instances of that time, because in real life the service time is never fixed.

One more important component that is missing is the set of variables for collecting performance measures. For example, we may want to compute the percentage of time that the car wash is busy, and the average delay experienced by a car, starting from the moment of its arrival until the completion of service.

1.2.3 Discrete Time

Time in SMURPH, as in many other simulators, is discrete, i.e., represented by integer numbers. This means that in any model there exists a time quantum (or a basic, indivisible time unit), and all time stamps and intervals are expressed as multiples of that quantum. The first question that usually comes up when designing a simulation model is how to choose the time unit, i.e., what interval of real time it should represent. At first sight it might seem that it would make better sense to express time as a real (floating point) number. We could then agree once for all that time is always expressed in some natural units (e.g., seconds) and any subtler intervals are simply represented by fractional numbers. There are a few problems with this simple idea.

First, on a computer, all numbers are finite sequences of bits and thus (de facto) integers. The floating-point arithmetic is an illusion cultivated for the somewhat deceiving convenience of programming mathematical calculations coming from the continuous domain of real (or possibly complex) numbers. Within the computer those calculations are implemented as series of operations on integers (or bits). For us, it basically means that there is no way around the discrete nature of numbers. The floating-point arithmetic doesn't eliminate the inherent granularity (discreteness) of representing sets of values, but only makes that granularity more complicated by automatically trading it for the range. This in a sense makes things worse, because one never knows whether adding something small (albeit still nonzero) to a number will in fact make that number different from what it was before the addition. There exists a whole discipline, numerical analysis, dealing with the rules relating the magnitude and accuracy of floating point numbers in advanced calculations.

In discrete-time, event-driven simulation, time arithmetic is apparently very simple. We basically want to be able to compare time values and increment them by time intervals. Sometimes we may also want to sum a few time intervals and divide the result by their (integer) count to calculate the average delay separating a pair of (repetitive) events. So where is the problem? Why is the issue of representation and granularity of time so important?

Note that the main loop of our simulator (in Sect. 1.2.2), only makes sense, if we can guarantee that the model never gets stuck in an infinite loop without making progress. This means that we want to know that the modeled time will advance whichever way the model's execution proceeds. The issue is delicate, because we admit time standing still, which should be OK, as long as we know that it cannot stand still forever. To see this within the context of the car wash model, suppose that a refinement of our simple simulator from Sect. 1.2.2 includes a detergent dispenser model. The device consists of two components, *dispenser_1* and *dispenser_2*, which alternately release some small amounts of two different detergents, e.g., to make sure that they properly mix. The release sequence consists of a series of alternating runs of the two dispensers carried out for some specific (total release) time, e.g., like this:

```
process dispenser_1 (state where) {
        switch (where) {
                case START_RELEASE:
                        start_time = now;
                        total_ time =  det_release_time (car);
                case GO_ON:
                        if (now —start_time >= total_time) {
                                schedule (START_WASH, wash, now);
                        } else {
                                TIME delta = det_1_release_time ( );
                                schedule (NONE, dispenser_2, now + delta);
                        }
        }
}
process dispenser_2 (action dummy) {
        TIME delta = det_2_release_time ( );
        schedule (GO_ON, dispenser_1, now + delta);
}
```

We can easily envision the wider context of the above fragment, so there is no need to obscure it with non-essential details. The first dispenser process is initially run at *START_RELEASE*, e.g., by the wash process as the starting step of a new car service. When the detergent release is complete, which event is detected by *dispenser_1* in state *GO_ON*, control is passed back to the *wash* process.

The transitions are illustrated in Fig. 1.2. Before *dispenser_1* schedules the next action of *dispenser_2* (in state *GO_ON*), it checks whether the difference between current time (*now*) and the time when the dispensing procedure started (*start_time*) has reached (or possibly exceeded) *total_time* which is the target duration of the combined dispensing operation. The necessary condition for this to ever happen is that at least some values of *delta* added to *now* in the above pseudocode are nonzero (it goes without saying that they are never less than zero). Strictly speaking, accounting for the discrete character of *now* and *delta*, we should rather say that when *now* is augmented by *delta* (in the two invocations of *schedule*), it ends up bigger than it was before the augmentation. This is needed so *now* can increase, eventually reaching the target time.

If time is represented as an integer number, and the assumed time unit turns out to be coarser than the magnitude of the intervals generated by *det_1_release_time* and *det_2_release_time*, then *now* will never increase and the simulator will be stuck in an infinite loop with time (apparently) standing still. If time is represented as a floating-point number, then the problem will occur when the magnitude of *now*

Fig. 1.2 The transition
diagram for the detergent
dispenser

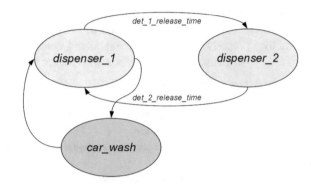

exceeds the threshold after which the accuracy of representation becomes less than
the magnitude of *delta*.

The problem may look exotic at first sight, but, with event-driven simulation,
long (or even extremely long) intervals of *real* time in the real system are trivially
easy to model (and cost no computer time), for as long as not too much happens
during that *real* time. Then, we are often tempted to model tiny actions (ones
taking very short intervals of time), lest we miss some important detail affecting
the accuracy of the model. This is particularly important in low-level network
protocols, ones dealing directly with hardware, where race conditions occurring
within picoseconds may affect decisions directly related to the formal correctness of
the protocol implementation [12]. Thus, there is no natural safeguard against huge
and tiny values coexisting within the same expression, something that computer
(floating point) arithmetic rightfully abhors.

Suppose that time is represented as an integer number of full seconds, i.e., the
grain of time is 1 s. Adding anything less than one second to *now* effectively adds
zero and amounts to no advancement of time in the model. Thus, we should make
sure that the smallest interval *delta* corresponding to some action that takes time in
the real system is never less than 1 s. Otherwise, the time grain is too coarse.

Suppose, for example, that the "actual" release time of the first detergent is
between 0.5 and 1 s, i.e., it can be any real number between 0.5 and 1.0. Then
we will expect *det_1_release_time* to return a random number in this range. If we
know something about the statistics of those values in the real system, then we will
try to make that function approximate those statistics through some smart algorithm.
But before the value returned by the function can be turned into *delta* and added to
now, it must become an integer number of seconds. If we just truncate it, i.e., take its
integer part, then it will always or almost always end up as zero (depending on the
tricky things that happen at the upper boundary of the interval). This way, *now* will
not be advanced when the first kind of detergent is released in our model. If a similar
mishap occurs with the second dispenser, then the detergent release time will never
reach its target.

On top of the obvious problem with potentially stalling the execution of the simulator, a situation like the one described above raises questions about the model's

fidelity. Probably, instead of truncating the floating-point value returned by the generator of the detergent release time, we would round it instead. That makes better sense whenever resolution is lost in a measurement. Then, the value of *delta* produced by *det_1_release_time* will be (almost) always 1. This is much better, because the time will begin to flow, but the statistics of the release time (and its stochastic nature) will be completely lost in the model. For example, assuming that, in its accurate representation, that time is uniformly distributed between 0.5 s and 1.0 s, the average observed over *n* consecutive actions of the dispenser (as seen in real life) should be about 0.75 s. But in the model, the average will end up as 1 s, because this is the only duration that may happen in the model.

The obvious implication is that the grain of time in the model should be (significantly) less than the shortest duration of action (*delta*) in the model that is supposed to be nonzero. With our approach, the model formally admits zero *delta*, but this feature should be restricted to those situations where the events in fact should (or harmlessly can) coincide in time in the real system. So how much smaller should the grain be in our model? There is no simple answer to this question. If the tiny delays are important, and their diversity is likely to translate into the diversity of paths in the model's execution, which can bring us important insights into the system, then we should worry about their accurate rendering, including the statistics. Then, the time unit should be at least one order of magnitude below the shortest relevant delay. And if the distribution happens to be exotic, e.g., characterized by a long tail, then one order of magnitude may not be enough. We simply have to understand what we are up to. Sometimes capturing the statistics of those tiny delays is not important, but it is still important to see their diversity, so the model can be exposed to all the interesting conditions that can occur in the real system without necessarily matching the frequency distribution of those conditions. This is important when the model is used for conformance/correctness tests: we want to catch abnormal situations (the interesting question is whether they occur at all) without worrying too much about their frequency. We may even want to bias the distribution to increase the rate of rare (but not impossible) occurrences, if we consider them interesting. Finally, there are situations where it doesn't matter. For example, if the detergent release operation amounts to a tiny fraction of the car service time in our model, and we only care about the statistics of that time, then the inaccurate statistics of the detergent release time can be probably ignored. But the fact that we have decided to include the detergent dispenser in the model means that we consider it important for some reason, so we should ask ourselves what that reason really is. Being able to correctly answer such questions belongs to the art of simulation. The fact that such questions can and should be asked, and that there are no blanket answers to them, is what makes simulation an art.

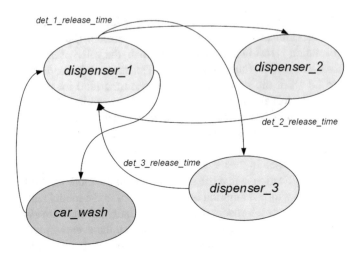

Fig. 1.3 A dispenser model with three units

1.2.4 Nondeterminism and the Time Grain

One more issue closely related to the discrete nature of modeled time is nondeterminism in the model. With time organized into little slots, it is not unlikely that two or more events become scheduled into the same slot. This may mean either that the two events should in fact occur concurrently in the real system, or that they would occur at different times, but the difference between the moments of their occurrence is less than the resolution of time in the model. Note that the first case is formally flaky, because there is no such thing as perfect simultaneity in distributed systems. Without invoking the authority of Albert Einstein, we can simply say that the fact that two events in the model fall into the same time slot means that the model cannot tell which of them should *really* go first.

OK, but is it important which of those events is *actually* run first by the simulator's main loop? Formally, there is no difference from the viewpoint of the modeled timing of those events. Well, we have agreed on a time unit, possibly after some deliberation inspired by the preceding section, so now we better be prepared to live with it. However, the order of the actions associated with those events, as they are carried out by the model, may have its implications. Suppose that the detergent dispenser in our model has been extended by one more unit, as illustrated in Fig. 1.3. The idea is that *dispenser_1*, having released a portion of its detergent, schedules two other dispensers that release their specific detergents simultaneously. The related piece of pseudocode may look like this:

delta = det_1_release_time ( );
schedule (NONE, dispenser_2, now + delta);
schedule (NONE, dispenser_3, now + delta);

The two dispensers may finish their operation at different times. Each of them may notify *dispenser_1* independently of the other. To make sense of this collaboration, *dispenser_1* will postpone its next action until both *dispenser_2* and *dispenser_3* are done, e.g., by using a counter.

Note that *dispenser_2* and *dispenser_3* are both scheduled to run at the same discrete time. We may suspect that in the real system the order in which they are *in fact* started is unknown and, as far as we can tell, unless there is some mechanism enforcing a specific order, both scenarios (*dispenser_2* starting first or second) are possible. Perhaps we don't really care (again, it depends on what the model is used for), but if we happen to care for some reason, then the model should include both paths in its evolution. Generally, we should follow the principle of minimal assumptions: if a sequence of actions is not explicitly forbidden by the model's rules, then it is allowed and should be explorable in a long-enough experiment.

One can easily think of situations where the exploration of all not-otherwise-prohibited paths is important. For example, both dispensers may share some valve, e.g., providing air under pressure to a pump. The setting of that valve is determined by the first dispenser that opens it, so when the second one is a bit late, it may fail to dispense the right amount of its detergent. Note that we cannot say what happens in the model without looking at the implementation. The simulator will pick events from the queue and run them in the order in which they have been found there. The sequence of *schedule* operations (see above) suggests that the event for *dispenser_2* is put into the queue first. Does it mean that it will precede the event for *dispenser_3* in the queue? Note that both events are marked with the same time, so both orderings are formally legal. If the ordering of events with the same time stamps is determined by the respective sequence of *schedule* statements, then they will be always run in the same order. This does not agree with the philosophy of discrete time which says that events falling within the same time slot are not pre-ordered in real time, but merely grouped within a certain interval with their actual order not being determined by the model. Thus, to be fair, the simulator should not presume any deterministic order, e.g., one that may transpire from a simple and natural implementation of the event queue, but rather randomize that order whenever two or more events in front of the queue happen to be marked with the same time. For example, the order of the *schedule* statements for the two dispensers should not be treated by the simulator as a hint regarding the processing order of the resulting events. The first obvious criterion of that order is the event's time, and this part is always clear. Then the fact that several events have been marked with the same (discrete) time can only mean that their *actual* ordering is unknown, so a fair execution of them, one that provides for the most complete exploration of the system's states reachable during a simulation experiment, is the one where such events are run in a random order. Note that a similar problem may occur when the two dispensers operating in parallel complete their actions, i.e., when their completion times fall within the same slot of discrete time.

Note that there are obvious cases when two events scheduled within the same time slot are in fact ordered quite deterministically: this happens when one event is the cause of the other, so it in fact *precedes* the other event (its effect), and the reason why both events fall into the same time slot is the (possibly excessive) coarseness of the time grain. As far as causes and their effects, we can talk about two categories of scenarios. In the first category, the effect is separated from the cause by an actual, perhaps quite pronounced and relevant, time gap which the model just fails to capture. This may be a problem, like the one responsible for the destroyed statistics of detergent release time in Sect. 1.2.3. Generally, in such scenarios, we should make sure that the cause-effect delays nullified by the finite time grain have no negative impact on the model's fidelity (which must be always considered in the context of its application). If they do, it means that the grain is too coarse and has to be reduced. The second category of scenarios is when the cause-effect mechanism is used to emulate true parallelism, in which case it is mostly always OK. It may be an artefact of the way things are programmed, and the zero delays may not affect the model's fidelity at all, regardless of the time grain. This will be the case, e.g., when a process schedules an event at zero delay to start another process supposed to take care of some action that must be performed in parallel with the subsequent action of the triggering process (also in the real system).[9] Then, even though formally one event is the cause of the other, insisting on their precise nonzero separation in time makes no sense (and may in fact be detrimental to the model's fidelity).

In many cases the *actual* ordering of events scheduled close in time may be deterministic and known, the requisite knowledge resulting from a more careful (more detailed) description of the real system in the model. If the ordering is known, and it matters for the model, then the unit of time should be adjusted (reduced) to make it possible to express the short intervals that get into the picture and tell them apart. One moral from this discussion is that in many cases we may want to use a very short time grain, especially when the impact of the simplifications implicitly introduced by a coarse grain is difficult to assess before the simulation study. Historically, SMURPH (or rather its predecessor LANSF) was first used for investigating synchronized, contention-based protocols for shared media networks where the (correct) operation of a protocol was based on distributed (albeit synchronized) perception of time by the different nodes of the network inferred from various signals detected in the shared medium. Via detailed simulation properly accounting for the nondeterminism of races and the inherent inaccuracies of independent clocks we were able to spot many problems in protocols formulated in idealistic (unrealistic) conditions and suggest proper ways of implementing them [4, 5, 13, 14].

Of course, the finer the time grain, the wider range is needed to represent the modeled time. For illustration, assuming 1 s as the time unit and 32-bit unsigned representation of time, we can model a car wash for over 100 years, which would probably suffice. Funny (and reassuring) as this illustration may sound, 1 s would be considered a ludicrously coarse grain from the viewpoint of a contemporary communication network where the amount of time needed to insert a bit into the medium may be of order 10^{-10} s. Then, for a truly accurate model of the physical layer, we may want to go well below that time, e.g., two orders of magnitude, to 10^{-12} s, i.e., a picosecond.

---

[9]We witness this kind of scenario in the *entrance* process in Sect. 1.2.2, which schedules an event at time *now* to notify the *wash* process about the new car arrival.

Then the 32-bit arithmetic of time will only let us go for about 4 milliseconds. With the 64-bit floating point representation, assuming 53 significant digits (the IEEE 754 standard [36]), the granularity of representation will be good for about 40 min, and then it will decay (increase by the factor of 2 every 40 min, or so). That would be confusing, so using floating point representation for time is not a good idea. With 64-bit unsigned integers, we can run the model for almost a year (in modeled time).

The need for an extremely fine time grain can, sometimes surprisingly, surface in models of apparently slow and unchallenging networks, depending on the kind of issues one may be forced to study. In one of my recent industrial projects I had to model a (some would call it legacy) wireless network consisting of a number of independent clusters, each cluster consisting of a base station (or an access point) and one or more client nodes communicating with the base in a synchronized (slotted) manner. The transmission rate was about 1 Mb/s, i.e., nothing special by contemporary standards. The predominant issue turned out to be the interference from neighboring clusters whose slots would slowly "slide in" at the drift rate of the otherwise accurate and precise, albeit uncoordinated, clocks of the different base stations. To accurately model those phenomena, I had to make sure that all delays issued by the model, including the smallest ones, were not only slightly randomized (as usual), but also included a persistent drift of order 1 ppm (part per million), or possibly even less. That meant, for example, that a delay of one-bit time had to include room for a tiny deviation of order 10^{-6} of its length, which called for the time grain corresponding to at most 10^{-7} of one bit insertion time, i.e., 10^{-13} s. With the 64-bit precision of time, I could only run the network for about 250 h of its virtual life. For a "more contemporary" bit rate, say 1 Gb/s, the time limit would become completely impractical.

1.3 An Overview of SMURPH

While SMURPH has been created to model communication protocols and networks, it has all the features of general purpose simulators. Indeed, the package has been applied to studying systems not quite resembling communication networks, e.g., disk controllers. Without unnecessarily narrowing the focus at this stage of the presentation, let us start by saying that SMURPH is a programming language for describing reactive systems (e.g., communication networks) and specifying event-driven programs (e.g., network protocols) in a reasonably natural and convenient way. Programs in SMURPH are run on *virtual hardware* configured by the user. Thus, SMURPH can be viewed both as an implementation of a certain (protocol specification) language and a run-time system for executing programs expressed in that language.

1.3.1 Reactive Systems: Execution Modes

The SMURPH kernel can execute in one of three modes. One (shall we say "traditional") way to run a SMURPH model is to simply execute it in the *simulation mode*, as a classical, event-driven, discrete-time simulator, basically cycling through a loop like that shown in Sect. 1.2.1. The execution time of the program bears no relationship to the flow of virtual time in the model. One way of looking at this kind of run, especially from the perspective of the other two modes, is to view it as off-line (or batch) execution of the model, where an input data set describes the system (network) layout, its load, and specifies the termination condition. The model executes unsupervised, on its own, with no contact with the outside world. Eventually, when the termination condition is met, the simulator will stop and produce a set of results. Those results are our only concern: they include all the answers that we wanted to obtain from the model and represent the entire interaction of the model with its user.

The simulation mode is intended for performance and conformance studies of communication networks and protocols. Typically, the purpose of a single run (experiment) is to identify one point of a curve that shows how some measure (or set of measures) depends on some parameter of the network load. Many experiments may be required to identify sufficiently many points of that curve, e.g., when comparing different (alternative) solutions with respect to their performance.

One handy feature of SMURPH, introduced primarily to facilitate the virtual execution of WSN applications, is the so-called *visualization mode* whereby SMURPH tries to match the intervals of virtual time flowing in the model to the corresponding intervals of real time. This way, the model poses for the real system, akin to the real-time simulators discussed in Sect. 1.2. To accomplish that, SMURPH doesn't change its modus operandi, say, in a way that would get it close to a flight simulator. It still operates in the same event-driven fashion as before, except that, whenever the time of the new event extracted from the queue is a jump into the future, SMURPH will delay that jump in *real* time until the corresponding interval has elapsed in the real world outside the model.

Note that this may not always be possible. A detailed model of a complex system may have problems catching up with its execution to the real time. Such a system can still be run in the visualization mode, albeit with some slow-motion factor. The mapping of virtual time to real time must be explicitly defined (see Sect. 1.2.3).

If the visualization mode can be so painlessly and seamlessly imposed on the regular simulation mode, one may want to ask what is the difference between *true* real-time simulators (as mentioned in Sect. 1.2) and an event-driven simulator (like SMURPH). Or, put differently, could a flight simulator be constructed within SMURPH's framework? While the difference need not be clearly delineated in all cases, one can attempt to divide physical systems into two classes based on which kind of approach to simulating them is likely to fare better. The key word is *reactive*. We say that a system is reactive, if its behavior mostly consists in responding to events. This definition is wide enough to encompass practically everything that exhibits a coordinated behavior; however, some systems are more reactive than others, because the sets of events describing their dynamics are smaller and better contained: it in fact makes sense to treat them as individual, discrete events.

Take an aircraft as an example. During a flight, its surfaces are subjected to varying pressure resulting from the air rushing in, every which way, mostly from the front. The lift, as well as lots of other forces collectively acting on the aircraft are the combined result of different vectors applied to the different parts of the wings, the fuselage, and so on. At the bottom of things, we do have events, e.g., the air molecules interacting with the solid structure. A level down, we have atomic forces and electromagnetic interactions maintaining that solid structure and, for example, determining its elasticity and strength. In principle, the model can be expressed using such events, because they underlay all the relevant physical processes; however, nobody in their sane mind would simulate an aircraft in flight by running an event queue of subatomic interactions.

Thus, for an aircraft in flight, it makes much better sense to model the relevant forces as the outcome of a continuous process described by differential equations which conveniently average out the zillions of elementary events into the expected perception of dealing with a solid object responding (mostly) smoothly to (mostly) smoothly varying conditions. Thus, an aircraft in flight is *not* a reactive system from our point of view. Its evolution is modeled by solving essentially continuous equations determining the system's complete state in the next while, given its state at the current moment. Since these equations are being solved on a computer, the solution consists in partitioning time into (usually small, ideally infinitesimal) intervals and using them as steps for approximate solutions of the differential equations. If we insisted on viewing this simulation as event-driven, then the events would occur whenever the time has advanced to the next step and there would be a single process with a single state responding to all events. The action carried out by that process would consist in the next iteration of the evaluation procedure for the aircraft's state.

On the other hand, a car wash (in our view from Sect. 1.2.2), or a telecommunication network, are good examples of reactive systems, because their behaviors are well captured by interesting events of different kinds whose timing is the essential element of description. When a car has been admitted to the wash, the most natural view on the evolution of the system is to ask how much time the different stages of that service will take, and turn directly to those moments, rather than, like a little kid stuck in a car on a weekend trip, keep asking incessantly "are we there yet?" Similarly, in a telecommunication network, the only important moments in the life of, say, a piece of wire are those where that wire carries a signal with some definite properties. It is more natural (and cost effective, from the viewpoint of simulation) to determine the timing of those relevant occurrences from other (previous) occurrences (e.g., based on the propagation time) than to constantly examine the wire for a possible status change. This is because the events are relatively sparse and their sequence amounts to a descriptive representation of the system's evolution. The overall complexity of the system results from a large number of such events, and, more importantly, from the different ways they can intertwine in time, but each of the individual events appears as a well-defined episode.

The third, and probably the most exotic, mode of the simulator is called the *control mode*. In this guise, the program does not really simulate anything (so it is not a simulator any more), but instead *controls* a real-life reactive system. From this angle, by a reactive system we understand something that can be represented for the control program in SMURPH via a collection of *sensors* and *actuators*, i.e., interfaces whereby the control program can receive events from the outside (possibly accompanied by data) and effect actions on its environment.

In the control mode, the programming language of SMURPH becomes a "regular" programming language without the implied embedding of the program into a simulation-oriented run-time system. The SMURPH tools for building processes of simulation models are used to create event-driven (reactive) threads of the control program. The program's (reactive) interface to the world is provided by SMURPH mailboxes which can be tied to devices, files, or TCP/IP ports.

1.3.2 Organization of the Package

SMURPH has been programmed in C++ and its protocol specification language is an extension of C++. We assume that the reader is familiar with C++ and the general concepts of object-oriented programming.

A schematic view of the package is presented in Fig. 1.4. A program in SMURPH is first preprocessed by *smpp* (a preprocessor) to become a C++ program. The code in C++ is then compiled and linked with the SMURPH run-time library. These operations are organized by *mks*, a generator script (see Sect. B.5) which accepts as arguments a collection of program files and options and creates a stand-alone, executable simulator (or a control program) for the system described by the source files and the input data. If we are interested exclusively in simulation, the resultant program can be fed with some input data and run "as is" to produce results.

SMURPH is equipped with many standard features of simulators, like random number generators and tools for collecting performance data. The program in SMURPH runs under the supervision of the SMURPH kernel, which coordinates the multiple reactive threads (processes, see Sect. 5.1) of the program, schedules events, and communicates the program with the outside world. The program can be monitored on-line via DSD,[10] a Java applet that can be invoked, e.g., from a web browser (Sect. C.1). This communication is carried out through a monitor server which typically runs on a dedicated host visible by both parties (i.e., the program and DSD). DSD is useful for debugging and peeking at the partial results of a poten-

---

[10]The acronym stands for Dynamic Status Display.

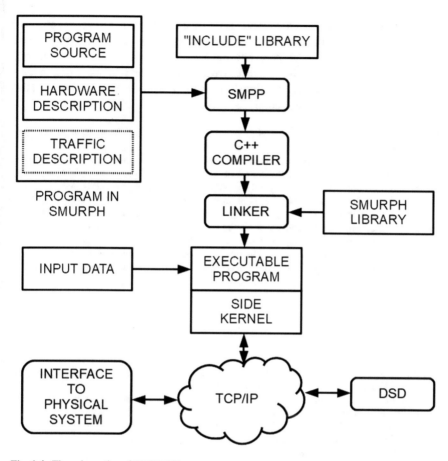

Fig. 1.4 The schematics of SMURPH

tially long simulation experiment. SMURPH also offers tools for protocol testing. These tools, the so-called *observers* (Sect. 11.3), are intended for verifying the protocol expressed in SMURPH in terms of its adherence to the specification. Observers look like programmable (dynamic) assertions describing the expected sequences of protocol actions and thus providing an alternative specification of the protocol. The simulator checks in run time whether the two specifications agree.

The source program in SMURPH is logically divided into three parts (see Fig. 1.4). The "program source" part describes the dynamics of the modeled system. The "hardware description" configures the hardware (network) on which the protocol program is expected to run. Finally, the "traffic description" part defines the load of the modeled system, i.e., the events arriving from the "virtual outside" (the network's "application") and their timing.

The partitioning of a SMURPH program into the three components need not be prominent. In a simple case, all three parts can be contained in the same, single source file. The words "protocol," "network," and "traffic" reflect the primary (original) purpose of SMURPH, but they should be viewed figuratively: a SMURPH program need not be a network model, need not execute a "protocol," and does not have to deal with "network traffic." Nonetheless, we shall use in the remainder of this book the convenient term "protocol program" to refer to the program written for and run by SMURPH, even if the program doesn't relate to a "protocol."

Acknowledgements I have collaborated with several people on research and industrial projects where SMURPH proved useful, or even essential. All those people have been credited in various places, more pertinent to the context of their contribution, so there is no need to mention them in this book, which is solely about SMURPH, for which I am modestly prepared to take full credit (and assume full responsibility). One of my collaborators, however, stands out in this respect. I mean my friend and first serious collaborator on protocol design, Piotr Rudnicki, who once, in 1986, drew on a piece of paper what must be seen today as the skeletal layout of the first ever SMURPH process, the transmitter of Ethernet. That picture forced me to discard my obscure collection of messy simulation programs and replace them with a unified simulation system good for them all. Piotr is no longer with us, as they say. His much premature departure has been a serious blow to reason and common sense. Too bad, because these days they need all the support they can possibly get.

References

1. R.M. Metcalfe, D.R. Boggs, Ethernet: Distributed Packet Switching for Local Computer Networks. Commun. ACM **19**(7), 395–404 (1976)
2. D.R. Boggs, J.C. Mogul, C.A. Kent, *Measured Capacity of an Ethernet: Myths and Reality*. Digital Equipment Corporation, Western Research Laboratory, Palo Alto, California (1988)
3. P. Gburzyński, and P. Rudnicki, A better-than-token protocol with bounded packet delay time for Ethernet-type LAN's, in *Proceedings of Symposium on the Simulation of Computer Networks*, Colorado Springs, Co., 1987
4. P. Gburzyński, P. Rudnicki, Using time to synchronize a token Ethernet, in *Proceedings of CIPS Edmonton '87*, 1987
5. P. Gburzyński, P. Rudnicki, A note on the performance of ENET II. IEEE J. Sel. Areas Commun. **7**(4), 424–427 (1989)
6. W. Dobosiewicz, P. Gburzyński, P. Rudnicki, An Ethernet-like CSMA/CD protocol for high speed bus LANs, in *Proceedings of IEEE INFOCOM '90*, 1990
7. W. Dobosiewicz, P. Gburzyński, Improving fairness in CSMA/CD networks, in *Proceedings of IEEE SICON '89*, Singapore, 1989
8. P. Gburzyński, X. Zhou, Ethernet for short packets. Int. J. Model. Simul. (1993)
9. W. Dobosiewicz, P. Gburzyński, Performance of Piggyback Ethernet, in *Proceedings of IEEE IPCCC*, Scottsdale, 1990
10. W. Dobosiewicz, P. Gburzyński, Ethernet with segmented carrier, in *Proceedings of IEEE Computer Networking Symposium*, Washington, 1988
11. W. Dobosiewicz, P. Gburzyński, On two modified Ethernets. Comput. Networks ISDN Syst., 1545–1564 (1995)
12. P. Gburzyński, P. Rudnicki, A virtual token protocol for bus networks: correctness and performance. INFOR **27**, 183–205 (1989)
13. W. Dobosiewicz, P. Gburzyński, P. Rudnicki, Dynamic recognition of the configuration of bus networks. Comput. Commun. **14**(5), 216–222 (1991)
14. W. Dobosiewicz, P. Gburzyński, P. Rudnicki, On two collision protocols for high speed bus LANs. Comput. Networks ISDN Syst. **25**(6), 1205–1225 (1993)

15. P. Gburzyński, P. Rudnicki, LANSF—a modular system for modelling low-level communication protocols, in *Modeling Techniques and Tools for Computer Performance Evaluation*, ed. by R. Puigjaner, D. Potier (Plenum Publishing Company, 1989), pp. 77–93

16. P. Gburzyński, P. Rudnicki, LANSF: a protocol modelling environment and its implementation. Softw. Pract. Experience **21**(1), 51–76 (1991)

17. P. Gburzyński, P. Rudnicki, Bounded packet delay protocols for CSMA/CD bus: modeling in LANSF, in *Proceedings of the 19th Annual Pittsburgh Conference on Modeling and Simulation*, 1988

18. P. Gburzyński, P. Rudnicki, On formal modelling of communication channels, in *Proceedings of IEEE INFOCOM '89*, 1989

19. M. Berard, P. Gburzyński, P. Rudnicki, Developing MAC protocols with global observers, in *Proceedings of Computer Networks '91*, 1991

20. P. Gburzyński, P. Rudnicki, Modeling low-level communication protocols: a modular approach, in *Proceedings of the 4th International Conference on Modeling Techniques and Tools for Computer Performance Evaluation*, Palma de Mallorca, Spain, 1988

21. P. Gburzyński, P. Rudnicki, On executable specifications, validation, and testing of MAC-level protocols, in *Proceedings of the 9th IFIP WG 6.1 International Symposium on Protocol Specification, Testing, and Verification*, June 1989, Enschede, The Netherlands, 1990

22. P. Gburzyński, J. Maitan, Deflection routing in regular MNA topologies. J. High Speed Networks **2**, 99–131 (1993)

23. P. Gburzyński, J. Maitan, Simulation and control of reactive systems, in *Proceedings of Winter Simulation Conference WSC '97*, Atlanta, 1997

24. W. Dobosiewicz, P. Gburzyński, Protocol design in SMURPH, in *State of the art in Performance Modeling and Simulation*, ed. by J. Walrand and K. Bagchi (Gordon and Breach, 1997), pp. 255–274

25. W. Dobosiewicz, P. Gburzyński, SMURPH: an object oriented simulator for communication networks and protocols, in *Proceedings of MASCOTS '93, Tools Fair Presentation*, 1993

26. P. Gburzyński, P. Rudnicki, Object-oriented simulation in SMURPH: a case study of DQDB protocol, in *Proceedings of 1991 Western Multi Conference on Object-Oriented Simulation*, Anaheim, 1991

27. P. Gburzyński, *Protocol Design for Local and Metropolitan Area Networks* (Prentice-Hall, 1996)

28. P. Gburzyński, J. Maitan, Specifying control programs for reactive systems, in *Proceedings of the 1998 International Conference on Parallel and Distributed Processing Techniques and Applications PDPTA '98*, Las Vegas, 1998

29. P. Gburzyński, J. Maitan, L. Hillyer, Virtual prototyping of reactive systems in SIDE, in *Proceedings of the 5th European Concurrent Engineering Conference ECEC '98*, Erlangen-Nuremberg, Germany, 1998

30. P. Gburzyński, P. Rudnicki, Modelling of reactive systems in SMURPH, in *Proceedings of the European Simulation Multiconference*, Erlangen-Nuremberg, W-Germany, 1990

31. E. Akhmetshina, P. Gburzyński, F. Vizeacoumar, PicOS: a tiny operating system for extremely small embedded platforms, in *Proceedings of ESA '03*, Las, 2003

32. W. Dobosiewicz, P. Gburzyński, From simulation to execution: on a certain programming paradigm for reactive systems, in *Proceedings of the First International Multiconference on Computer Science and Information Technology (FIMCSIT '06)*, Wisla, 2006

33. P. Gburzyński, I. Nikolaidis, Wireless network simulation extensions in SMURPH/SIDE, in *Proceedings of the 2006 Winter Simulation Conference (WSC '06)*, Monetery, 2006

34. P. Gburzyński, W. Olesiński, On a practical approach to low-cost ad hoc wireless networking. J. Telecommun. Inf. Technol. **2008**(1), 29–42 (2008)

35. N.M. Boers, P. Gburzyński, I. Nikolaidis, W. Olesiński, Developing wireless sensor network applications in a virtual environment. Telecommun. Syst. **45**, 165–176 (2010)

36. M. Cowlishaw, Decimal arithmetic encodings, *Strawman 4d. Draft version 0.96*. IBM UK Laboratories, 2003

Chapter 2
Examples

In this chapter we shall inspect four examples of programs in SMURPH illustrating its three modes introduced in Sect. 1.3.1. We shall try to present reasonably complete programs, including sample data sets and outputs, even though the precise definitions of the requisite programming constructs will come later. All four programs are included with the SMURPH package (see subdirectory *Examples*), and they can be played with as we proceed through their discussion.

2.1 The Car Wash Redux

Let us start from the car wash model, as introduced in Sect. 1.2.2, which we shall now turn into a complete program. This exercise will let us see the essential part of the simulation environment provided by SMURPH, the part that is independent of the package's intended, official application (simulating communication networks and specifying communication protocols) and underlines its general simulation features.

The program can be found in *Examples/CarWash*. For simplicity, it has been put into a single file, *carwash.cc*. The other file present in that directory, *data.txt*, contains a sample data set for the simulator.

The program file starts with this statement:

identify (Car Wash);

whose (rather transparent) role is to identify the program. The statement proved useful in the days when I was running lots of simulation experiments concurrently, on whatever workstations in the department I could lay my hands on. The program

© Springer Nature Switzerland AG 2019
P. Gburzyński, *Modeling Communication Networks and Protocols*,
Lecture Notes in Networks and Systems 61,
https://doi.org/10.1007/978-3-030-15391-5_2

identifier, appearing prominently in the output file (and a few other places), would let me quickly tell which program the (otherwise poorly discernible) sets of numbers were coming from.

Then we see a declaration of three constants:

const int standard = 0, extra = 1, deluxe = 2;

representing three service classes that a car arriving at the wash may ask for. As the programming language of SMURPH is C++, with a few extra keywords and constructs thrown in, a declaration like the one above needs no further comments regarding its syntax and semantics. Generally, we will not elaborate in this book on straightforward C++ statements, assuming that the reader is reasonably well versed in C++. If this is not the case, then other sources, e.g., [1, 2], will be of better assistance than any incidental elucidations I might be tempted to insert into this text.

Instead of discussing the contents of the program as it unveils sequentially in the file, it makes more sense to proceed in a methodological way, based on the concepts and tools around which the model is constructed. The first thing that we see when trying to comprehend the program is its organization into *processes*, so this is where we start.

A SMURPH process amounts to a more formal embodiment of the entity with this name introduced in Sect. 1.2.2. Thus, a process is a function invoked in response to an event. The body of such a function is typically partitioned into *states*, to account for the different kinds of actions that the process may want to carry out. The state at which the process function is called (entered) is determined by an attribute of the *waking* event. SMURPH conveniently hides the implementation of the event queue relatively deep in its interiors. Also, we never directly see the main loop where the events are extracted from the queue and distributed to processes.

Processes are created dynamically, and they can be dynamically destroyed. If a SMURPH process represents an active component of the physical system modeled in the simulator (as we informally introduced processes in Sect. 1.2.2), then it must be created as part of the procedure of setting the model up, i.e., building the modeled counterparts of the relevant components of the physical system. This procedure is explicit in SMURPH.

2.1.1   The Washer Process

The essence of our model is the process delivering the service to the cars. It is simple enough, so we can see that process at once in its entirety:

```
process washer {
        car_t *the_car;
        states { WaitingForCar, Washing, DoneWashing };
        double service_time (int service) {
                return dRndTolerance (
                        service_time_bounds [service] . min,
                        service_time_bounds [service] . max,
                        3);
        };
        perform {
                state WaitingForCar:
                        if (the_lineup->empty ( )) {
                                the_lineup->wait (NONEMPTY, WaitingForCar);
                                sleep;
                        }
                        the_car = the_lineup->get ( );
                transient Washing:
                        double st;
                        st = service_time (the_car->service);
                        Timer->delay (st, DoneWashing);
                        TotalServiceTime += st;
                        NumberOfProcessedCars++;
                state DoneWashing:
                        delete the_car;
                        sameas WaitingForCar;
        };
};
```

While we will often informally say that what we see above is a process declaration, it is in fact a declaration of a process type, which corresponds to a C++ class
declaration. This is essentially what such a process declaration amounts to after the
program in SMURPH has been preprocessed into C++ (Sect. 1.3.2). A true actual
process is an object (instance) of the class brought in by the above declaration. Such
an object must be *created* to become an actual process. In our car wash example, the
shortcut may be warranted by the fact that the car wash model runs a single instance
of each of its process types.

As SMURPH introduces several internal class types of the same syntactic flavor
as *process*, it may make sense to digress a bit into a friendly introduction to this
mechanism (which will be explained in detail in Sect. 3.3). At the first level of
approximation (and without losing too much from the overall picture), we can assume
that there exists a predefined class type representing a generic process, declared as:

class Process {...};

and a SMURPH declaration (declarator construct) looking like this:

process NewProcessType {...};

which is informally equivalent to:

class NewProcess: Process { ...};

i.e., it extends the standard, built-in *Process* class. The equivalence is not that much informal (and mostly boils down to the syntax), because there does exist a class type named *Process*, and it does define some standard attributes and methods that specific processes declared with the *process* declarator inherit, override, extend. You can even reference the *Process* type directly. For example, you can declare a generic process pointer:

*Process *p;*

assign it:

p = create washer;

and reference it, e.g.:

p->terminate ();

   While you cannot write something like this:

class washer: Process { ... };

literally and by hand, because some not-extremely-relevant extra magic is involved with the SMURPH *process* declaration, a syntactically correct SMURPH process declaration in fact does something very similar to that.

   Typically, a *process* defines a list of *states* and the so-called *code method* beginning with the keyword *perform*. It can also declare regular C++ methods, e.g., *service_time* above. The code method is organized into a sequence of states (like the informal processes introduced in Sect. 1.2.2). A process declaration can be split according to the standard rules applicable to C++ classes, so we can put the declaration in one place, e.g.:

```
process washer {
        car_t *the_car;
        states { WaitingForCar, Washing, DoneWashing };
        double service_time (int);
        perform;
};
```

and specify the methods, including the code method, elsewhere (perhaps in a different file), like this:

```
double washer::service_time (int service) {

    ...

};
washer::perform {
        state WaitingForCar:

            ...

        transient Washing:

            ...

        state DoneWashing:

            ...

};
```

Of course, the header must be visible in the place where the methods are specified, as for any regular C++ class.

The washer process has three states. The first state (*WaitingForCar*) is assumed when the process is created. It can also be viewed as the default top point of the main loop where the process is waiting for a customer (car) to show up. Having found itself there, the process checks if the car queue (variable *the_lineup*) is empty. If it happens to be the case, the process must wait until a car arrives. It looks like the requisite event will be delivered by *the_lineup* (whatever it is), because the process indicates its willingness to respond to that event by executing the *wait* method of the *the_lineup* object. We can also safely bet that the second argument of *wait* is the state in which the process wants to be awakened when the event occurs. Note that it is the very same state in which the process executes the present action, so the process means "taking it again from the top."

The simple idea is that every object capable of triggering events defines a *wait* method with two arguments: the first one identifies an event that the object can possibly generate, the other indicates a state of the invoking process (the third implicit argument is the process invoking the method). Everything in SMURPH happens within the context of some process, so this implicit argument always makes sense. The (invoking) process declares this way that it wants to respond to the indicated event. An object capable of triggering events, and defining a *wait* method is called an *activity interpreter* (or an *AI*, for short) (Sect. 5.1.1).

The next statement is *sleep*, which is the straightforward way to terminate the current action of the process and exit the process, or rather put it to sleep. The difference is subtle. The process is put to sleep, if there is at least one event that the process is waiting for (we say that there is at least one outstanding *wait request* issued by the process). Otherwise, the process can never possibly be run again, so it is in fact terminated.

Normally, when in the course of its current action a process exhausts the list of statements at its current state, it effectively executes *sleep*, i.e., state boundaries are hard (unlike, e.g., *cases* in a *switch* statement). Sometimes it is convenient to be able to cross a state boundary without clumsily jumping around it. This is illustrated in the above code by *transient* which is the same as *state*, except that it does not stop the sequence of statements from the preceding state. Thus, if *the_lineup* is not empty, the process will smoothly fall through to the next state (*Washing*). In that state, the process generates a service time for the car (method *service_time*) and invokes the *delay* method of *Timer* specifying that time as the first argument. The second argument points to the last state of *washer*. Clearly, this way the process indicates that it wants to be awakened in state *DoneWashing* when the service time is up.

Note that neither *wait* nor *delay* takes effect immediately, i.e., the methods return and the process is allowed to continue its action in the present state until it executes *sleep* or hits a (non-transient) state boundary. Methods declaring future events are just that, i.e., declarations, and their effects become meaningful when the process completes its present action (state). More than one invocation of a *wait* method may occur within one action and the effect of those multiple invocations is cumulative, i.e., the process ends up simultaneously waiting for *all* the declared events. The earliest of them (the first one to occur) will be the one to *actually* wake the process up.

Another important thing to remember is that whenever a process is awakened, before it starts executing the statements in its just-entered state, its list of declared future waking events is empty, regardless of how many events the process awaited before getting into the present state. It may have been waiting for a dozen different events, but it has been awakened by exactly one of them (the earliest one) at the state associated with that event (with the respective *wait* or *delay* call). All the remaining declarations, which have proved ineffective, have been ignored and forgotten.

Note that in state *Washing*, besides setting up an alarm clock for the time required to wash the current car, the process updates two variables: *TotalServiceTime* and *NumberOfProcessedCars*. This way it keeps track of some simple performance measures for the system. For example, at the end of the simulation experiment, the ratio of *TotalServiceTime* to the total time that has elapsed in the model will yield the normalized throughput, i.e., the fraction of time spent by the system on doing useful work.

When the process wakes up in state *DoneWashing*, the service has been completed. Thus, the process deallocates the data structure representing the car that was previously extracted from *the_lineup* and loops back to *WaitingForCar* to see if there is another car to be taken care of. Statement *sameas* acts as a direct "go to" the indicated state.

One may note that the *Washing* state is effectively superfluous, because it is never mentioned as the argument of a *wait* (or *delay*) operation, i.e., the *washer* process can never "wake up" in that state (it can only be fallen through from *WaitingForCar*). Arguably, the state is convenient as a comment: it is natural to have a formal state of the *washer* process indicating that the wash is currently processing a car. It costs nothing to keep it around.

2.1.2 The Entrance Process

The second process of the model is responsible for car arrivals. It operates according to the simple scheme discussed in Sect. 1.2.2, and its declaration looks like this:

```
process entrance {
        double IAT;
        void setup (double iat) {
                IAT = iat;
        };
        states { NothingHappening, CarArriving };
        perform {
                state NothingHappening:
                        Timer->delay (dRndPoisson (IAT) , CarArriving);
                state CarArriving:
                        int s;
                        double d = dRndUniform (0.0, 1.0);
                        for (s = standard; s < deluxe; s++)
                                if (d <= service_percentage [s])
                                        break;
                        the_lineup->put (new car_t (s));
                        sameas NothingHappening;
        };
};
```

The *double* attribute of the *entrance* process is the mean interarrival time[1] between a pair of cars consecutively arriving at the wash. That attribute is set by the *setup* method of the process which plays a role similar to a standard C++ constructor (in that it is invoked automatically when the process is created, Sect. 2.1.4). The process starts in its first state, *NothingHappening*, where it delays for a random amount of time before transiting to *CarArriving*. As all simulators, SMURPH comes equipped with functions for generating pseudo-random numbers drawn from various useful distributions. For example, *dRndPoisson* generates a *double* pseudo-random number according to the Poisson (exponential) distribution [3] with the mean specified as the function's argument.

Having awakened in state *CarArriving*, the process generates another pseudo-random number to determine the type of service for the car. The uniformly distributed real value between 0 and 1 is compared against this table declared earlier in the program file:

---

[1] The reader may be concerned that having argued at length for a discrete, integer representation of time (in Sect. 1.2.3) we now use a floating point (*double*) variable to store something that evidently looks like a time interval. In Sect. 2.1.4 we explain why this is OK.

const double service_percentage [ ] = { 0.8, 0.92, 1.00};

where the three numbers stand for the cumulative probabilities of the three service types: *standard*, *extra*, and *deluxe*. The way to read them is that 80% of all cars ask for standard service, $92 - 80 = 12\%$ of the cars ask for extra service, and the remaining 8% opt for deluxe service. A car is represented by this C++ class:

```
class car_t {
        public:
        int service;
        car_t (int s) {
                service = s;
        };
};
```

whose single attribute describes the service type requested. Now we are ready to have another look at the *service_time* method of the *washer* process (Sect. 2.1.1) which generates the randomized service time for the car. That is accomplished by invoking yet another built-in random number generator of SMURPH, *dRndTolerance*, taking three arguments: the minimum, the maximum, and something interpreted as the distribution's "quality" (see Sect. 3.2.1 for details). The first two arguments are extracted from this array:

```
typedef struct {
        double min, max;
} service_time_bounds_t;
. . .
const service_time_bounds_t service_time_bounds [ ] = {
        { 4.0,  6.0 },
        { 6.0,  9.0 },
        { 9.0, 12.0 }
};
```

consisting of three pairs of numbers describing the bounds of the service time for the three service types. We may suspect that those bounds are in minutes. The actual randomized service time, as generated by *dRndTolerance*, is a β-distributed number guaranteed to fall between the bounds (see Sect. 3.2.1).

The car queue is described by a SMURPH *mailbox* (Sect. 9.2) which is an object of a special class declared as follows:

```
mailbox car_queue_t (car_t*) {
        void setup ( ) {
                setLimit (MAX_Long);
        };
};
. . .
car_queue_t *the_lineup;
```

Similar to a SMURPH *process* type declaration, a *mailbox* declaration, like the one above, defines a (special) class type that can be used as the basis for creating objects (instances). Note that *the_lineup*, as declared above, is merely a pointer and the actual object must be created elsewhere (Sect. 2.1.4).

In a nutshell, a mailbox is a queue of objects of a simple type specified in the first line of the declaration. What makes it different from a straightforward list is the ability to accept *wait* requests from processes and trigger events. Objects with this property are called *activity interpreters* (Sect. 5.1.1). In our present context, when a car is put into *the_lineup* (at state *CarArriving* of the *entrance* process), the *NONEMPTY* event awaited by the *washer* process on the same mailbox will be triggered, and the *washer* process will be awakened.

2.1.3   Completing the Program

A SMURPH model must be built before it can be run. The act of building is usually quite dynamic (involving execution of SMURPH code), because most systems are naturally parameterizable by input data; thus, many details in the program are flexible. For example, in a network model, the number of nodes, the number of their interconnecting channels, their geographical distribution, and so on, can be (and usually are) flexible parameters passed to the model in the input data. From this angle, the car wash model looks pathetically static. But even in this case there is a healthy chunk of dynamic initialization to take care of.

For one thing, regardless of any other initialization issues, all processes in the model must be explicitly created. On the other hand, as everything that executes any code in the simulator must be a process, the previous sentence cannot be entirely true (if it were, there would be no way to start). Therefore, one special process is created automatically at the beginning. This is akin to the *main* function in a standard (C, or C++) program, which must be called implicitly (by the system) to start things running. This special process in SMURPH is called *Root*. Any other process of the model will be created from the code run by *Root* or, possibly, by its descendants.

Here is how the *Root* process is declared in our car wash model (for now, we focus on the structure, so the details are skipped):

```
process Root {
        states { Start, Stop };
        perform {
                state Start:
                        // Read input data, initialize things

                        ...

                        Kernel->wait (DEATH, Stop);
                state Stop:
                        // Write output results

                        ...

                        Kernel->terminate ( );
        };
};
```

Of course, what we have above is a process *class* declaration, so when we say that the *Root* process is automatically created on startup, we really mean that an *object* of the above type is instantiated. This is exactly what happens: a single instance of the *Root* process class is created and, as is usual for a newly created process, run at its first state.

Most *Root* processes (at least in true SMURPH simulators) have the same general structure consisting of two states: the first one representing the action to be carried out at the beginning of simulation (to start things up), the other to finalize things upon termination. The termination state is only needed, if the run ever ends, i.e., the program is in fact a simulator expected to produce some "final" results. In such a case, there is usually some termination condition, i.e., a formula prescribing when SMURPH should conclude that the experiment has reached its goal. The termination condition can be awaited by *Root* (or by any other process) as the *DEATH* event on the *Kernel* object (Sect. 5.1.8). In response to that event (in state *Stop*), the *Root* process will produce some output and terminate the whole thing (exit the simulator). The last action is officially accomplished by terminating the *Kernel* object.

Sometimes a third state is inserted in between *Start* and *Stop*. In serious simulation studies we often worry about the model reaching a steady state, i.e., a stable behavior, following its initialization, at which point it makes sense to begin collecting performance data. For a network, this typically means that the nodes have created enough traffic to spread into the channels, fill in the buffers, queues, and so on, and any changes in the system state that we are going to see from this time on will be representative of its "normal" behavior in the middle of some typical "production" day. To that end, the third state of *Root* will be triggered by some condition (e.g., a timer) and its action will be to initialize the procedure of collecting the performance measures (Sect. 10.2.4).

2.1.4 About Time

In the *washer* process (Sect. 2.1.1), the service time for a car is generated as a pseudo-random number, based on a set of prescribed bounds. Looking at those bounds, we guessed that they were expressed in minutes. How can we know for sure what they mean?

Internally, at the very bottom, when events are stored in the event queue and their timing is determined, SMURPH represents the time stamps of events as integer numbers. It doesn't care what those numbers correspond to in terms of real time. It is up to the model's creator to assign a meaning to them, and, at least in the "normal" simulation mode (Sect. 1.3.1), that meaning need not leak beyond the creator's mind.

In any case, the internal unit of time inside the simulator is fixed and indivisible. It is called an *ITU* (for Indivisible Time Unit). Its correspondence to some specific interval of real time is usually relevant, but, as we said above, not important to the simulator. For example, 1 *ITU* may correspond to 1 ns in the model of some telecommunication network. If we assume (just assume, without changing anything in the model) that instead of 1 ns, the *ITU* corresponds to 10 ns, that would be equivalent to re-mapping the model to a slightly different network where all distances have been multiplied by 10 and all bit rates have been divided by the same number. In most cases, the *ITU* is chosen from the viewpoint of the model's accuracy: we want it to be fine enough, so nothing important is lost. This seldom coincides with naturalness and convenience from the viewpoint of everyday's units, like seconds, hours, days. In a telecommunication network, the *ITU* tends to be comparable to a single-bit insertion time or the signal propagation time across some minimum relevant distance (say 10 cm). If the network is heterogeneous and the model is meant to be accurate, the *ITU* will tend to be even finer (e.g., corresponding to the greatest common divisor of several low-level units popping up in the system). Notably, there is little penalty in SMURPH for making the *ITU* extremely fine, so one is inclined to be generous, just in case.

This kind of detachment of the *ITU* from natural units calls for a parameter that would map the simulator's time to those natural units for the sake of convenience in specifying the input data and interpreting the output results. This brings in a second unit, called *ETU*, for Experimenter Time Unit. About the first thing that needs to be set up before starting the model is the correspondence between the two units. Here is the first state of the *Root* process from the car wash model:

state Start:

    double d;

    setEtu (60.0);

    readIn (d);

    the_lineup = create car_queue_t ( );

    create entrance (d);

    create washer ( );

    readIn (d);

    setLimit (0, d);

    Kernel->wait (DEATH, Stop);

The first statement executed in the above sequence, *setEtu (60.0)*, sets up the conversion factor between the two units. In this case, it says that 1 *ETU* corresponds to 60 *ITU*s. Note that we still don't know what those units are, but we agree to express (some) data and view (some) results in units that are multiples of 60 internal units of the simulator. For example, we may want to assume that 1 *ITU* corresponds to 1 s and 1 *ETU* represents 1 min. This may not appear as a tremendous advantage for our car wash model, but the simple feature is certainly useful when the *ITU* is not as natural as it happens to be in our present simple setup.

By default, if *setEtu* is never called, 1 *ITU* is equal to 1 *ETU*. But even then, the two units are different in a sense, because different numerical types are used to represent intervals expressed in *ITU*s and in *ETU*s. While the former are nonnegative integer numbers (of the appropriately large range), the latter are *double* (floating point) values. Internally, the simulator must keep time at a fixed grain (see Sect. 1.2.3), but for its presentation to the user (experimenter) the so-called "scientific" notation is more meaningful and appropriate.

With a fine time grain, the absolute time in a non-trivial model can legitimately grow to a huge (or humongous) integer number; however, most time intervals (delays) can still be accurately representable with much smaller values. So it may be OK (and occasionally more convenient) to specify them in *ETU*s, or still in *ITU*s, but as floating point (*double*) numbers. This appears especially congenial when such delays are generated as (pseudo) random numbers, which naturally come from distributions formally described using floating point arithmetics. In such cases, we understand that the values are not cumulative, usually quite well bounded, and the problems discussed in Sect. 1.2.3 do not apply to them.

At any moment during the model's execution the amount of simulated time that has passed since the execution was started can be told by looking at two global variables named *Time* and *ETime*. The first one expresses it in *ITU*s, the other in *ETU*s.

Two more numerical parameters are set by *Root* in its first state. When the *entrance* process is created (the first *create* operation), it is passed a *double* value read from the input data file (the first call to *readIn*). As we remember, that number determines the mean interarrival time between cars (Sect. 2.1.2). That time is expressed in *ETU*s, i.e., in minutes in our present case. The last *time* parameter, also expressed in *ETU*s, is the time limit, i.e., the virtual-time duration of the simulation experiment. It is one of the simplest and often most natural termination condition for an experiment: the *DEATH* event on *Kernel* (Sect. 2.1.3) will be triggered when the simulated time (as expressed in *ETUs*) reaches or exceeds the specified bound. See Sect. 10.5 for a detailed description of *setLimit*.

2.1.5 Presenting the Results

The performance measures collected in our model amount to two numbers: the total service time and the total number of processed cars. Those numbers are output in the second state of *Root* which looks as follows:

```
state Stop:
        print (TotalServiceTime, "Busy time:", 10, 26);
        print ((TotalServiceTime / ETime) * 100.0, "Normalized throughput:", 10, 26);
        print (NumberOfProcessedCars, "Cars washed:", 10, 26);
        print (the_lineup->getCount ( ), "Cars queued:", 10, 26);
        Kernel->terminate ( );
```

The detailed meaning of the arguments of *print* is not important at this stage (it will be explained in Sect. 3.2.2). It is rather obvious what we want to accomplish with the above sequence of commands. In addition to the total service time (in *ETUs*, i.e., in minutes), we also show the percentage of service time in total time, which we call the "normalized throughput" (terminology resulting from our addiction to simulating networks). We also take the final snapshot of the car queue, taking advantage of a standard method of the SMURPH *mailbox* class (Sect. 9.2.3).

2.1.6 Running the Model

One little detail that we still need to account for to make our program complete is the declaration of three global variables:

```
car_queue_t *the_lineup;
double TotalServiceTime = 0.0;
Long NumberOfProcessedCars = 0;
```

The initialization of the two simple performance measures is superfluous (this is how uninitialized global variables are automatically initialized in C++), but it emphasizes their role as accumulating counters.

The model accepts a simple input data set consisting of two numbers (see the first state of *Root* in Sect. 2.1.4). Directory *Examples/CarWash* of the package contains the model (file *carwash.cc*) and a sample data set (file *data.txt*) with these contents:

Interarrival time   10.0 min
Simulated time   10080 min = one week

Any non-numbers are ignored by the *readIn* function used to input the data to the model (see Sect. 3.2.2) and treated as comments. The data set amounts to two values: the mean car interarrival time (10 min) and the limit on simulated time (10,080 min).

To compile the simulator, execute this program (see Sect. B.6) in the example's directory:

mks

After a few seconds the model will be compiled into an executable program and put into file *side* (*side.exe* on Windows). Run it this way:

./side data.txt output.txt

After a second or so the program will stop having simulated a week of operation of our car wash. Here is what is written to the output file (*output.txt*):

SMURPH Version 3.26-OA   Car Wash   Mon May 4 20:10:24 2019

Call arguments: data.txt output.txt

Busy time:	5736.8
Normalized throughput:	56.899
Cars washed:	999
Cars queued:	0

@@@ End of output

The performance measures returned by the simulator can be quickly verified for sanity. Of course, the simple model can be easily analyzed by hand and the simulation experiment is not extremely interesting to a car wash designer. While there is little to learn from it about car washes, it does illustrate a few points about SMURPH.

Within about 10,080 min of activity the car wash has serviced 999 cars. This is more or less what we would expect at the mean spacing (interarrival time) of 10 min between cars. The exact expectancy is 1008, so the figure is a bit on the low side, but well within the reasonable statistical error margin (less than 1%). When we increase the simulated time to 1000 weeks (by appending three zeros to the time limit), the execution time of the simulator increases slightly (to about three seconds on my laptop) yielding 1,008,616 cars, with less than 0.1% deviation from the exact expected value.

Needless to say, this is all nonsense from the viewpoint of any realistic setup. Even if the car arrival process can be sensibly modeled by the Poisson distribution at some local time scales, the arrival rate is obviously not the same at all moments. For one thing, the car wash is probably going to be closed at some hours. Besides, there are holidays. Even if the car wash is going to be open 24/7, the arrival rate at 3 a.m. on a Sunday is probably going to be drastically different from that on a sunny Friday afternoon before a long weekend and following a thaw. Obviously, the most interesting aspect of the model would be a car arrival process well capturing all those little details. And a good model for weather prediction would probably be of assistance as well.

Let us have a look at *busy time* and *normalized throughput*. Can we calculate these numbers by hand, without running the model? Of course, we can. There are three types of service (Sect. 2.1.2) with the mean service times 5, 7.5, and 10.5 min. Their probabilities are, respectively, 0.8, 0.12, and 0.08. Thus, the average service time of a car is $5 \times 0.8 + 7.5 \times 0.12 + 10.5 \times 0.08 = 5.74$ min. Thus, every 10 min the car wash gets enough work to last for 5.74 min, so the expected figure for normalized throughput is 0.574.

We see that there are no cars queued at the wash at the end of simulation. No wonder, because the car wash is somewhat underutilized (operating at less than 60% of its capacity). We can push the load by reducing the mean interarrival time. Setting it to 5.74 min, i.e., to match the average service time (with the original time limit of one week) yields these numbers:

Busy time   *9858.3*
Normalized throughput   *97.77*
Cars washed   *1724*
Cars queued   *15*

This time we do see some lineup, and the car is utilized close to 100%. Again, it is probably not very realistic to see that many cars lined up for a car wash, but this is not the least realistic piece of our naive model.

2.1.7   In Real Time

It is easy to run the model in visualization mode (Sect. 1.3.1). There will be no problem catching up with the virtual time to the real time, because the model is simple, and the computation effort of the simulator is truly minuscule: running the model for 1000 weeks (19 years) took my (rather pathetic) laptop less than 3 s.

The visualization mode is primarily intended for cooperation with some GUI, so the model can be truly *visualized*. In our simple case, we will content ourselves with printing some lines of text to the output file. Suppose that we want to see these events:

- a new car arriving at the wash
- service commencement for a car
- service termination

To be able to trace the individual cars as they are queued and then serviced, we need to identify them somehow. To this end, we may want to add an attribute to the *car_t* class (Sect. 2.1.2), e.g., assigning a ticket number to every car entering the wash. The first interesting event (car arrival) is marked by state *CarArriving* in the *entrance* process (Sect. 2.1.2). In that state, a new *car_t* object is created and *put* into the queue (mailbox) identified by variable *the_lineup*. Thus, it makes sense to print the arrival message from the *car_t* constructor. Here is how the updated version of the class may look:

```
int Ticket = 0;
...
class car_t {
        public:
        int service, number;
        car_t (int s) {
                service = s;
                number = Ticket++;
                trace ("car %1d arriving for %s service", number, service_name [service]);
        };
};
```

The standard SMURPH function *trace* is intended for debugging (see Sect. 11.1), but it will nicely do for our little extension of the model. Its convenient feature is that it automatically shows the current virtual time (in *ITUs*). To make the service type more legible in the printed lines, the *service_name* array maps the integer enumeration ordinals into strings:

const char *service_name [ ] = { "standard", "extra", "deluxe"};

The remaining two events are identified in the washer process whose last two states have been modified as follows:

```
transient Washing:
        double st;
        st = service_time (the_car->service);
        trace ("car %1d service %s start", the_car->number, service_name [the_car->service]);
        Timer->delay (st, DoneWashing);
        TotalServiceTime += st;
        NumberOfProcessedCars++;
state DoneWashing:
        trace ("car %1d service %s end", the_car->number, service_name [the_car->service]);
        delete the_car;
        sameas WaitingForCar;
```

i.e., by inserting two more invocations of *trace*. When we recompile and run the modified model, we will see in the output file lines like these:

```
...
Time: 22 car 0 arriving for standard service
Time: 22 car 0 service standard start
Time: 186 car 1 arriving for standard service
Time: 326 car 0 service standard end
Time: 326 car 1 service standard start
```

Time: 620 car 1 service standard end
Time: 1564 car 2 arriving for standard service
Time: 1564 car 2 service standard start
Time: 1886 car 2 service standard end
Time: 2489 car 3 arriving for standard service
Time: 2489 car 3 service standard start
Time: 2823 car 3 service standard end
...

and so on. Note that time is expressed in seconds. All those lines will be written almost instantaneously, because we haven't changed the execution mode of the simulator. To run the model in real time, we must do these additional tweaks:

1. add this statement:

   setResync (1000, 1.0/60.0);

   into the *Start* state of *Root* (Sect. 2.1.4), right after *setEtu (60.0);*
2. recompile the model using this command:

   mks −W

   to enable the visualization mode
3. run the model directing the output to the terminal, so the events are immediately seen as they happen:

   ./side data.txt

We must wait patiently, because now the model pretends to run the car wash under a real-life timing. The output file header shows up immediately, but it takes a while for the first event line to materialize: the first car arrives after 22 s (at least on my system with the standard seeding of the random number generator). This is rather a lucky fluke, because (with the original data set, Sect. 2.1.6) the expected waiting time for a car is 10 min.

Enabling the visualization mode consists of two steps. The simulator must explicitly declare the real-time duration of its time unit (by calling *setResync*). As we said in Sect. 2.1.4, the time unit used by the simulator is purely virtual and its meaning is solely in the experimenter's head. The function (which is described in more detail in Sect. 5.3.4) fulfills a dual purpose. Its first argument is the real-time length (in milliseconds) of the so called *resync interval*, i.e., the window of real time determining the grain of synchronization. Intuitively, at the boundary of this window the simulator will wait until its virtual time has caught up with real time. Note that any events occurring within the window may not be visualized correctly. The second argument tells how many *ETUs* correspond to the resync interval. In our case, the resync interval is set to 1000 ms, i.e., 1 s, and 1 *ETU* = 60 *ITUs*; thus, 1 s of real time is set to correspond to 1 *ITU* in the model.

The second step is to recompile the model with –W which enables the code for visualization. That code is not compiled in by default to avoiding unnecessarily slowing down the simulator by checks that will never evaluate to true. The *setResync* function is only available, if the simulator has been compiled with –W.

Looking at the car wash model operating in this kind of visualization mode may become boring after a while: it takes long minutes to see any progress at all. We can easily speed up the model, e.g., by changing the second argument of *setResync* to 1.0 (so one second of real time corresponds to one minute in the model). This is one of the rare cases when a model visualization must be sped up to become entertaining.

2.2   The Alternating-Bit Protocol

For the next example, we shall look at something closer to what SMURPH was designed for, namely, a simple communication network. The network consists of two nodes connected via two separate channels, as shown in Fig. 2.1. The functional responsibility of our network is to send data, continuously, in one direction, i.e., from the *sender* to the *recipient*. One of the two channels is used by the sender to send the actual data packets. The other channel is used by the recipient to send acknowledgment packets indicating that the data packets have been received. As in all telecommunication setups, we worry about the correct reception of data in the face of faulty channels, interference, and any kinds of errors that the people involved in telecommunication are well familiar with.

We assume that the sender has a continuous supply of data packets to be sent to the other party. They do not have to be all available immediately: the node gets them from some "application" that creates them as part of some stream of data. The very same stream must arrive (and be properly recognized) at the recipient, even though some packets may be lost (or received erroneously, which basically means the same thing). Both channels are faulty. Whichever way a channel can fail, it can only damage a packet in transit. Such a packet may fail to be recognized by the recipient, or it may be recognized as a packet, but its contents will be damaged. The latter situation is detected by the recipient via a checksum (or a CRC code), so we may assume that when a packet is formally *received*, then we know that it has been received correctly. Thus, the problem is basically reduced to dealing with lost packets. The same applies to the acknowledgment packets with the roles of sender and recipient reversed.

Our channels are simple, e.g., implemented as cables or optical fibers. For example, they cannot misorder packets, i.e., absorb several packets from the sender and

Fig. 2.1 A simple one-way communication setup

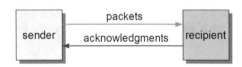

present them to the recipient as correct packets arriving in an order different from that in which they were sent. Also, they cannot create packets that were never sent. I mention this, because some theoretical problems in reliable telecommunication deal with advanced malicious scenarios like those.

Our goal is to implement a reliable communication scheme for the two nodes, so the recipient can be sure to be receiving the same stream of data as that being transmitted by the sender, with the simplest possible means. For every packet that the sender transmits and considers done, it must also make sure that the packet has made it to the other party in a good shape. The sender wants to be sure of that before dispatching the next packet.

The example can be found in directory *AB* in *Examples*. The entire program has been put into a single file named *prot.cc*. A sample data file, *data.txt*, is included.

2.2.1   The Protocol

At first sight, the problem doesn't look too complicated. The sender transmits a packet and waits until it receives an acknowledgment from the recipient. If the acknowledgment doesn't arrive within some time interval, the sender will conclude that the packet hasn't made it and will try again. Only when the current packet has been acknowledged can the sender discard it and take care of the next packet. Note, however, that acknowledgments are also packets, and they can be lost. Thus, confusing scenarios are possible. For example:

1. The sender sends its packet, the packet arrives at the recipient and is received correctly.
2. The recipient sends an acknowledgment, the acknowledgement is damaged and doesn't make it to the sender.
3. After the timeout, the sender concludes that its packet has been lost, so it retransmits the packet which arrives undamaged at the recipient and is received as a new packet.

Note that without some special markers, the recipient will not be able to tell a new packet from the last-received one. Thus, it may erroneously insert the same packet twice into the received stream of data passed to the application. An obvious way to solve this problem is to tag packets dispatched by the sender with serial numbers. In our simple setup we don't really need full-fledged numbers. As the sender is retransmitting the same packet until it receives an acknowledgment, we never get into a situation where two retransmissions of the same packet are separated by a different packet. Thus, the recipient must only be able to tell "new" from "old." That information is completely conveyed in a modulo-2, single-bit "counter" which alternates with every new packet to be transmitted.

Note that the recipient should apply a similar scheme to acknowledgments. First, if it receives no new data packet for a while, it may assume that its last acknowledgment has been lost, so it can re-send the previous acknowledgment, to make sure it has been conveyed to the other party. We can easily see that acknowledgment packets must also be tagged with alternating bits referring to the packets they acknowledge. Otherwise, the following scenario would be possible:

1. The sender transmits a packet, call it p_0 (assume that its alternating bit is zero) which is correctly received by the recipient. The recipient sends an acknowledgment which makes it to the sender.
2. The sender assumes (quite correctly) that its last packet has been has been received and transmits the next packet p_1.
3. Packet p_1 is lost. The recipient sends another acknowledgment for p_0 (after a timeout). That acknowledgment arrives at the sender in a good shape.
4. The sender concludes that p_1 has been acknowledged and proceeds with the next packet p_2.

As we can see, detecting duplicate acknowledgments is no less important than detecting duplicate packets. We are now ready to formulate the protocol in plain language.

The sender

The node maintains a binary flag denoted *LastSent* reflecting the contents of the alternating bit in the current outgoing packet. The flag is initialized to zero.

1. When the node acquires a new packet for transmission, it inserts the contents of *LastSent* into the alternating-bit field of the packet. Having completed the transmission, the node sets up a timer and starts waiting for an acknowledgment.
2. Upon the arrival of an acknowledgment, the node looks at its alternating-bit field. If it matches *LastSent*, then the current packet is assumed to have been acknowledged. *LastSent* is flipped and the node tries to acquire a new packet for transmission. Otherwise, the acknowledgment is ignored, and the node continues waiting.
3. If the timer goes off, the node retransmits the current packet and continues waiting for its acknowledgment.

The recipient

The node maintains a binary flag denoted *Expected* indicating the contents of the alternating bit in the expected packet. The flag is initialized to zero.

1. The node sets up a timer while waiting for a data packet from the sender.
2. If the timer goes off, the recipient sends an acknowledgment packet with the alternating-bit field set to the reverse of *Expected*. Then it continues at 1.
3. When a data packet arrives from the sender, the node checks whether the alternating bit of that packet matches *Expected*. If this is the case, the packet is received and passed to the application. The recipient immediately sends an acknowledgment packet with the alternating bit set to *Expected*. Then *Expected* is flipped and the node continues from 1.

The above protocol is known in the literature as the Alternating Bit protocol. It was formally proposed in 1968 [4, 5].

2.2.2 Stations and Packet Buffers

We want to implement the simple network shown in Fig. 2.1, as well as the protocol described in Sect. 2.2.1, in SMURPH. One difference in comparison to the car wash from Sect. 2.1 is that a network naturally consists of number of somewhat independent components, e.g., corresponding to the different nodes. For example, it is quite natural for two or more nodes to run the same program (or similar programs, understood as collections of SMURPH processes) and it is thus natural to see the node as an encapsulation for a bunch of related activities. In other words, we would like to perceive the node as a virtual computer running the set of SMURPH processes describing the node's behavior. The car wash, in contrast, appeared to us as a single place where all the interesting activities were happening, so we did not have to worry about the encapsulation hierarchy for those activities.

All the network components of a SMURPH model, as well as some components of the network environment, must be built explicitly, which usually happens at the very beginning, i.e., in the first state of *Root* (Sect. 2.1.3). The network backbone is described by a configuration of objects of some standard types: *station*, *port*, *link*, *transceiver*, and *rfchannel*. We can view them as the basic building blocks of the virtual hardware on which the protocol program will be run. For the alternating-bit example, we need the first three items from this list: the remaining two are useful in models of wireless networks.

Three more built-in SMURPH types are needed to represent some soft components of the model: *traffic*, *message*, and *packet*. The last one is intuitively clear. The first two take part in modeling the network's "application" or "user."

A *station* object represents a network node, called a station in the SMURPH lingo. Here we have the type of the sender station:

```
station SenderType {
        PacketType PacketBuffer;
        Port *IncomingPort, *OutgoingPort;
        Mailbox *AlertMailbox;
        int LastSent;
        void setup ( );
};
```

A *SenderType* object consists of a packet buffer, holding the next outgoing packet, and two ports interfacing the station to the two channels. The packet buffer is a class variable (of type *PacketType*), while *IncomingPort* and *OutgoingPort* are pointers, suggesting that the respective actual objects will be created dynamically. Along with the mailbox pointed to by *AlertMailbox* they are created by the station's *setup* method (which we will see a bit later).

As we explained in Sect. 2.1.1, a built-in SMURPH class type comes with a declarator keyword used to declare extensions of that type, e.g., *process*, *station*, *mailbox*, and an actual class type name representing generic objects of that type, e.g., *Process, Station, Mailbox*. Thus, *Port* and *Mailbox* refer to the generic types for SMURPH ports and mailboxes.

PacketType is not a built-in type, but one derived from the built-in type *Packet* in the following way:

```
packet PacketType {
        int SequenceBit;
};
```

On top of the standard set of attributes of the built-in type, *PacketType* declares a single additional attribute representing the alternating bit carried by the packet. This is the protocol-specific extension of the standard *Packet* type.

The mailbox pointed to by *AlertMailbox* will provide an interface between the processes run by the station. As we shall see, the sender's part of the alternating-bit protocol will be implemented by two processes. These processes will synchronize by exchanging notifications via the mailbox. The role of *LastSent* is clear from the protocol description in Sect. 2.2.1.

The recipient station is very similar:

```
station RecipientType {
        AckType AckBuffer;
        Port *IncomingPort, *OutgoingPort;
        Mailbox *AlertMailbox;
        int Expected;
        void setup ( );
};
```

with two minor differences: the packet buffer has been replaced by an acknowledgment buffer, and the integer variable has been renamed to *Expected*. The first of the two differences is rather illusory, because *AckType* is declared as:

```
packet AckType {
        int SequenceBit;
};
```

i.e., in the same way as *PacketType*. Why do we need the separate declarations? Well, we just want to make a formal distinction between the data and acknowledgment packets (basically as a comment) without introducing unnecessary complication beyond the minimum needed by the scheme. Note that in a real network there probably would be bits or codes in the packet headers identifying packets of different types. Such an attribute, a packet type, is in fact a standard component of the built-in type *Packet* (Sect. 6.2), so a distinction is possible, but the protocol doesn't need it, because only one type of packets can be sent on a given channel.

One can claim that in a blown-up implementation of the protocol, the two station types in fact should look identical. The traffic between a pair of connected nodes is typically bidirectional. In such a case, it still might be better to retain the two separate station types and build a third type combining their functionality (see Sect. 3.3.4).

2.2.3 The Sender's Protocol

The program executed by the sender (a station of class *SenderType*) consists of two processes, each process taking care of one port. This idea can be recommended as a rule of thumb wherever the division of the protocol responsibilities among processes is not obvious. In our case, one of the processes, called the *transmitter*, will be responsible for transmitting packets to the recipient, while the other process, the acknowledgment receiver, will be handling the recipient's acknowledgments and notifying the transmitter about the relevant ones. The type declaration for the transmitter process is as follows:

```
process TransmitterType (SenderType) {
        Port *Channel;
        PacketType *Buffer;
        Mailbox *Alert;
        TIME Timeout;
        states { NextPacket, Retransmit, EndXmit, Acked };
        void setup (TIME);
        perform;
};
```

Note a new syntactic feature in the first line of the above declaration: the process type name is followed by a station type name enclosed in parentheses. We didn't do anything like that in Sect. 2.1, for the car wash model, because we needed no stations there. The station type name mentioned in a process type declaration indicates the type of the station at which the process will be running. More precisely, we say that all instances of the declared process type will be running at stations belonging to the specified station type. Here comes another blanket rule: every process in SMURPH runs in the context of some station. This also means that while any piece of code belonging to the model is always run as part of some process, it is also run within the context of some specific station.

So, what was going on in the car wash model where no stations were mentioned at all? If not indicated otherwise, a newly created process is associated with one special, built-in station, which always exists and can be referenced via the global pointer variable *System*. The station is called *the system station*, and its actual class type is not directly visible (and not relevant). Since that that type is not extensible and offers no *public* attributes[2] (plus the system station occurs in exactly one copy), it is never needed and can be assumed to be just *Station*.

On the other hand, the system station happens to be quite important, because it is a natural placeholder for running all those processes that don't naturally belong to regular stations. It did save us tacitly in the car wash case, because, owing to its quiet

---

[2]Except its specific exposure methods (see Sect. 12.5.12).

existence, we didn't have to talk about stations while building the model. But even in network models, there often are special processes (operating behind the scenes), like custom traffic generators, performance data collectors, mobility drivers (in wireless networks) that most comfortably belong to the system station.

Here is the *setup* method of the transmitter process:

```
void TransmitterType::setup (TIME tmout) {
        Channel = S->OutgoingPort;
        Buffer = &(S->PacketBuffer);
        Alert = S->AlertMailbox;
        Timeout = tmout;
};
```

which sets up its attributes. *Channel*, *Buffer*, and *Alert* are set to point to the objects belonging to the station running the process. Variable *S*, acting as an implicit process attribute, refers to that station. Its type is "a pointer to the station type indicated with the first line of the process type declaration." Thus, the above *setup* method copies the pointers *OutgoingPort* and *PacketBuffer* from the station running the process and sets up a local pointer to the packet buffer declared (as a class variable) at the station. We can guess that the process wants to have personal (direct) attributes to reference those objects (instead of having to apply remote access, via *S->*, at every reference).

The transmission end of the protocol is described by the code method of *TransmitterType* which looks like this:

```
TransmitterType::perform {
        state NextPacket:
                if (!Client->getPacket (Buffer, MinPktLength, MaxPktLength, HeaderLength)) {
                        Client->wait (ARRIVAL, NextPacket);
                        sleep;
                }
                Buffer->SequenceBit = S->LastSent;
        transient Retransmit:
                Channel->transmit (Buffer, EndXmit);
        state EndXmit:
                Channel->stop ( );
                Alert->wait (RECEIVE, Acked);
                Timer->wait (Timeout, Retransmit);
        state Acked:
                Buffer->release ( );
                S->LastSent = 1 — S->LastSent;
                proceed NextPacket;
};
```

The code method is partitioned into four states. Immediately after the process is created, it enters the first state labeled *NextPacket*. In this state, the process attempts to acquire a packet for transmission and store it in the station's packet buffer (pointed to by the process's attribute *Buffer*). *Client* is a (global) pointer to an external agent maintained by the SMURPH kernel whose role is to simulate the network's application (or user). It acts as the supplier of data packets (the session) for transmission. From this angle, *Client* is visible to the station as a queue of *messages* awaiting transmission and dynamically filled according to the prescribed, automatic, external from the viewpoint of the protocol, arrival scheme.

The *getPacket* method of the *Client* examines the message queue at the *current* station. Note that the notion of "current station" is always relevant in SMURPH: it means "the station that owns the process running the current piece of code." If *getPacket* fails, returning *false*, which means that there is nothing to send, the process executes the *wait* method of the *Client*. As we remember from Sect. 2.1.1, this is the standard way to wait for some event for which the method's owner is responsible. We can easily deduce that the *ARRIVAL* event consists in a new message showing up in the station's queue. When that happens, the process wants to be resumed in its first state, *NextPacket*, to have a next go at packet acquisition.

If *getPacket* succeeds, meaning there is a message queued at the station for transmission, it fills the packet buffer (passed to it as the first argument) with a new packet to transmit. The process then sets the alternating bit (aka *SequenceBit*) of the packet to *LastSent* and falls through to state *Retransmit*. There it initiates the packet's transmission by calling the *transmit* method of the station's output port (pointed to by *Channel*).

The *transmit* method accepts two arguments. The first one points to the packet buffer containing the packet to be transmitted. The second argument identifies the process's state to be assumed when the packet has been transmitted. Thus, the transmitter will be resumed in state *EndXmit* when the packet's transmission is complete. Note that the transmitter is supposed to explicitly terminate the transmission by invoking the *stop* method of the port (*Channel*). The way we should look at it is that the *transmit* method fires up the transmitter hardware, which will now strobe the packet's bits onto the channel at the prescribed rate, and (at the same time) sets up an alarm clock for the time needed to process all the packet's bits. When the timer goes off, the transmitter hardware is switched off with the *stop* method.

According to the protocol, having completed a packet transmission, the transmitter will wait for an acknowledgment from the recipient. This is accomplished by invoking the *wait* method of the station's mailbox pointed to by the process's *Alert* attribute. Acknowledgments are received by the other process running at the station (the acknowledgment receiver) which will deposit an alert into the mailbox whenever an ACK packet with the right value of the alternating bit has been received. The transmitter will be resumed in state *Acked* when the mailbox becomes nonempty, which moment is marked by the *RECEIVE* event triggered on the mailbox.

While expecting the ACK alert, the transmitter process is also waiting for a time-out. We are already familiar with the *Timer* object from Sect. 2.1.1, where its *delay* method was used to set up an alarm clock for the specified number of *ETU*s. Here, the process takes advantage of another (more elementary) method of the *Timer*, *wait*, whose first argument is interpreted as the delay interval in *ITU*s. Thus, *delay* and *wait* do essentially the same job, except that the former method accepts the time interval in *ETU*s, while the latter in *ITU*s(Sect. 5.3.1).

Note that only one of the two events awaited by the process in state *EndXmit* will wake it up, so the process will continue either in state *Ack* or *Retransmit*. When both events occur within the same *ITU*, which is not impossible owing to the discrete nature of time, SMURPH will pick one of them at random as the one to which the process will in fact respond. This agrees with the interpretation of the *ITU* as the time quantum: when two events are separated by less than the quantum, then it may become impossible to order them chronologically. In such a case, the simulator should statistically explore both options (execution paths) of running the actions triggered by them (see Sect. 1.2.4).

If the timer goes off before the acknowledgment has been received, the process will find itself in state *Retransmit*, where it transmits the same packet again. In state *Acked*, where the transmitter gets upon the reception of the acknowledgment, it releases the packet buffer, flips the *LastSent* bit, and moves to state *NextPacket*, to acquire a new packet for transmission. The action of releasing the packet buffer (the *release* method of *Packet*) can be viewed as emptying the buffer and making it ready to accommodate a new packet.

Here is the class type of the second process run by the station:

```
process AckReceiverType (SenderType) {
        Port *Channel;
        Mailbox *Alert;
        states { WaitAck, AckBegin, AckArrival };
        void setup ( );
        perform;
};
```

This process is simpler, because it only receives acknowledgments on its port (*Channel*) and notifies the other process of their arrival. The *setup* method:

```
void AckReceiverType::setup ( ) {
        Channel = S->IncomingPort;
        Alert = S->AlertMailbox;
};
```

just copies two pointers from the station to the process's private attributes for easier access. The code method looks like this:

```
AckReceiverType::perform {
        state WaitAck:
                Channel->wait (BMP, AckBegin);
        state AckBegin:
                Channel->wait (EMP, AckArrival);
                Channel->wait (SILENCE, WaitAck);
        state AckArrival:
                if (((AckType*)ThePacket)->SequenceBit == S->LastSent) Alert->put ( );
                skipto WaitAck;
};
```

In its first state, or at the top of its event loop, the process issues a wait request to its port for the *BMP* event. The acronym stands for "Beginning of My Packet" and marks the moment when the port begins to receive a packet addressed to *this* station, i.e., the station running the process. When that happens, the process transits to state *AckBegin* where it awaits two possible events that can ensue. One possibility is that the packet will be received completely, to the very end, thus triggering the *EMP* event (End of My Packet). That would mean that the acknowledgment packet has been received without bit errors. The other possible outcome is that the packet will dissolve into silence, i.e., it will end without its terminating mark being recognized as such. In our present model, that would mean that there were errors in the packet's reception.

We need to digress into two issues to make the meaning of the two *wait* requests in state *AckBegin* clear. As the purpose of the Alternating Bit protocol is to provide for reliable communication in the face of errors, our channel (link) models should implement some kind of erroneous behavior. In simple terms, that behavior will boil down to a nonzero probability that a random bit of a packet transmitted over a channel is received in error. SMURPH offers enough tools to implement practically arbitrarily intricate models of such phenomena, but, in this example, we take advantage of a certain (rather natural and mostly innocent) shortcut. We assume that if a packet contains one or more bit errors, then its terminating marker will not be recognized by the receiver. In real life that would correspond to the situation where the receiver evaluates some checksum (or CRC code) on the packet and, at the end of its reception, determines whether the packet is valid or not. The port model of SMURPH offers a shortcut for this operation. The default behavior of a faulty channel is that a nonzero number of bit errors in a packet will inhibit the packet from triggering packet-end events on ports, i.e., events normally caused by the reception of the packet's end marker.

All this must be interpreted in the context of the more general propagation model of packets in links). A transmitted packet is started (operation *transmit*) and stopped (operation *stop*) on the source port. Both moments translate into potential events to be triggered on all ports connected to the link, according to the propagation time needed for the original signals to reach those ports from the source. When the first signal (corresponding to starting the packet transmission) reaches a port on which some process is waiting for the packet's beginning, the event will be triggered, and the process will be awakened. And, of course, it is quite similar with the second signal, except that the corresponding events have different names.

A packet (or rather its SMURPH model) carries a few standard attributes. Two of those attributes identify the packet's sender and the receiver (a pair of stations). They can be set directly by the protocol program, or automatically, in some cases. When the packet's receiver attribute is set to a specific station, that station can claim the packet as *mine*, which means that the packet's boundaries will trigger some events only at the rightful recipient. In real life, this corresponds to (low-level) packet addressing, where the receiver hardware only tunes up to packets carrying some marks identifying the receiving station.

A packet passing by a port can trigger several events, some of them simultaneously. For example, the *EMP* event is triggered when the last bit of a qualifying packet has passed through the port. Formally speaking, the event falls on the first *ITU* following the last bit of the packet's activity perceived by the port. Thus, it coincides with *SILENCE*, which event falls on the first *ITU* in which the port perceives no activity. According to what we said earlier about the nondeterministic way of presenting multiple events occurring within the same *ITU*, it appears that the *SILENCE* event competes with *EMP* and can be presented as *the one*, even though an *EMP* event may be pending at the same time. Obviously, this is not what we want, as it would cause the acknowledgment to be lost. To make it easy (and to account for all the sane cases of handling packet events), SMURPH (exceptionally) prioritizes multiple events caused by the same packet (and falling onto the same *ITU*) such that, for example, *EMP* takes priority over *SILENCE* (see Sect. 7.1.5 for more details). Thus, the *SILENCE* event will only be triggered, if no *EMP* event is pending within the same *ITU*.

Thus, there are two possible continuations from state *AckBegin*. If the *SILENCE* event is triggered, which means that the acknowledgment packet hasn't been received correctly, the packet is ignored, and the process resumes from *WaitAck*, to wait for a new acknowledgment packet. Otherwise, the process continues at *AckArrival*. If the alternating bit of the ACK packet agrees with *LastSent*, the other process (the transmitter) is notified via a signal deposited in the mailbox.

An event caused by a packet, like *BMP* and *EMP* in the above code method, makes the packet causing the event accessible via variable *ThePacket* available to the process. As that variable is generic, i.e., it cannot assume a single, specific packet type (because the same port and the same process may perceive packets of various kinds), its type is a pointer to the generic type *Packet*. Thus, it often must be cast to the actual packet type, especially when the process needs access to an attribute defined in a non-standard *Packet* extension, which happens in our case.

Another important feature of packet events on ports, and generally all events caused by activities in the underlying link, is their persistent nature. What this means is that such an event is defined by the *ITU* in which it occurs, and it doesn't auto- matically go away when triggered and accepted. Say, the *EMP* event forcing the process to state *AckArrival* occurs because the packet that has been passing through the port has reached its terminating mark. The time being discrete, this happens within some specific (*ITU*-sized) time slot and, more notably, lasts for that entire slot. Put differently, the condition will remain pending until the time is advanced. Put yet differently, if you issue a wait request to the port for the same event within the same *ITU*, the event will be triggered (and the process will be awakened again) within the same *ITU*, i.e., immediately.

This explains why, at the end of state *AckArrival*, the process executes *skipto* to return to its "ground" state, instead of *proceed*. The operation behaves almost exactly as *proceed*, except that the transition is delayed by 1 *ITU*, to make sure that the persistent port event has gone. It is a recommended way out of a port event for which there exists a possibility that it may be awaited again without advancing the virtual time, which would cause an infinite, no-progress event loop (Sect. 2.1.4). At first sight, the situation may appear safe in our present case, because the process transits to *WaitAck*, where it issues a wait request for a different event (*BMP*). However, it is not formally impossible that the last *EMP* event coincides with a *BMP* event triggered by the next ACK packet.[3] Even though we will never send acknowledgments in such a way (back-to-back), it is better to safeguard the ACK receiver process against whatever malicious configurations of events are not logically impossible.

2.2.4 The Recipient's Protocol

Similar to the sender, the recipient station runs two processes. One of them, the receiver, takes care of incoming packets. The other process, the acknowledger, sends acknowledgments to the sender. Each of the two processes handles one of the two station ports. The type declaration of the receiver process is listed below:

```
process ReceiverType (RecipientType) {
        Port *Channel;
        Mailbox *Alert;
        TIME Timeout;
        states { WaitPacket, BeginPacket, PacketArrival, TimeOut };
        void setup (TIME tmout) {
                Channel = S->IncomingPort;
                Alert = S->AlertMailbox;
                Timeout = tmout;
        };
        perform;
};
```

---

[3]This is because *BMP* falls on the first bit of the packet (the first *ITU* where a packet signal is perceived), while *EMP* falls on the first *ITU* behind the packet (which, in principle, may be filled with the beginning of the next packet sent back-to-back).

Like the sender, the receiver uses a timeout to detect missing packets. Here is the process's code method:

```
ReceiverType::perform {
        state WaitPacket:
                Channel->wait (BMP, BeginPacket);
                Timer->wait (Timeout, TimeOut);
        state BeginPacket:
                Channel->wait (EMP, PacketArrival);
                Channel->wait (SILENCE, WaitPacket);
        state PacketArrival:
                if (((PacketType*)ThePacket)->SequenceBit == S->Expected) {
                        Client->receive (ThePacket, Channel);
                        S->Expected = 1 - S->Expected;
                }
                Alert->put ( );
                skipto WaitPacket;
        state TimeOut:
                Alert->put ( );
                sameas WaitPacket;
};
```

The reception drill for a data packet is the same as for an acknowledgment (Sect. 2.2.3). In state *WaitPacket*, the process waits for the beginning of a packet addressed to *this* station and for a timeout, whichever happens first. If the beginning of a packet is sensed, the process transits to state *BeginPacket* where it awaits the *EMP* event (marking a correct reception of the packet) or *SILENCE* (indicating that the reception has been erroneous). When a packet is correctly received (state *PacketArrival*), and its alternating bit is as expected, the process formally absorbs the packet, by calling the *receive* method of the *Client*, flips the *Expected* bit, and puts a signal into the mailbox to notify the other process that an acknowledgment should be sent the other way. Note that an acknowledgment is also sent if the packet's alternating bit is wrong, but then it relates to the previous (old) packet.

Recall that *Client* represents the network's application Sect. 2.2.3. When a packet that has originated in the network's application reaches the point of its (final) destination in the network, it should be passed back to the application, so the application "knows" that the network has fulfilled its task. Usually, that means tallying up the bits, calculating performance measures (delays), and so on. In general, the application model can be quite elaborate, but the reception link to it is simple and consists solely of the *receive* method of the *Client*, which should be invoked for an application packet when its peregrination through the network has come to an end.

Having responded to the packet's reception, and signaled the other process that an acknowledgment packet should be sent towards the sender node, the process transits back to its top state via *skipto* (see Sect. 2.2.3), to make sure that time has been advanced and the packet event has disappeared from the port. In state *TimeOut*, the transition is accomplished via a different operation, *sameas*, which we remember from Sect. 2.1.1. In summary, there are three ways to carry out a direct transition to a specific state: *proceed*, *skipto*, and *sameas*. See Sect. 5.3.1 for a detailed discussion of the differences among them.

The last process of the set is the one responsible for sending acknowledgments from the recipient to the sender. Here is its type declaration:

```
process AcknowledgerType (RecipientType) {
        Port *Channel;
        AckType *Ack;
        Mailbox *Alert;
        states { WaitAlert, SendAck, EndXmit };
        void setup ( ) {
                Channel = S->OutgoingPort;
                Ack = &(S->AckBuffer);
                Alert = S->AlertMailbox;
        };
        perform;
};
```

and the code:

```
AcknowledgerType::perform {
        state WaitAlert:
                Alert->wait (RECEIVE, SendAck);
        state SendAck:
                Ack->SequenceBit = 1 − S->Expected;
                Channel->transmit (Ack, EndXmit);
        state EndXmit:
                Channel->stop ( );
                proceed WaitAlert;
};
```

There is nothing new in the above code method that would call for an explanation. Note that the alternating bit in the outgoing acknowledgment packet is set to the reverse of *Expected*, because the bit is flipped by the receiver process after receiving a data packet and before notifying the acknowledger. If the action is in response to a timeout, then the acknowledgment should be for the previous packet, not the expected one.

2.2.5   Wrapping It Up

Whatever we have seen so far (in Sects. 2.2.2–2.2.4) is merely a bunch of declarations, so a *spiritus movens* is needed to turn them into a program and breathe life into it. Specifically, the following loose ends must be taken care of:

1. Input data parameterizing the experiment must be read in. These data should specify the numerical parameters related to the network configuration (e.g., the lengths of channels) and traffic conditions (the behavior of the network's application).
2. The network must be built, i.e., the stations and the channels must be created and configured.
3. Traffic conditions in the network (the message arrival process) must be described.
4. The protocol processes must be created and started.
5. If we want to see some results from running the model, we should make sure they are produced when the simulation ends.

The task of coordinating these activities is delegated to the *Root* process (Sect. 2.1.3) declared as follows:

```
process Root {
        void readData ( ), buildNetwork ( ), defineTraffic ( ), startProtocol ( );
        void printResults ( );
        states { Start, Stop };
        perform;
};
```

The five methods mark the four initialization stages (to be carried out in state *Start*) and one final stage of producing the output results performed in state *Stop*, at the end of the simulation experiment. The code method joins all these stages together:

```
Root::perform {
        state Start:
                readData ( );
                buildNetwork ( );
                defineTraffic ( );
                startProtocol ( );
                Kernel->wait (DEATH, Stop);
        state Stop:
                printResults ( );
};
```

Besides the five methods of *Root*, three other code fragments are still missing: the *setup* methods for the two station types and the list of global variables used by the program. Note that no such variables were explicitly needed by the protocol, barring items like *Client*, *ThePacket*, *Timer*, which identify elements of the local environment of the current process or station rather than conceptually global objects or parameters. This is not surprising: by definition, the protocol program is local, and its behavior can only be influenced by what happens at a given station. User-introduced global variables are needed, however, to organize the protocol into a stand-alone executable program. These "variables" are in fact flexible constants parameterizing the protocol and the experiment:

```
SenderType      *Sender;
RecipientType   *Recipient;
Link            *STRLink, *RTSLink;
double          FaultRate;
double          MessageLength, MeanMessageInterarrivalTime;
int             HeaderLength, AckLength, MinPktLength, MaxPktLength;
TIME            TransmissionRate, SenderTimeout, RecipientTimeout;
double          Distance;
long            MessageNumberLimit;
```

Sender and *Recipient* will be set to point to the two stations of our network. Similarly, *STRLink* (sender-to-recipient) and *RTSLink* (recipient-to-sender) will point to objects representing the two links (channels). Strictly speaking, we don't absolutely need global variables to point to these objects, but there is no harm to keep them this way, if only to better illustrate the explicit way the network "hardware" is constructed.

FaultRate is a global parameter that will affect the quality of the links. This will be the probability that a single bit transmitted over the link is flipped (or received in error). From the viewpoint of the model, the only thing that matters is the probability that an entire packet is received correctly, which is the simple product of the probability of a correct reception of all its bits.

The next two parameters, *MessageLength* and *MeanMessageInterarrivalTime*, describe the supply end of the network's application, i.e., the network's *Client*. The simple arrival process is characterized by a single "frequency" (inter-arrival) parameter specified as the (average) number of time units (*ETUs*) separating two consecutive arrivals of a message, at the sender station, to be transmitted over the network to the recipient station. The other parameter describes the message length. The actual distribution is specified when the traffic pattern is defined (we shall see it shortly).

The next four integer values apply to packets. *HeaderLength* is the fixed number of bits in every packet representing the combined header and trailer, i.e., the part of the packet that carries no "useful" information (from the viewpoint of the application). In SMURPH, we call that part the packet's *frame* content, as opposed to the (proper) information content (aka the payload). Unless the model needs it for some reason, SMURPH doesn't care how the frame content is physically laid out inside a packet; what only matters is the total number of bits contributed to the packet's length, which affects the transmission/reception time. Then, *AckLength* will determine the information content (payload) of an acknowledgment packet, which we assume to be fixed. *MinPktLength* and *MaxPktLength* describe the minimum and maximum length of a (data) packet's payload. All these values are in bits.

When we look at the invocation of the *getPacket* method of the *Client* in the transmitter process in Sect. 2.2.3, we see three of the above parameters mentioned in the arguments. Recall that the method extracts (into a packet buffer) a packet from a message queued at the station for transmission (Sect. 6.7.1). The packet is constructed in such a way that the length of its payload is at most *MaxPktLength* (if that many bits are available in the message), and at least *MinPktLength*. If the message (whatever is left of it at the time of the packet's acquisition) is shorter than *MinPktLength* bits, then the payload is *inflated* with dummy bits to *MinPktLength*. Then, *HeaderLength* frame content bits are added to the packet to make it complete. The inflated bits, if any, do not count to the payload, but are treated as extra frame bits. This is important, when the packet is tallied up on reception, for the assessment of its effective contribution to the application.

Next, we see three parameters representing time intervals, three of them of type *TIME* plus one *double*. The first of them, *TransmissionRate*, determines the transmission rate of our simple network (which is the same for both ends). The last one, *Distance*, is the length of the links expressed as the amount of time required for a signal to propagate between the two stations. The remaining two are the timeouts used by the transmitter and the recipient.

The last parameter, *MessageNumberLimit*, will let us declare a simple termination condition for the experiment. The model will stop, and the *DEATH* event on *Kernel* will be triggered (thus making the *Root* process transit to its second state) when the total number of messages received at the recipient reaches the specified boundary value.

Those of the above variables that represent numerical parameters of the model are read from the input data set. This is accomplished by the *readData* method of *Root* in this way:

```
void Root::readData ( ) {

        readIn (HeaderLength);
        readIn (AckLength);
        readIn (MinPktLength);
        readIn (MaxPktLength);
        readIn (TransmissionRate);
        readIn (SenderTimeout);
        readIn (RecipientTimeout);
        readIn (Distance);
        readIn (MessageLength);
        readIn (MeanMessageInterarrivalTime);
        readIn (FaultRate);
        readIn (MessageNumberLimit);
};
```

This is just a dull series of calls to *readIn* to read the numerical parameters of our model from the input file. The function correctly handles arguments of different numerical types.

Before we discuss the next step carried out by *Root*, which is the creation of the network's virtual hardware, we should look at the *setup* methods of the two stations. We have postponed their introduction until now, because they carry out functions belonging to the network creation stage.

```
void SenderType::setup ( ) {
        IncomingPort = create Port;
        OutgoingPort = create Port (TransmissionRate);
        AlertMailbox = create Mailbox (1);
        LastSent = 0;
};

void RecipientType::setup ( ) {
        IncomingPort = create Port;
        OutgoingPort = create Port (TransmissionRate);
        AlertMailbox = create Mailbox (1);
        AckBuffer.fill (this, Sender, AckLength + HeaderLength, AckLength);
        Expected = 0;
};
```

Note that the methods can only be called (i.e., the stations can be created) after the parameters referenced in them have been read in. They set up the ports and mailboxes, and initialize the alternating bits. The *setup* method of the recipient station also pre-fills the acknowledgement buffer with the fixed part of its content.

The *create* operation for *Port* accepts an optional argument defining the transmission rate, or rather its reciprocal[4] specifying the amount of time (expressed in *ITU*s) needed to insert a single bit into the port. Note that only those ports that are used to transmit packets (via the *transmit* method of *Port*, see Sects. 2.2.3 and 2.2.4) need a transmission rate. It is needed as the multiplier of the total packet length to produce the amount of time needed by the method to complete the packet's transmission, i.e., to correctly set the timer for a transition to the state specified as its second argument.

Both mailboxes are created with capacity 1 (the argument of the standard *setup* method for *Mailbox*). This is the maximum number of (pending) signals that can be stored in the mailbox (by *put*) before being accepted (when the *RECEIVE* event is triggered). Both mailboxes are used to implement a simple handshake mechanism between a pair of processes where the presence of a single item represents a pending condition.

The acknowledgement packet buffer at the recipient station is filled by calling the *fill* method belonging to the standard *Packet* type. The arguments correspond to the following attributes of the packet:

---

[4]The term "transmission rate" is used inconsistently in SMURPH, because the parameter, as an argument of the *Port* setup method, in fact stands for the reciprocal of what is normally considered the transmission rate. Normally, transmission rates are expressed in bits per second, or, more generally in inverse time units, whereas in SMURPH they are expressed in time units. We shall live with this inconsistency for the remainder of this book.

1. the packet's sender station (note that this argument points to the station whose *setup* method is being executed)
2. the recipient station
3. the total length of the packet (this is the length that, together with the port's transmission rate, determines the packet's transmission time)
4. the "information content" of the packet, i.e., the number of bits that should count as the packet's payload.

The primary reason why we fill the recipient attribute is to enable events like *BMP* and *EMP* (see the receiving processes in Sects. 2.2.3 and 2.2.4). We do not have to do this for the data packet buffer (at the sender station), because that buffer is filled by *getPacket* (see the transmitter process in Sect. 2.2.3) which takes care of all the relevant attributes.

The network is put together by *buildNetwork* in the following way:

```
void Root::buildNetwork ( ) {
        Port *from, *to;
        Sender = create SenderType;
        Recipient = create RecipientType;
        STRLink = create Link (2);
        RTSLink = create Link (2);
        (from = Sender->OutgoingPort)->connect (STRLink);
        (to = Recipient->IncomingPort)->connect (STRLink);
        from->setDTo (to, Distance);
        RTSLink->setFaultRate (FaultRate);
        (from = Recipient->OutgoingPort)->connect (RTSLink);
        (to = Sender->IncomingPort)->connect (RTSLink);
        from->setDTo (to, Distance);
        STRLink->setFaultRate (FaultRate);
};
```

First, the two stations are created (note that their *setup* methods are automatically invoked, which results in their ports being built and set up). Next, the two links come into existence; the *setup* argument indicates the number of ports that will be connected to the link. The connection is accomplished with the *connect* method of *Port*. Then, the distance between the pair of ports on the same link is set with *setDTo* (set distance to). Note that *setDTo* is a port method. It takes two arguments: the other port (the one to which the distance is being defined) and the distance itself, i.e., the propagation time. The reason why *Distance* is of type *double* (rather than *TIME*) is that the propagation time for *setDTo* is expressed in the so-called distance units (whose formal type is *double*), rather than in *ITUs*, like the remaining three time intervals. In absolute numerical terms, all those units are the same in our case (the difference will be explained in Sect. 3.1.2). The standard type *Link* represents symmetric (simple, uniform) links where the distance is the same in both directions. Thus, defining the distance in (any) one direction (say from *Sender->OutgoingPort* to *Recipient->IncomingPort*) automatically sets the (same) distance in the opposite direction (i.e., from *Recipient->IncomingPort* to *Sender->OutgoingPort*). Incidentally, it doesn't really matter in our case, as each link is used to transmit packets in one direction only. For each link, the fault rate (the probability of any single bit being received in error) is set to the same value stored in *FaultRate*.

Traffic conditions in the modeled network (the behavior of the network's *Client*) are described as a collection of *traffic patterns*. Only one traffic pattern is used in our program. Its type is declared in the following way:

traffic TrafficType (Message, PacketType) {};

The body of the declaration is empty, which is not uncommon for a traffic pattern. The only reason why we need to define our private traffic pattern, instead of relying on the built-in type *Traffic*, is that we must indicate to the *Client* that the packets generated by the network's application should be of type *PacketType* (which is introduced by the protocol program), rather than the standard type *Packet* (assumed by *Traffic* as the default). This is accomplished by the part in parentheses. The two identifiers appearing there specify the type of messages (in this case we get away with the standard type *Message*) and packets handled by the traffic pattern.

The traffic pattern is created and parameterized in this method of *Root*:

```
void Root::defineTraffic ( ) {
        TrafficType *tp;
        tp = create TrafficType (MIT_exp + MLE_fix,
                MeanMessageInterarrivalTime, MessageLength);
        tp->addSender (Sender);
        tp->addReceiver (Recipient);
        setLimit (MessageNumberLimit);
};
```

In general, the list of *setup* arguments for a *Traffic* can be long and complicated. The complete description of a traffic pattern consists of the arrival process (telling how and when messages arrive to the network) and the distribution of senders and receivers (telling at which stations the arriving messages are queued for transmission and where they are destined). In our simple case, the three arguments describe a Poisson arrival process [3] with an exponentially distributed message interarrival time and with fixed-length messages. The mean message interarrival time and the fixed message length are specified as the last two arguments; their values have been read from the input data set.

The trivial configuration of senders and receivers is described by the next two statements. The *addSender* method (of *Traffic*) adds the indicated station to the population of senders for the traffic pattern. Similarly, *addReceiver* includes the specified station in the set of recipients. In our case, there is just one sender and one recipient, so the procedure is straightforward and devoid of options. Every message arriving from the *Client* to the network is queued at one possible sender and addressed to one possible recipient.

The call to the global function *setLimit* (the last statement in the above method) sets the termination condition. Formally, the condition will be met when the specified number of messages have been received at their destinations. In our case, it means that the recipient station has received that many messages. The act of reception of a message is defined as receiving the last packet of that message, as it was queued at the sender by the *Client*. By "receiving" we mean, of course, executing *Client->receive* (Sect. 2.2.4) for the packet.

The last stage of the startup is the sequence to create the processes:

```
void Root::startProtocol ( ) {
        create (Sender) TransmitterType (SenderTimeout);
        create (Sender) AckReceiverType;
        create (Recipient) ReceiverType (RecipientTimeout);
        create (Recipient) AcknowledgerType;
};
```

Note the new syntax of the *create* operation, one we haven't seen yet. The optional item in the first set of parentheses immediately following the *create* keyword is an expression identifying the station (a single, specific station, not a station type) to which the created process is supposed to belong. As we said in Sect. 2.2.3, every process must belong to some station, and that station is determined at the time when the process is created, once in its lifetime. Sometimes this can be done implicitly (Sect. 5.1.3), and the (somewhat clumsy) variant of *create* exemplified above is seldom used. The above sequence may not be the most elegant way to start the processes, but it well illustrates the point that those processes must run at definite stations.

Following the invocation of *startProtocol* in the first state of *Root*, the model is up and running. The *Root* process will hit the end of the list of statements in its first state (*Start*) and become dormant until the declared limit on the number of received messages is reached (the *DEATH* event on *Kernel* is triggered).

2.2.6 Running the Model

We run the model in the simulation mode where we submit a data set parameterizing the experiment, execute the program on that data set, and then, when the program terminates (as it is expected to) we look at its output.

Here is the sample data set from the protocol's directory (*AB*) in *Examples*:

Header length	*256 bits*
ACK length	*64 bits*
Min packet length	*32 bits*
Max packet length	*8192 bits*
Transmission rate	*1 (ITUs per bit)*
Sender timeout	*512 ITUs*

Recipient timeout	*1024 ITUs*
Distance	*24 ITUs*
Message length	*1024 bits (fixed)*
Mean message interarrival time	*4096 ITUs (bits)*
Fault rate	*0.0001 (one bit per ten thousand)*
Message number limit	*200000*

Recall that only the numbers from the above sequence are interpreted when the data set is read in by the simulator (the *Root* method *readData* in Sect. 2.2.5). They should be matched to the calls to *readIn* in *readData* in the order of their occurrence.

In our data set, we see values of five kinds: bits (referring to packets and messages and denoting information length), time (the two timeouts and the mean message interarrival time), distance (the separation between the sender and the recipient), probability (the likelihood that a single bit of a packet will be damaged), and one simple (unitless) number at the very end. Ignoring bits and the last two kinds (whose interpretation is always clear), we must come to terms with time and distance. Note that we have never defined the *ETU* (there is no call to *setEtu* in our present program, unlike in the car wash model in Sect. 2.1.4). This means that 1 *ETU* = 1 *ITU*, and in all cases where we mention time we mean *ITU*, i.e., the smallest indivisible time grain used by the simulator.

We shall see later, in Sect. 3.1.2, that it is possible to use units for expressing distance that are more convenient than the *ITU*. But ultimately, the only way for the model to interpret distance is as the propagation time, i.e., the amount of time needed for a signal to propagate from point *A* to point *B*. This is only needed for channels (wired or RF). A single network may include channels with different properties, so the model's measure of "distance" need not unambiguously and uniformly translate into the same notion of geographical distance for all channels.

The only place in our model where the *ITU* is correlated with something other than time is where we assign the bit rate to the (output) ports (see the station *setup* methods in Sect. 2.2.5). Note that the corresponding parameter in the data set (*Transmission rate*) is 1. This way bits become synonymous with *ITUs*. To interpret them in real-life time units, we must agree on a transmission rate expressed in the "normal" units of bits per seconds. For example, if the rate is 100 Mb/s, then one bit (and also one *ITU*) equals 10^{-8} s. Then, the distance of 24 bits will translate into about 48 m (i.e., 1 bit = 2 m, assuming the signal propagation speed of 2/3 of the speed of light in vacuum).

The model is compiled by running:

mks −z

The −z parameter is needed to enable faulty links (see Sects. 7.3, B.5). It takes a few seconds to run the above data set through the simulator. The command line is:

./side data.txt output.txt

assuming that the file *data.txt* contains the data. The results are written to *output.txt*. The output file consists of whatever the model decides to write to it. Typically, if the purpose of the model is to study the performance of some system (network), the relevant data are produced at the end, when the model terminates on its stop condition. Sometimes, especially during testing and debugging, the output may contain other data reflecting the model's intermittent state.

In SMURPH, there is a standard set of operations (methods) associated with objects that collect performance-related data to write those data to the output file. More generally, objects of some types can be *exposed*, possibly in several different ways (see Sect. 12.1), which means producing some standard information, specific to the object, on the object's content. We see some of those operations in the *printResults* method of *Root* which looks like this:

```
void Root::printResults ( ) {
        Client->printPfm ( );
        STRLink->printPfm ( );
        RTSLink->printPfm ( );
};
```

It calls a method with the same name, *printPfm*, for three objects: the *Client* (representing the network's application) and the two links. The method is the standard way to ask an object (capable of delivering this kind of information) to write its standard set of *performance measures* to the output file.

The standard set of *Client* performance measures consists of the contents of six *random variables* and a few counters (see Sect. 10.2 for details). A random variable (an object of type *RVariable*, Sect. 10.1) collects samples of a numerical measure and keeps track of a few standard parameters of its distribution (the minimum, the maximum, the average, the standard deviation, and possibly more).

The roles of the different random variables of the *Client* are explained in Sect. 10.2.1. Below we show an initial fragment of the *Client* output, including the contents of the first two random variables:

Time: 820660064 (Client) Global performance measures:
AMD—Absolute message delay:

Number of samples	*200000*
Minimum value	*1304*
Maximum value	*28169*
Mean value	*2590.26234*
Variance	*3696834.001*
Standard deviation	*1922.715268*
Confidence int @95%	*0.003253211487*
Confidence int @99%	*0.004273989581*

APD—Absolute packet delay:

Number of samples	*200000*
Minimum value	*1304*
Maximum value	*12056*

Mean value	*1547.6224*
Variance	*496100.4603*
Standard deviation	*704.3439929*
Confidence int @95%	*0.001994623271*
Confidence int @99%	*0.002620487205*

A single sample of *absolute message delay* is defined as the amount of time elapsing from the moment a message arrives from the *Client* at a sender station for transmission, until the last packet of that message is delivered to its destination. By the latter we understand the packet's formal reception, i.e., the execution of *Client-> receive* (Sect. 2.2.4) for the packet. Note that generally a message can be transmitted in several packets (if its length is greater than the maximum packet length). In our case, looking at the data set, we see that the maximum packet length (understood as the maximum length of the packet's payload) is 8192 bits, which is more than the (fixed) message length of 1024 bits; thus, every message is always transmitted in one packet. This tallies up with the printed contents of the second random variable, which applies to individual packets (the *number of samples* is the same in both cases). The absolute packet delay counts from the moment a packet becomes ready for transmission (it is put into a buffer via *Client->getPacket*, Sect. 6.7.1) until the packet is received (*Client->receive*) at the destination.

Both message and packet delays are expressed in *ETUs*, which in our case are the same as *ITUs* (which in turn are the same as the transmission rate, i.e., bit slots). We see, for example, that the average packet delay is 1547.6224 bits. The absolute lower bound on that number is 1024 bits, because that much time is needed by the transmitter to insert the packet into the output port, and it certainly cannot be received sooner than the transmitter is done with this job. Then we must throw in the propagation delay across the channel (24 bits). Anything more than the sum of those two (1048) should be attributed to delays incurred by the protocol. In our case, this means waiting for acknowledgments and retransmissions.

Note that the average message delay is higher than the average packet delay, even though every message is transmitted in one packet. This is because the message delay, in addition to the delay experienced by the packet, includes the time spent by the message in the queue at the sender.

Here is the list of counters appearing in the output file after the random variables:

Number of generated messages	*200001*
Number of queued messages	*2*
Number of transmitted messages	*199999*
Number of received messages	*200000*
Number of transmitted packets	*199999*
Number of received packets	*200000*
Number of queued bits	*2048*
Number of transmitted bits	*204798976*
Number of received bits	*204800000*
Measurement start time	*0*
Throughput	*0.2495552166*

Most of these numbers are trivially easy to interpret. The fact that the number of received messages (packets) is higher than the number of transmitted ones may raise brows, but it is perfectly OK, if we realize that to be deemed *transmitted*, a packet must be acknowledged, and its buffer must be released at the sender (operation *Buffer->release* in the transmitter process in Sect. 2.2.3). Thus, the reception of the last packet (triggering the termination of the experiment) precedes the moment when the packet can be formally tallied as transmitted.

The throughput is calculated as the ratio of the number of all received bits to the total simulated time in *ETUs*. In our case, it is in bits per bit time, so it can be viewed as being normalized to the transmission rate, with 1 (corresponding to 100% channel utilization) being the (unreachable) upper bound. Note that the 25% throughput that we witness correlates with the message interarrival time of 4096 bits. As the fixed message length is 1024 bits, we would expect the effective throughput to be about $1024/4096 = 1/4$, at least as long as all the arriving messages are eventually transmitted and received (the network can cope with the offered load). This is clearly the case, as the message queue at the end of simulation is almost empty.

The *Client* performance measures do not let us see acknowledgments and retransmissions, because those measures only relate to the messages and packets that arrive from the *Client* and are formally transmitted and received as useful data. Thus, the same packet retransmitted a few times only counts once: transmitted once at the sender (operation *Client->release*) and received once at the recipient (operation *Client->receive*). Also, the *Client* doesn't care about acknowledgments (and is not even aware of them). Some insight into the internal workings of the protocol can be obtained from the link output listed below:

Time: 820660064 (Link 0) Performance measures:

Number of jamming signals	*0*
Number of transmission attempts	*234533*
Number of transmitted packets	*234533*
Number of transmitted bits	*240161792*
Number of received packets	*200000*
Number of received bits	*204800000*
Number of transmitted messages	*234533*
Number of received messages	*200000*
Number of damaged packets	*28058*
Number of h-damaged packets	*5941*
Number of damaged bits	*28731392*
Throughput (by received bits)	*0.2495552166*
Throughput (by trnsmtd bits)	*0.292644668*

(Link 0) End of list
Time: 820660064 (Link 1) Performance measures:

Number of jamming signals:	*0*
Number of transmission attempts:	*621134*
Number of transmitted packets:	*621134*

Number of transmitted bits:	*39725576*
Number of received packets:	*0*
Number of received bits:	*0*
Number of transmitted messages:	*0*
Number of received messages:	*0*
Number of damaged packets:	*19698*
Number of h-damaged packets:	*15844*
Number of damaged bits:	*1260672*
Throughput (by received bits):	*0*
Throughput (by trnsmtd bits):	*0.04843975934*

(Link 1) End of list

We can ignore for now the (irrelevant for our present model) *jamming signals* (leaving them until Sect. 7.1). The number of received packet bits in *Link* 0 (which is the data link from Sender to Recipient) equals the number of bits counted as received by the *Client*. The respective counters have been updated in exactly the same way, i.e., whenever *Client->receive* was executed. Note that the method's second argument points to a port and thus identifies a link. The number of transmitted bits in a link is calculated differently and reflects the total number of bits *ever* inserted into the link via the *transmit* method of *Port*. This way, the difference between the number of transmitted packets, as shown by the link, (234,533) and the number of received packets as shown by the *Client* (200,000) amounts to the number of retransmissions (34,533).

The number of transmission attempts (for both links) is the same as the number of transmitted packets, because we never *abort* a packet transmission, i.e., every *transmit* operation is matched by *stop* (see Sect. 7.1.2). The number of acknowledgment packets (transmitted in*Link* 1) is much larger than the number of transmitted (and received) packets, because acknowledgments are also sent by the recipient (periodically on a timeout) when no data packets arrive from the sender.

Generally, a packet damaged during its passage through a link can be damaged in one of two ways (see Sect. 7.3 for details). Our model does not distinguish between them, so the first value (*Number of damaged packets*) is the one that matters. Note that the number of damaged data packets (28,058) is smaller than the number of retransmitted data packets (34,533) which is explained by the fact that acknowledgments can also be damaged, so a correctly received (and acknowledged) packet may also have to be retransmitted. No packet is ever received on the second link, because acknowledgments are never formally "received" (by the *Client*).

2.3 A PID Controller

In this section we introduce a SMURPH program controlling a hypothetical plant. In control theory, a *plant* denotes a controlled, dynamic object producing some quantifiable output depending on some quantifiable input. We shall implement a generic PID controller and see how it can be interfaced to real devices representing real plants via sensors and actuators. Then we will apply the controller to drive a fake plant whose virtual behavior will be programmed in SMURPH (what else?).

2.3.1   PID Control in a Nutshell

About the simplest (while still non-trivial) case of system control involves adjusting a knob on some device to bring its output close to some target value. Imagine a robotic arm driven by a motor. The arm has to reach a specific position. The motor has a range of settings, say between -1 and $+1$, with $+1$ meaning full forward, -1 meaning full reverse, and zero meaning off. The arm has a non-trivial inertia. The problem is how to adjust the motor setting as the arm keeps moving, to make it smoothly stop at the target position without too much overshooting and backtracking, preferably in the shortest possible time.

Assume that a measure of distance of the arm from its target position is available, call it the *error*, and can be used as an input parameter to calculate the momentary setting of the motor. Let that measure have the property that zero means "on target," a positive value means "below target" (connoting with the forward action of the motor required for a correction), and a negative value means "above target" (i.e., the motor should be put in reverse). The problem is how to dynamically transform the error indications into adjusted settings of the motor to reach the target without undue jerks.

The above system fits a more general pattern shown in Fig. 2.2. By tradition, the target value is called the *setpoint*, and the controlled system is called the *plant*. Our problem is to construct the *compensator* and fit it into the overall scheme.

The compensator takes as its input the error, i.e., the difference between the setpoint and the current output of the plant (the position of our robotic arm), and produces on its output the new value of the plant's input (the motor setting). This procedure is carried out in cycles, strobed by some clock, which also accounts for the system's inertia, i.e., the delay separating a modification of the input variable from the plant's response. Following a new setting of the plant's input, the next output value (to be compared against the setpoint) will be produced at the beginning of the next cycle.

PID stands for Proportional-Integral-Derivative, which adjectives apply to the three components of the compensator. The theory behind PID controllers is described

Fig. 2.2 A PID-controlled system

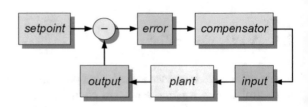

in many easily available books and papers, e.g., [6, 7]. Here we shall focus on the simple mechanism of the PID approach.

One obvious idea would be to adjust the plant's input in proportion to the error assuming that a more drastic measure is required when the error is big. This is what we call proportional control. For the robotic arm, the idea boils down to setting the motor value to:

$$M = K_p \times e$$

where e is the error and K_p is a tunable factor. Normally, we would make sure never to exceed the allowable range for M, so the value would be forced to be between the minimum and the maximum allowed. To keep things independent of the scales and units of M and e, K_p is often expressed (or thought of) as the fraction (or percentage) of the M range per unit of e. In a program, we would write:

```
v = Kp * error;
if (v > VMax)
        M = VMax;
else if (v < VMin)
        M = VMin;
else
        M = v;
```

Depending on K_p, a proportional controller may tend to oscillate (if K_p is large and the controller is aggressive), or be slow to converge to the target (if K_p is small). When the target is moving, a proportional controller will often tend to under-reach the target, because it needs a nonzero error to actually change the plant's input, and that error diminishes as the plant gets close to the target.[5] A way to improve its behavior is to add a term encoding the history of error. The idea is that if the error has been nonzero (or large) for some time, then it makes sense to push the plant harder (even if the current momentary error appears low), reducing the input as the error begins to consistently get smaller. Formally, this means to augment the formula by an integral component, i.e.:

$$M = K_p \times e + K_i \times \int_{t_0}^{t_c} e\, dt$$

where the integral is taken over some time preceding the current moment t_c. K_i is another tuneable coefficient of the controller responsible for the integral part (with K_p factoring in the contribution from the proportional component). Quite often the integration time covers the entire control period (so $t_0 = 0$). In some control scenarios

---

[5]In many practical cases the allowable values of M and e are (often drastically) discretized which makes the problem more pronounced.

it makes sense to occasionally redefine that period, e.g., by resetting t_0 to current time.

The last addition to the formula is the derivative factor compensating for jumps in the momentary behavior of the error. When the difference in error indications between two consecutive steps is large, say the error drops considerably, this may be taken as a signal that the input should be reduced (relative to the outcome of the proportional and integral components), lest the plant overshoots the target. This yields the following formula:

$$
M = K_p \times e + K_i \times \int_{t_0}^{t_c} e\, dt + K_d \times \frac{de}{dt}
$$

parametrized by three coefficients and including all three parts.

The first two parts are used in most of practical controllers, with the differential part being optional and sometimes controversial. It may be confusing in situations where the error (or output) readings are tainted with non-trivial "errors" (especially if the time step is short), which may cause large and disruptive swings of the derivative having no grounds in the plant's reality.

As the time is in fact discrete, the integral boils down to a sum and the derivative becomes a difference. A code fragment implementing the controller may look like this:

```
lasterr = 0;
I = 0;
while (1) {
        output = wait_for_output ( );
        e = Setpoint − output;
        I += e;
        D = e − lasterr;
        v = Kp * e + Ki * I + Kd * D;
        if (v > VMax)
                v = VMax;
        else if (v < VMin)
                v = VMin;
        setinput (v);
        lasterr = e;
}
```

We assume that *wait_for_output*, in addition to producing the current output from the plant, also waits for that value. This way, each turn of the loop executes one cycle of the control loop at some regular intervals.

2.3.2   Building the Controller

Our objective is to program a generic PID controller in SMURPH. From the previous section we easily see that the structure of that program is going to be rather simple: the control loop can be handled by a trivial SMURPH process with a single state triggered by output readouts of the plant. So, the only interesting part of the control program is going to be its interface to the real world. That interface will include events triggered by the arrival of new plant output, receiving the output value, and passing a new input value to the plant.

   This kind of interface is implemented in SMURPH by associating mailboxes with devices or network ports. We have seen simple mailboxes in Sects. 2.1.2, 2.2.2, and 2.2.3, used for passing signals among processes. Generally, a mailbox can store a list of objects. It can also be *bound*, in which case its contents can arrive from the outside or be sent to the outside of the SMURPH program.

   The advantage of using mailboxes, instead of talking directly to the devices or network ports, is that the communication is then nicely encapsulated into the same event-driven framework as the rest of the SMURPH constructs. Otherwise, we couldn't really claim that the SMURPH program had things under its control, because the blocking inherent in reading and writing data from/to devices would get in the way of servicing events occurring during the blockage.

   In our PID controller, we shall use network mailboxes, which will turn the controller into a server. The plant will be represented by a single SMURPH station declared like this:

```
station Plant {
        int VMin, VMax, Setpoint, SMin, SMax;
        double Kp, Ki, Kd, integral, lasterr;
        Process *CP, *OP;
        Mailbox *CI, *OI;
        void setup ( );
        void control (int);
        void setpoint (int);
};
```

We assume, as it usually happens with real-life control, that the plant's input and output are integer values. The legitimate setting of input (i.e., the controlling knob) is between *SMin* and *SMax*, inclusively. Similarly, the output, as well as *Setpoint*, are between *VMin* and *VMax*.

At any moment, there may be up to two connections to the controller represented by two mailboxes and two processes, each process handling its dedicated mailbox. One of those processes, pointed to by *CP* (for *control process*) interfaces the controller to the plant and implements the actual control loop. The other process, pointed to by *OP* (for *operator's process*) will allow the operator to adjust the knob.

When the plant station is created, its attributes are initialized in this way:

```
void Plant::setup ( ) {
        readln (VMin);
        readln (VMax);
        readln (Setpoint);
        readln (SMin);
        readln (SMax);
        readln (Kp);
        readln (Ki);
        readln (Kd);
        lasterr = integral = 0.0;
        CP = OP = NULL;
        CI = OI = NULL;
}
```

The respective minimums and maximums, as well as the initial setpoint, plus the three PID factors are read from the input data file. All the remaining attributes are zeroed out. The fact that the mailbox and process pointers are *NULL* indicates that the objects are absent. They will be set up when the plant and the operator connect to the controller. The connection procedure follows the drill of a network client connecting to a server over a TCP socket [8, 9], except that the sockets are disguised by mailboxes. Here is the complete definition of the *Root* process implementing the setup stage of the server:

```
process Root {
        states { Start };
        perform {
                state Start:
                        Mailbox *ci, *oi;
                        int port_p, port_o;
                        Plant *p;
                        readln (port_p);
                        readln (port_o);
                        p = create Plant;
                        ci = create Mailbox;
                        oi = create Mailbox;
                        if (ci->connect (INTERNET + SERVER + MASTER, port_p) != OK)
                                excptn ("Cannot set up plant interface");
                        if (oi->connect (INTERNET + SERVER + MASTER, port_o) != OK)
                                excptn ("Cannot set up operator interface");
                        create (p) CConnector (ci);
                        create (p) OConnector (oi);
        };
};
```

At some disagreement with our earlier recommendations (Sect. 2.1.3), the above *Root* process only has one state. Nothing wrong with that. This is not a simulation program, which we would want to stop at some point and produce results, but a server that, once started, is expected to be there forever (or until something drastic happens). The process starts by reading two integer values from the input data set. These are two TCP port numbers for receiving connections from the plant and the operator. Then it creates the plant station and two mailboxes. The latter will act as server sockets making the controller accessible (at least in principle)[6] over the Internet.

At the risk of repeating myself, let me say once again that one advantage of representing sockets (and I/O devices in general) via the mailbox paradigm makes the data transfers natural from the viewpoint of SMURPH and compatible with the SMURPH *process* model. In that model, we are only allowed to execute something that will always immediately complete, without having to wait for something external to happen behind the (SMURPH) scenes. If that something cannot deliver its expected outcome right away, then there must exist a SMURPH event triggered when the outcome is available (or when it makes sense to ask again) and a SMURPH object (an activity interpreter, see Sect. 5.1.1) responsible for delivering that event. While direct non-blocking I/O with sockets and most other devices is of course possible [8], it needs to be encapsulated into the right kind of mechanism to be usable in SMURPH.

---

[6]These days, most systems that are not intended to act as servers run on VPNs and/or behind firewalls, which makes the task of turning them into servers visible across the Internet more involved.

Another advantage of that mechanism is that it is significantly simpler than the raw tools offered by the operating system.

The *connect* operations, which are identical for the two mailboxes, except for the difference in the port number, make them visible as server ports on the host running the program. The first argument is a collection of flags describing the nature of the "device" to which the mailbox is being bound. Thus, *INTERNET* means a socket (as opposed to an I/O device), and *SERVER+MASTER* indicate that the socket is a "master server" socket on which incoming connections can be accepted. The function returns *OK* when it succeeds. Practically the only reason for it to fail is that the specified port number is already taken.

It is (at least in principle) legal to execute *connect* for any mailbox (see Sect. 9.4.1 for details). The operation rather drastically changes the mailbox's nature, effectively transforming it from an internal repository for signals and simple objects into an interface to something located outside the SMURPH program. It also enables new functions and events.

To handle the incoming connections, *Root* creates two connector processes and then becomes dormant: the process issues no wait request in its only state, so it will never run again. The connector processes are almost identical. The idea is that when the plant or the operator connects to the controller, a new mailbox will be created, to provide for the direct link to the client, and a new process will be run to take care of the communication. This is, in a nutshell, the standard scheme of delivering a socket-based, network service, with the mailboxes acting as sockets in disguise.

The fact that the two connector processes are not completely identical, while the differences between them are rather minute, allows us to illustrate a certain convenient feature of the SMURPH process declaration. In a case like this, the most natural way to implement the processes is to have a single code method, possibly parameterized by some exchangeable functions (methods), and share it by multiple, formally different, process types. Our implementation of this idea may go like this (see Sect. 5.1.2):

```
process Connector abstract {
        Mailbox *M;
        int BSize;
        states { WaitClient };
        virtual void service (Mailbox*) = 0;
        perform;
};
```

The *Connector* process represents the shared part of the two *actual* connector processes. Its code method is listed below:

```
Connector::perform {
        state WaitClient:
                Mailbox *c;
                if (M->isPending ( )) {
                        c = create Mailbox;
                        if (c->connect (M, BSize) != OK) {
                                delete c;
                                proceed WaitClient;
                        }
                        service (c);
                }
                M->wait (UPDATE, WaitClient);
}
```

The mailbox pointed to by *M* is the respective master server mailbox (see the *setup* methods below). By executing *isPending* for such a mailbox the process can learn if there is an incoming connection awaiting acceptance (note that accepting connections is the process's only function). If this happens to be the case, the process creates a new mailbox and *connects* it to the master mailbox. The meaning of *connect* with a (server) mailbox pointer appearing as the first argument is to accept a client pending on the server mailbox and make the client's socket available via the connecting mailbox. The operation may fail, because nothing is taken for granted in networking, e.g., the client may have disappeared in the meantime. In that case the new mailbox is deallocated and the connection attempt is ignored. When the operation succeeds, the process invokes its *service* method. This method is *virtual*, and it will be defined by the actual (final) connector processes to implement their specific actions required to set up the connection. Here is the plant connector process:

```
process CConnector : Connector (Plant) {
        void service (Mailbox *c) {
                if (S->CP != NULL) {
                        delete c;
                        return;
                }
                S->CI = c;
                S->CP = create Control;
        };
        void setup (Mailbox *m) {
                M = m;
                BSize = 4;
        };
};
```

The first line declares *CConnector* as a process type descending from *Connector*. There is no code method (no *perform* declaration), which means that the process inherits the code method from the parent process class.

Note that the *BSize* attribute (set to 4 in the *setup* method of *CConnector*) is used in the (shared) code method as the second argument of the *connect* method of the new mailbox. This argument sets the buffer size for a data item sent or received over the mailbox. The plant mailbox will be acting as a virtual sensor/actuator pair handling fixed-size, binary data, i.e., 4-byte integer values.

The service method checks if the station's (*Plant*) attribute *CP* (pointing to the control process) is *NULL*; if not, it means that the plant is already connected, and the current connection attempt is ignored. Deleting the mailbox has the effect of closing the connection socket. Otherwise (the plant is not connected yet), the mailbox pointer is stored in the station's attribute *CI* and the control process is created.

The operator's connector process is similar:

```
process OConnector : Connector (Plant) {
        void service (Mailbox *c) {
                if (S->OP != NULL) {
                        delete c;
                        return;
                }
                S->OI = c;
                S->OP = create Operator;
                c->setSentinel ('\n');
        };
        void setup (Mailbox *m) {
                M = m;
                BSize = OBSIZE;
        };
};
```

The operator's mailbox will be used for receiving textual input, i.e., lines of text terminated by the newline character. The buffer size (*BSize*) is set to some constant (*OBSIZE* defined as 256) which, in this case, limits the length of a line received from the operator. The last statement of the *service* method sets the *sentinel* character for the client mailbox. There is a way of extracting (reading) data from a mailbox (see below) where the sentinel character delimits the chunk extracted in one step.

When the connector processes are created by the *Root* process they are explicitly assigned to the *Plant* station (see the syntax of the *create* operation, Sect. 3.3.8). The processes created by the connectors (*Control* and *Operator*) are also assigned to the *Plant* station, because it happens to be the *current* station at the time of their creation (the connector process belongs to it). In the first case, when the connectors are created, the current station is *System* (see Sects. 2.2.3 and 4.1.2), so the station assignment must be explicit.

From the viewpoint of its configuration of states and awaited events, the control process is going to be extremely simple, e.g., its code method may look like this:

```
perform {
        state Loop:
                if (plant_input_not_available) {
                        wait_for_plant_input (Loop);
                        sleep;
                }
                get_plant_input ( );
                do_control_loop ( );
                sameas Loop;
}
```

All we need is a single state and a single event amounting to the reception of a new reading from the plant. The rest is a straightforward execution of non-blocking code. One potentially problematic item, not immediately visible in the above pseudocode, is the operation of sending the input to the plant. We may assume that it is included in *do_control_loop* and that it will cause no blocking (and thus no need for introducing additional events). The clock cycling the control loop is provided by the input readings from the plant. Those readings will be arriving on the plant's mailbox, and their arrivals will provide the triggering events strobing the control loop.

The same simple structure also befits the operator process, which will wait for lines of text arriving from the operator and decode from them new values for the setpoint. The reason why we prefer this interface to be textual is the ease of implementing a trivial, external operator to test our controller, e.g., by entering data directly from a *telnet* of *putty* terminal connected to the appropriate port.

The situation is like the already handled case of two slightly different connector processes, which could be serviced by the same code method inherited by two formally separate process types. Thus, we shall once again declare an abstract process class providing the frame of a generic callback process with a single state:

```
process Callback abstract (Plant) {
        states { Loop };
        virtual void action (int) = 0;
        perform {
                state Loop:
                        action (Loop);
                        sameas Loop;
        };
};
```

The virtual method *action* captures the entire action that the process is supposed to carry out after receiving the triggering event. The single argument of the method identifies the *Loop* state, because the method must also issue the wait request for the trigger.

The actual processes are declared as follows:

```
process Control : Callback (Plant) {
        void action (int st) { S->control (st); };
};
process Operator : Callback (Plant) {
        void action (int st) { S->setpoint (st); };
};
```

and look even simpler, because their actions are delegated to the respective methods of the Plant station. Here is the control action:

```
void Plant::control (int st) {
        int32_t output, input;
        int nc;
        double derivative, error, setting;
        if ((nc = CI->read ((char*)(&output), 4)) == ERROR) {
                delete CI;
                CP->terminate ( );
                CP = NULL;
                sleep;
        }
        if (nc != ACCEPTED) {
                CI->wait (4, st);
                sleep;
        }
        error = (double) Setpoint – (double) bounded (output, VMin, VMax);
        integral += error;
        derivative = error – lasterr;
        lasterr = error;
        setting = Kp * error + Ki * integral + Kd * derivative;
        input = (int32_t) bounded (setting, SMin, SMax);
        CI->write ((char*)(&input), 4);
}
```

The key to understanding the triggering mechanism is in the operations on the plant mailbox pointed to by *CI*. The *read* method tries to extract from the mailbox the indicated number of bytes (4) and put them into the buffer pointed to by the first argument. Those bytes (if available in the mailbox) amount to a 4-byte integer value representing the new output of the plant. Note that *output* and *input* have been explicitly declared as 32-bit integers to make sure that they can always be safely handled as 4-byte structures.[7]

If the mailbox contains at least four bytes, the first four bytes are extracted from the it and stored in the *output* variable. The method then returns *ACCEPTED*. Otherwise, if the data is not available, *read* returns *REJECTED*. In the latter case, the process should wait for the so-called *count* event, i.e., the nearest moment when the number of bytes available in the mailbox reaches the specified threshold. Note that the buffer size of the plant mailbox is 4, so it can never contain more than 4 bytes at a time.

One exceptional situation, that needs to be taken care of, is when the *read* method returns *ERROR*, which means that the client (i.e., the plant) has disconnected from the controller. In such a case, the mailbox, as well as the *Control* process, are destroyed.

Having received a new output value, we are ready to run a control cycle and produce a new input to be passed to the plant. The role of *bounded* is to guarantee that the plant parameters are between their limits. The operation is a macro defined as:

#define bounded(v,mi,ma) ((v) > (ma) ? (ma) : (v) < (mi) ? (mi) : (v))

i.e., it returns the value of the first argument forced not to exceed the specified minimum and maximum.

The new input value is written to the mailbox as a binary sequence of four bytes. In principle, the *write* operation might be *REJECTED*, if the socket behind the mailbox couldn't immediately accept the four bytes (there are mailbox events to provide for that, see Sect. 9.4.6). In practice, this is not possible for as long as the other party (i.e., the plant) remains connected to the server. The input to the plant is only sent as often as the output is available, so the plant cannot be overwhelmed by an avalanche of input data that it couldn't absorb.

The operator's action (*setpoint*) illustrates a slightly different variant of data extraction from a mailbox:

---

[7] Strictly speaking, as the PID server can be made visible across the Internet (and different computer architectures) we should also account for the endianness of the numbers passed as sequences of bytes (operations *htonl* and *ntohl* [8]). We skip this step for clarity, assuming that the server as well as the plant execute on machines with the same endianness. Note that the operator is exempt from problems of this kind, because its interface is textual.

```
void Plant::setpoint (int st) {
        char buf [OBSIZE];
        int s, nc;
        if ((nc = OI->readToSentinel (buf, OBSIZE)) == ERROR) {
                delete OI;
                OP->terminate ( );
                OP = NULL;
                sleep;
        }
        if (nc == 0 || sscanf (buf, "%d", &s) != 1) {
                OI->wait (SENTINEL, st);
                sleep;
        }
        Setpoint = (int32_t) bounded (s, VMin, VMax);
}
```

The pattern of acquiring data from the mailbox is essentially the same, except that *readToSentinel* is used instead of *read*. The method extracts from the mailbox a sequence of bytes up to the first occurrence of the sentinel byte. Recall that the sentinel was declared when the mailbox was connected to the operator (see the *service* method of the *OConnector* process above). The second argument of *readToSentinel* limits the length of the extracted string to the buffer size. The method returns the number of extracted characters with zero meaning that no sentinel has been found in the mailbox (see Sect. 9.4.7). The sequence of bytes is assumed to be a character-string representation of the new value of *Setpoint*.

2.3.3 Playing with the Controller

To connect to the controller, the plant must become a TCP client. Note that the controller could present a simpler interface, e.g., via a serial port, which would probably make more sense in a realistic application. For verification, we can substitute a dummy plant following the simple format of conversation expected by the controller. It makes perfect sense to program such a plant in SMURPH. We don't have to be too precise, because the plant model doesn't have to be very realistic. This set of global variables captures the data needed by the model:

int ServerPort, VMin, VMax, Target, SMin, SMax;
int32_t Output, Setting;
Mailbox *CInt;

Their names are self-explanatory, especially with reference to the preceding section. *Target* stands for the "target" output that the plant will eventually reach, if its current input (variable *Setting*) persists for a sufficiently long time. *CInt* points to the client mailbox connecting the plant model to the controller. Here is the complete *Root* process:

```
process Root {
        states { Start, Sensor, Update };
        void setup ( ) {
                readln (ServerPort);
                readln (VMin);            // Read and ignored
                readln (VMin);
                readln (VMax);
                readln (Target);
                readln (SMin);
                readln (SMax);
                Output = (int32_t) VMin;
                Setting = (int32_t) SMin;
        };
        perform {
                state Start:
                        CInt = create Mailbox;
                        if (CInt->connect (INTERNET + CLIENT, SERVER_HOST,
                            ServerPort, 4) != OK)
                                excptn ("Cannot connect to controller");
                        create Actuator;
                transient Sensor:
                        CInt->write ((char*)(&Output), 4);
                        Timer->delay (DELTA_INTERVAL, Update);
                state Update:
                        update ( );
                        sameas Sensor;
        };
};
```

For simplicity, the model uses the same data set as the controller; the operator port, not needed by the model, is read and ignored. The plant is initialized at its minimum input and output.

In its first state, the process creates the mailbox and connects to the controller server. The configuration of arguments of the *connect* method used for this occasion includes *SERVER_HOST*, which is the Internet address of the machine running the server. We have set the constant to *"localhost"* as we intend to run both programs on the same system. The buffer size (the last argument) is 4, because the mailbox will be used to exchange 4-byte binary data (the plant's input and output) acting as a sensor/actuator combo. The role of the sensor process is assumed by *Root* (in states

Sensor and *Update*), and a separate process (*Actuator*) is created for the actuator role. The former consists in sending periodically over the mailbox the current output of the plant at *DELTA_INTERVAL* (set to 1.0 s).

Let us comment now on the interpretation of time in a control program. Note that both the PID controller and the plant model are formally control programs (executing in the control mode, Sect. 1.3.1), because they both interface to the outside world (via bound mailboxes). As it happens, there is no single *delay* (or timer *wait*) operation in the controller (the program only responds to mailbox events), so we only have to deal with the time issue in the plant model.

In contrast to the simulation mode, the relationship between the internal and external time in the control mode is necessarily explicit, because the program does run in real time. Thus, 1 *ETU* is pinned to 1 s. The default granularity of time, i.e., the setting of 1 *ITU* is 1 μs, i.e., 0.000001 s. It is legal to redefine that setting, with *setEtu* (Sects. 2.1.4 and 3.1.2), but one must remember that the *ITU* is defined with respect to 1 s of real time. The argument to the *delay* method of *Timer* is always expressed in straightforward seconds and internally converted to the closest (integer) number of *ITUs*.

The *Actuator* process looks like this:

```
process Actuator {
        states { Loop };
        perform {
                state Loop:
                        int32_t val;
                        int nc;
                        if ((nc = CInt->read ((char*)(&val), 4)) == ERROR)
                                excptn ("Connection lost");
                        if (nc != ACCEPTED) {
                                CInt->wait (4, Loop);
                                sleep;
                        }
                        val = bounded (val, SMin, SMax);
                        if (val != Setting) {
                                Setting = val;
                                set_target ( );
                        }
                        sameas Loop;
        };
};
```

In its single state, the process reads the plant input arriving from the controller (we saw similar code in Sect. 2.3.2), and when the plant input changes (the received value is different from the previous *Setting*), it invokes *set_target* to redefine the plant's target state. The plant's behavior is described by the functions *set_target* and *update*. The first function determines the output that the plant will (eventually) converge at for the given input (if it stays put long enough), the second function executes incremental steps towards the target, i.e., produces a sequence of output updates to report along the way.

We can program the two functions in many ways, possibly even to approximate the real-time behavior of some actual plant. Here is a naive, although not completely trivial idea for *set_target*:

```
void set_target ( ) {
        double s;
        s = (((double) Setting — (double) SMin) * 2.0 / (SMax — SMin)) — 1.0;
        if (s >= 0.0)
                s = pow (s, TTR_ROOT);
        else
                s = —pow (—s, TTR_ROOT);
        s = round ((double) VMin + ((VMax — VMin) * (s + 1.0) / 2.0));
        Target = (int32_t) bounded (s, VMin, VMax);
}
```

Note that the function is only executed when *Setting* (the plant's input) changes. The first statement transforms *Setting* linearly into a *double* value between -1 and 1. If the value ends up positive (falling into the upper half of its range understood as the interval [*SMin*, *SMax*], it is subsequently transformed (deformed from the linear shape) while retaining the endpoints (*TTR_ROOT* is defined as 0.25, i.e., we mean taking the 4-th root of the original value). For a negative value, the transformation is symmetrical w.r.t. the origin. The function looks as shown in Fig. 2.3, and represents the normalized relationship between the plant's input and output.

Here is how the plant converges to the target output:

```
void update ( ) {
        double DOutput, DTarget;
        DOutput = Output;
        DTarget = Target;
        if (DTarget > DOutput)
                DOutput += (DTarget — DOutput) * HEATING_RATE;
        else
                DOutput += (DTarget — DOutput) * COOLING_RATE;
        Output = (int32_t) round (DOutput);
}
```

Recall that *update* is invoked regularly at 1 s intervals. The function simply increments or decrements the output towards the target by a fixed fraction of the residual difference. Both *HEATING_RATE* and *COOLING_RATE* are constants defined as 0.05.

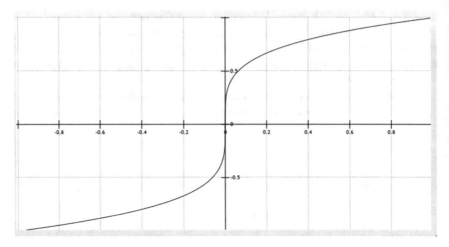

Fig. 2.3 The plant's target output versus *Setting*

The controller program can be found in directory *Examples/REALTIME/PID*, with the plant model in subdirectory *Plant*. The sample data set (shared by both programs) is in file *data.txt* and consists of these numbers:

Plant port	*3798*
Operator port	*3799*
VMin	*250*
Vmax	*5000*
Initial setpoint	*250*
SMin	*0*
SMax	*1023*
Kp	*2.0*
Ki	*0.1*
Kd	*0.2*

which are somewhat accidental, but can be used as a starting point for experiments. Both, the controller as well as the plant should be compiled with this command (see Sect. B.5):

mks –R

Then the controller should be run first and the plant second (on the same data). The simplest way to make the plant/controller state visible is to add a line like this:

trace ("O: %1d [S: %1d, E: %1.0f, I: %1.0f, D: %1.0f, V: %1d]",

output, Setpoint, error, integral, derivative, input);

as the last statement of the *control* method of the *Plant* station (Sect. 2.3.2). The line will show (at the control-loop rate, i.e., at 1 s intervals) the current output of the plant, the setpoint, the error, the accumulated integral of the error, the current derivative (i.e., the increment of error in the current step), and the new computed input for the plant.

An easy way to change the setpoint is to connect to the controller from a telnet terminal, e.g.,

telnet localhost 3799

and enter the new value, e.g., 3000 (note that the range is 250–5000, as per the data set above). With the *trace* line incorporated into the controller, we shall see a sequence of updates from which the plant's output is the most interesting item. The consecutive values illustrating the plant's response are plotted in Fig. 2.4. The controller overshoots a bit before converging on the setpoint. To reduce the overshot, we can play with the three factors. By reducing K_i to 0.05 and increasing K_d to 0.5 we get the behavior shown in Fig. 2.5.

Tuning the PID factors for best performance is generally a challenging goal, especially that real-life plants tend to exhibit inertia and noise in their behavior. The latter may have a serious impact on the differential component, which is based on the difference in error. If the error tends to be noisy, the difference can become random, especially as the plant approaches the setpoint [10].

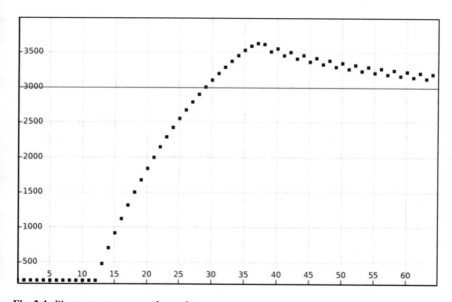

Fig. 2.4 Plant output versus cycle number

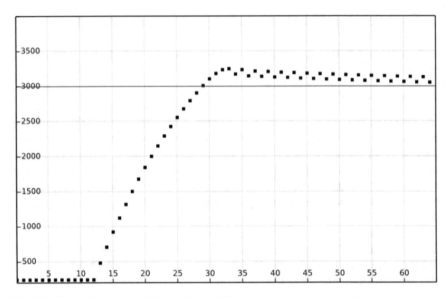

Fig. 2.5 Plant behavior after adjusting K_i and K_d

2.4   A Wireless Network Model

In this section, we illustrate how one can build models of wireless networks in SMURPH. Wireless networks are generally more difficult to model than wired ones for these main reasons:

1. The wireless medium is more temperamental than a piece of wire, because of attenuation, interference, and so on. One of the goals of a realistic model of a wireless network is to capture the releated problems, features, and phenomena.
2. Wireless links and networks (as opposed to wired ones) are often deployed for flexibility. This means that we may want node mobility to be an essential feature of the model.

In a nutshell, there are more features to model in a wireless network than in a wired one, and, consequently, the model is going to have a larger number of parameters. Some of those parameters, e.g., the channel description (the way bits and packets propagate and interfere, and how those phenomena impact their reception) cannot be turned into a small set of pre-programmed options adjustable with a few numbers. This is because, in many cases, the wireless medium is one of the essential components of the modeled network, and its behavior strongly depends on some idiosyncratic features of the environment, network hardware, etc. There exists a large number of wireless channel models, and new ones are created every day [11–13]. We just cannot anticipate them all and equip the simulator with their ready implementations that would be good once and for all.

2.4.1 The ALOHA Network

For our illustration, we shall use the ALOHA network deployed in the early 70's at the University of Hawaii. While it can hardly be considered an example of modern communication technology, it will do perfectly for our little exercise, because, to see the essential features of SMURPH supporting wireless network models, we need to focus just on the "wireless," without the numerous distracting features in which advanced contemporary networking solutions so painfully abound. The network seems to strike a good compromise between the simplicity of concept and the number of interesting details whose models will make for an educational illustration.

The ALOHA network [14–16] is of some historical importance, so, despite its age, it makes perfect sense to honor it in our introduction. It was the first wireless data network to speak of, and its simplistic (as some people would say) protocol paved the way for a multitude of future networking solution, not only in the wireless world [17].

The network's objective was to connect a number of terminals located on the Islands of Hawaii to a mainframe system on the Oahu Island (see Fig. 2.6). In those days (the year was 1970), it was a standard telecommunication problem to provide remote access to highly centralized computing facilities from (extremely slow by contemporary standards) terminals. Such terminals would typically offer batch capabilities (reading-in jobs on punched cards and receiving printouts on local printers) plus rudimentary, command-line, interactive, time-sharing access to the mainframe.

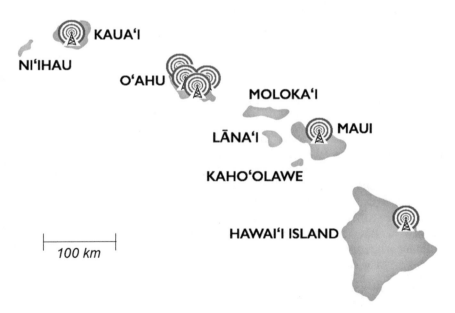

Fig. 2.6 The ALOHA network

The terminals did not communicate among themselves, at least not directly, mostly because there was no need for that. They were basically interface devices with little interesting features of their own. The traffic consisted of terminal data addressed to the mainframe, and mainframe data addressed to a terminal, one at a time. The two traffic types were assigned to two separate UHF channels.[8] From the viewpoint of networking, the mainframe system acted as a *hub*. While a terminal could cause a message to be sent to another terminal, such an action involved the hub, because the terminal could only talk directly to the hub.

The hub-to-terminal traffic incurred no channel multiplexing problems, because the hub was the sole node transmitting on the channel. On the other hand, the terminal traffic was prone to contention and collisions. When multiple terminals transmitted simultaneously, their packets interfered and stood a good chance of being destroyed. The network employed a naive scheme to cope with this problem, which nonetheless proved effective under the circumstances and seminal for a broad range of future solutions. When the terminal received a new packet to transmit to the hub, it would do so right away. If no acknowledgement packet arrived from the hub after the prescribed amount of time, the terminal would retransmit the packet after a randomized delay. This procedure would continue until an acknowledgement would eventually materialize.

A packet in ALOHA consisted of a receiver synchronization sequence, a 32-bit identification sequence, a payload of up to 80 bytes (640 bits), and a 32-bit error detection (CRC) code.[9] The bit rate was 9600 bps.[10] The receiver synchronization sequence, which today we would call a *preamble*, was equivalent to 90 bits. The transmission time of a complete packet, including the preamble, was $t = (90 + 64 + 640)/9600 \approx 83$ ms. The payload length of 80 bytes may appear exotic from the contemporary point of view, but it was extremely natural in those days: 80 bytes was the amount of information encoded on one punched card.

2.4.2   The Protocol Specification

In Sects. 2.2.2–2.2.5, we saw a model of a wired network consisting of two nodes (SMURPH *stations*). We learned that connections in wired networks are built of *links* and *ports*, the latter acting as taps into the former. The corresponding objects for wireless communication are *rfchannels* and *transceivers*. As we said earlier, the trickiest (and most controversial) part of a wireless network model is usually the model of the wireless medium. SMURPH does offer a small library of ready wireless channel models (see Appendix A), but we shall see here how one gets around to building such a model from scratch. But let us assume for now that the channel model has been built already: its actual construction can be comfortably postponed

---

[8]The frequencies were 407.350 and 413.475 MHz.

[9]That code was split into two 16-bit parts applicable separately to the header and to the payload.

[10]In [14], the rate of 24,000 bps is mentioned, but the later paper [15] gives the figure of 9600 bps.

until the very end. We shall focus now on a specification of the algorithms run by the ALOHA network, which would provide us with enough detail for a gratifying SMURPH model.

The model will consist of several stations. All but one of them will act as terminals, the remaining one will be the hub. This much is rather clear. What about the actual protocol?

Looking at the original, seminal paper [14], we only see there a rather informal description of the function carried out by the terminals, giving us little more than the contents of the last two paragraphs from the preceding Sect. 2.4.1. Based on it, we can arrive at the following outline of the terminal transmission protocol:

1. Acquire a packet for transmission (from the network's application, i.e., the *Client*). Wait until a packet is available.
2. Transmit the packet.
3. Wait for an acknowledgment and also for a randomized timeout.
4. If the acknowledgment arrives first, clear the timeout and resume from 1.
5. If the timer goes off instead, continue at 2.

Then, the terminal also runs a receiver which handles two types of packets addressed to the node and arriving from the hub. A regular data packet is passed to the *Client* while an acknowledgment is treated as an event awaited by the transmitter in step 3.

Here is how much we can infer about the behavior of the hub transmitter (which seems to be considerably simpler):

1. Acquire a packet for transmission. This can be an application (*Client*) packet or an acknowledgment triggered by the receiver. Both packet types are always addressed to one specific terminal. If no packet is ready for transmission, wait until one becomes available.
2. Transmit the packet, then continue at 1.

The hub also runs a receiver accepting packets arriving from the terminals and passing them to the *Client*. Those packets trigger acknowledgments which are queued for transmission.

The above description has the flavor of a typical protocol specification one can find in a research paper. It may suffice for building a simple (toy) simulation model, but it will not do as a blueprint of the real-life network. Our goal is somewhere in between, but probably closer to the latter. This is what makes SMURPH different from other network simulators. While we do not intend to put together actual hardware compatible with the original equipment of ALOHA, we want to try to come close to its authoritative virtual replica. Here are some of the relevant questions that the above specification leaves unanswered:

1. What is the distribution of retransmission delays in step 3 of the terminal transmitter?
2. How are the acknowledgments matched to the packets being acknowledged?
3. How does the receiver/application detect retransmitted duplicates of previously received packets?
4. How is the cooperation between the transmitter and the receiver organized? Do they look like separate processes operating in parallel and synchronizing at one point?
5. Does the terminal acknowledge received hub packets? What is the mechanism of those acknowledgments?

Looking up detailed answers to the above questions is not easy, and we won't find them in [14]. This is understandable, because the purpose of that paper was not to offer a precise specification of the protocol, but rather to focus, in its limited space of a conference paper, on one novel and theoretically intriguing aspect of the scheme, viz. the random-access, shared, terminal-to-hub channel, and estimate analytically its maximum effective throughput.

As a side note, collecting enough details to build a high-fidelity simulation model, or, put differently, obtaining an adequate specification of the system for its authoritative virtual implementation, is often difficult in simulation studies of telecommunication systems. In those few cases where the system has been implemented in real life, one can try to get hold of (and then plough through) the technical documentation. Probably, if we tried hard, we could procure the original specs of the ALOHA hardware and software; however, we won't go that far, because the effort would be substantial and, with all our excitement about ALOHA, we do not care *this* much. After all, this is merely an exercise of no direct, material consequences. But, generally, the problem is quite serious, as we argued in [18, 19], and in several other papers. Especially for academic (theoretical) protocol concepts, which are often "specified" in research papers at a level of detail comparable to that of [14], the descriptions leave a lot of room for guessing and "creative" misrepresentations of the authors' ideas. This often translates into sloppy simulation models whose role becomes to support (simplified) analytical studies, rather than investigate how a realistic embodiment of the abstract idea would fare in the real world. The next tempting step is to extrapolate that sloppiness onto the models of known, well-established, well-documented, real-life systems, whose simulated performance is compared to that of their theoretically proposed improvements. Performance studies of such a stature can be confusing and damaging to the reputability of the academic community in practical telecommunication [20]. As the ALOHA network provides a never-ending inspiration for student assignments and projects, one can find on the Internet numerous student papers investigating its performance by simulation. Virtually all of them are naively superficial contenting themselves with idealistic models of the terminal-to-hub traffic, usually assuming (to name just one popular fallacy) that the retransmission delays (step 3 of the terminal transmitter above) are exponentially distributed. Exponential distribution in something supposedly realistic always smells of a detachment from reality. People use it in analytical models, because it makes the models tractable (or just plain easy), but the distribution is impractical, because of its infinite tail. Even if we didn't know how exactly the retransmission delays were generated in the real-life ALOHA network, we could safely bet that their distribution was not exponential. Abramson used that assumption in [14] to arrive at his analytical formula bounding the maximum throughput of the scheme. It is one of those (mostly innocent) simplifications that people who are into analytical performance studies make without as much as a blink. But it has no implications for practical designs, and it should not impact simulation studies, which need not be constrained by the limitations of analytical models.

The later paper [15] brings in more implementation details also covering the evolution of the original network since its inception. The distribution of retransmission delays for the terminal node is described there as uniform between 0.2 and 1.5 s. The numerical parameters for our virtual implementation are taken from [15], and they only partially agree with those given in [14]. Still not everything is 100% clear. The maximum packet transmission time is listed as 73 ms which seems to exclude the preamble. The authors argue that the preamble can be reduced from 90 bits to 10 bits, or even eliminated, without impairing packet synchronization, but the status of those suggestions is not clear. Apparently, as the network evolved, new ideas were developed and tried resulting in some diversification of the equipment and the algorithms. In particular, repeaters and concentrators were developed to extend the network's range over all the islands of Hawaii. We shall not concern ourselves with those enhancements: the ALOHA network is history now, and it would be pointless to turn our simulation exercise into a study in archaeology.

While it was obvious from the very beginning that the terminal-to-hub traffic had to be acknowledged (the acknowledgments were part of the scheme's very definition), the issue of acknowledging the hub-to-terminal traffic was considered secondary, because the hub-to-terminal channel was contention-free and (theoretically) interference-free. In consequence, acknowledgments on that channel were basically delegated to the application, and the exact procedures were complicated and messy. To make it easier on ourselves, we shall thus ignore the issue, assuming that the hub-to-terminal traffic is not acknowledged in our model. We couldn't really do much about it without introducing multiple traffic classes and applications, which would complicate things for no good reason.

The terminal devices were half duplex, meaning that the terminal transmitter and receiver were not independent processes. In other words, when the transmitter was active, the receiver was not. There was a switchover delay of ca. 10 ms. Following a packet transmission, the activation of the receiver (e.g., to receive the acknowledgment for that packet) took that much time.[11]

Note that a packet in ALOHA only needs one address. If the packet is sent on the hub-to-terminal channel, the address identifies the terminal to receive the packet (the hub being the only possible sender). Conversely, for a packet transmitted on the terminal-to-hub channel, the same address identifies the sender (with the hub

---

[11]Originally, in the first version of the terminal equipment, the delay was 100 ms, because the switchover was handled by electro-mechanical relays.

being the only possible recipient). The maximum number of terminals in ALOHA was limited to 256, so the address field was one byte [15]. While we don't have to concern ourselves with the exact layout of the packet header (it is never mentioned in the papers), we see that all the requisite header information can be comfortably encoded on 32-bits (the address field only takes one byte). Another byte was used to specify the payload length (0–80). Packet duplicates were detected with a single (alternating) bit in the header. The same bit was used to match acknowledgments to packets (note that a terminal is not supposed to send a new packet until the previous one has been acknowledged). Throwing in one additional bit to tell acknowledgments from data packets, we use up 18 bits of the header.

The hub, in contrast, operates in a fully duplex manner. Its transmitter must deal with regular (*Client*) packets addressed to the terminals as well as the acknowledgments sent in response to packets received from the terminals. The acknowledgments have priority in the sense that if there is an acknowledgment awaiting transmission, it will be sent before the next outgoing data packet, but without disrupting a possible data transmission in progress. To make it a bit more difficult, the hub makes sure to delay the acknowledgment by at least 10 ms following the reception of the data packet being acknowledged. The delay is needed to account for the terminal's switchover to receive after the transmission. As explained in [15], the outgoing acknowledgments formed a queue which was examined by the transmitter before the *Client* queue. It very seldom happened that the queue contained more than one pending acknowledgment at a time.

2.4.3   The Hub Station

Our model of the ALOHA network can be found in subdirectory *Aloha* of *Examples/WIRELESS*. It makes sense to assume that the hard, numerical parameters of the protocol are constants, so they don't have to be read from the input file. Therefore, the header file starts with these definitions:

```
#define DATA_RATE            9600
#define PACKET_LENGTH        (80 * 8)
#define ACK_LENGTH           0
#define FRAME_LENGTH         (32 + 32)
#define PREAMBLE_LENGTH      90
#define MIN_BACKOFF          0.2
#define MAX_BACKOFF          1.5
#define SWITCH_DELAY         0.01
#define ACK_DELAY            0.02
```

Note that the timing parameters are specified in seconds, so we shall make sure to define the *ETU* to correspond to one second of real time. *ACK_DELAY* represents the delay preceding the acknowledgment sent by the hub after a data packet reception, compensating for the switchover delay of the terminal from transmit to receive. We don't know that delay exactly, but considering that the switchover delay (*SWITCH_DELAY*) is 0.01 s (10 ms), it probably makes sense to set it to twice that value, to play it safe.

We shall use these packet types:

packet DataPacket { unsigned int AB;};
packet AckPacket { unsigned int AB;};

which look identical except for the type name. It will be easy in the model to tell one packet type from another (given a generic packet pointer) by its association with the application traffic, i.e., the *Client* (we will see that shortly), so the alternating bit is the only extra attribute needed by the model.

Let us start from the hub station. One non-trivial data structure needed by the hub is the queue of outgoing acknowledgments. We shall not store in that queue the actual acknowledgment packets, but rather the specific information to be inserted into a preformatted acknowledgment packet before it is transmitted, i.e., the recipient (terminal) address and the alternating bit setting. Here is the layout of an item stored in the acknowledgment queue:

```
typedef struct {
        unsigned int AB;
        Long Terminal;
        TIME when;
} ackrq_t;
```

Station identifiers (useful as network node addresses) are *Long* integers in SMURPH.[12] Attribute *when* will be used to implement the minimum delay separating the acknowledgment from the reception of the acknowledged packet. It will be set to the earliest time when the acknowledgment can be expedited.

The most natural way to implement the queue is to use a SMURPH *mailbox*. This way we can automatically trigger and perceive events when the queue becomes nonempty. The mailbox is defined this way:

---

[12]This does not conflict with the 8-bit terminal address in ALOHA. Data representation in a SMURPH structure and the formal length of the packet carrying it are two different things.

```
mailbox AckQueue (ackrq_t*) {
        void newAck (Long rcv, unsigned int ab) {
                ackrq_t *a = new ackrq_t;
                a->Terminal = rcv;
                a->AB = ab;
                a->when = Time + etuToItu (ACK_DELAY);
                put (a);
        };
        Boolean getAck (AckPacket *p, int st) {
                ackrq_t *a;
                TIME w;
                if (empty ( )) {
                        wait (NONEMPTY, st);
                        return NO;
                }
                if ((w = first ( ) -> when) > Time) {
                        Timer->wait (w – Time, st);
                        return NO;
                }
                a = get ( );
                p->Receiver = a->Terminal;
                p->AB = a->AB;
                delete a;
                return YES;
        };
};
```

The two methods provide the interface to the mailbox. The first one (*newAck*) is invoked to add an acknowledgment to the queue. It allocates the requisite record, fills it in, and deposits its pointer into the mailbox, which has the effect of appending the new acknowledgement at the end of the queue. The earliest transmission time for the acknowledgement is calculated by adding *ACK_DELAY* (transformed into *ITUs*) to the current time.

With *getAck*, the caller may attempt to retrieve the first acknowledgment from the queue. The method takes two arguments: a pointer to the packet buffer containing the actual, vanilla acknowledgment packet, whose *Receiver* and *AB* attributes are to be filled, plus a process state identifier. If the queue is empty, the method issues a wait request to the mailbox for a *NONEMPTY* event to awake the invoking process in the specified state. Otherwise, the first entry is examined. If the *when* attribute is greater than *Time* (meaning that the acknowledgment cannot be transmitted right away, because of the need to obey the terminal switchover delay), the method issues a *Timer* wait request for the difference. Otherwise, the acknowledgment packet is filled with the target values of *Receiver* and *AB*, and the queue entry is removed and deallocated. The method returns *YES* on success (when the acknowledgment is ready) and *NO* on failure (when the caller must wait).

The two station types are derived from the same base type defined as follows:

```
station ALOHADevice abstract {
        DataPacket Buffer;
        Transceiver HTI, THI;
        void setup ( ) {
                double x, y;
                readln (x);
                readln (y);
                HTChannel -> connect (&HTI);
                THChannel -> connect (&THI);
                setLocation (x, y);
        };
};
```

which contains all the common components, i.e., a data packet buffer and two *transceivers* interfacing the station to the two RF channels (*HTI* and *THI* stand for hub-to-terminal and terminal-to-hub interface). Note that the transceivers are declared as *Transceiver* class variables within the station class (the variables are not pointers). This is an option, also available for ports (Sects. 4.3.1, 4.5.1). A port or a transceiver can be created explicitly (e.g., as we did in Sect. 2.2.2), or declared as a class object (like above), in which case it will be created together with the station. In the latter case, there is no way to pass parameters to the port/transceiver upon creation, but they can be set individually (by invoking the respective methods of the objects, see Sect. 4.5.2) or inherited from the underlying link or rfchannel (Sects. 4.2.2 and 4.4.2). In some cases, this is more natural than creating (and parameterizing) the objects explicitly.

Like ports and links, transceivers must be *connected* to (associated with) rfchannels. *HTChannel* and *THChannel* are pointers to two objects of type *ALOHARF* which descends from the standard type *RFChannel*. We shall see in Sect. 2.4.5 how the wireless channels are set up in our model.

The *setup* method of the base station type reads two numbers from the input data file interpreted as the planar coordinates[13] of the station expressed in the assumed units of distance. The station's method *setLocation* assigns these coordinates to all the transceivers owned by the station. This is a shortcut for invoking *setLocation* of every transceiver. Strictly speaking, location coordinates are attributable to transceivers rather than stations: it is not illegal to assign different locations to different transceivers of the same station.

---

[13]The model assumes planar geometry with two location coordinates. It is possible to use three coordinates for a 3-d model (Sect. 4.5.1).

The hub station extends the base station type with a few more attributes and methods:

```
station Hub : ALOHADevice {
        unsigned int *AB;
        AckQueue AQ;
        AckPacket ABuffer;
        void setup ( );
        Packet *getPacket (int st)  {
                if (AQ.getAck (&ABuffer, st))
                        return &ABuffer;
                if (Client->getPacket (Buffer, 0, PACKET_LENGTH, FRAME_LENGTH)) {
                        unwait ( );
                        return &Buffer;
                }
                Client->wait (ARRIVAL, st);
                sleep;
        };
        void receive (DataPacket *p) {
                Long sn;
                unsigned int ab;
                sn = p->Sender;
                ab = p->AB;
                AQ.newAck (sn, ab);
                if (ab == AB [sn]) {
                        Client->receive (p, &THI);
                        AB [sn] = 1 − AB [sn];
                }
        };
};
```

The role of *AB* (which is an array of *unsigned int*) is to store the expected values of alternating bits for all the terminals (so the station can identify and discard duplicates of the received data packets on the per-terminal basis) . The array is initialized in the *setup* method whose definition must be postponed (see below), because of its need to reference process types defined later. Method *getPacket* implements the operation of acquiring a packet for transmission in a way that gives acknowledgments priority over data packets. The method returns the pointer to the packet buffer, which can be either the acknowledgment buffer or the data packet buffer. If neither packet is available, the method does not return: the process is put to sleep waiting for two events, the arrival of an acknowledgment (see *getAck* in *AckQueue*) or a *Client* packet (see Sects. 2.2.3 and 6.8), whichever comes first. Both events will wake the process up in the state passed as the argument to *getPacket*.

Incidentally, the role of *unwait* invoked in *getPacket* is to deactivate all the wait requests that have been issued by the process so far (Sect. 5.1.4). It is needed, because, by the time we realize that the outcome of *Client->getPacket* is successful, the *getAck* method of *AckQueue* has already issued a wait request. That request must be retracted now, because it turns out that the process is not going to sleep after all, at least not for that reason.

The second method, *receive*, will be invoked after the reception of a data packet. It triggers an acknowledgment (appended to the queue with *newAck*) and passes the packet to the *Client* (Sects. 2.2.4 and 8.2.3), but only if the alternating bit of the received data packet matches the expected value for the terminal. In that case, the expected alternating bit is also flipped.

We are now ready to see the two processes run by the hub station. Here is the transmitter:

```
process HTransmitter (Hub) {
        Packet *Pkt;
        Transceiver *Up;
        states { NPacket, PDone };
        void setup ( ) {
                Up = &(S->HTI);
        };
        perform {
                state NPacket:
                        Pkt = S->getPacket (NPacket);
                        Up -> transmit (Pkt, PDone);
                state PDone:
                        Up->stop ( );
                        if (Pkt == &(S->Buffer))
                                S->Buffer.release ( );
                        skipto NPacket;
        };
};
```

The process is almost trivial. Confronting it with the transmitter from Sect. 2.2.3, we see that the way to transmit a packet over a wireless interface closely resembles the way it is done on a wired port. The process blindly transmits (on the *HTI transceiver*) whatever packet it manages to acquire (by invoking the *getPacket* method of the hub station) and then *releases* the packet buffer, to tell the *Client* that the packet has been expedited, if it happens to be a data packet (as opposed to an acknowledgment). Note that hub packets are not acknowledged, so the process fulfills its obligation to the *Client* as soon as it finishes the packet's transmission.

In a realistic implementation of the protocol, there should be some minimum spacing between packets transmitted back-to-back, i.e., some delay separating the end of the previous packet from the beginning of the next one. We do not know this delay, so the process assumes "the next thing to zero," i.e., the delay of 1 *ITU* provided by *skipto*.

The receiver is no more complicated, although its brief code method calls for a bit of discussion:

```
process HReceiver (Hub) {
        Transceiver *Down;
        states { Listen, GotBMP, GotEMP };
        void setup ( ) {
                Down = &(S->THI);
        };
        perform {
                state Listen:
                        Down->wait (BMP, GotBMP);
                        Down->wait (EMP, GotEMP);
                state GotEMP:
                        S->receive  ((DataPacket*)ThePacket);
                transient GotBMP:
                        skipto Listen;
        };
};
```

The *transceiver* events *BMP* and *EMP* are similar to their *port* counterparts (Sects. 2.2.3 and 8.2.1). So, *BMP* is triggered when the transceiver sees the beginning of a packet addressed to *this* station, and *EMP* marks the end of such a packet. In contrast to the wired case, the circumstances of the occurrence of those events can be qualified with potentially complex conditions in the (user) description of the wireless channel model. As we shall see in Sect. 8.2.1, one necessary (although not sufficient) condition for *EMP* is that the packet potentially triggering that event was earlier marked with *BMP*. The above code should be interpreted as follows.

In state *Listen*, the process waits for *BMP*. Even though it also seems to be waiting for *EMP*, that wait is only effective if a *BMP* was detected first. When a *BMP* event is triggered (meaning that the transceiver sees the beginning of a packet addressed to this station), the process transits to *GotBMP*. Nothing interesting seems to be happening in that state. The transition to *GotBMP* only marks the fact that the beginning of a packet (addressed to the station) has been spotted and that packet has been internally flagged as one that (potentially) can trigger *EMP* (when its end arrives at the transceiver). The process *skips* back to *Listen*. As we remember from Sect. 2.2.3, the transition with *skipto* makes sure that time has been advanced, so the *BMP* event (that triggered the previous transition to *Listen*) is not active any more. So, when the process gets back to *Listen*, it will be waiting for *EMP* and another possible *BMP* event. The only way for *EMP* to be triggered now is when the end of the packet that previously triggered *BMP* is detected. When that happens, the process will transit to *GotEMP*. If a *BMP* event occurs instead, it will mean that the first packet didn't make it, e.g., because of an interference, and the game commences from the top with another packet.

This procedure in fact closely resembles how a packet reception is carried out by many real-life receivers. First, the receiver tries to tune itself to the beginning of a packet, typically by expecting a preamble sequence followed by some delimiter. That event will mark the beginning of the packet's reception. Then the receiver will try to collect all the bits of the packet up to some ending delimiter, or until it has collected the prescribed number of bytes based on the contents of some field in the packet header. That procedure may end prematurely (e.g., because the signal becomes unrecognizable). Also, at the formal end of the packet's reception, the receiver may conclude (e.g., based on the CRC code in the packet's trailer) that the packet contains bit errors (and its reception has failed). Both cases amount to the same conclusion: the end of packet event is not triggered. Of course, the receiver will also check if the packet is addressed to its node before marking it as received. Technically, this check can be carried out at the end of reception, or maybe earlier, after the reception of a sufficient fragment of the packet's header. Its implementation bears no impact on the outcome or, what is important from the viewpoint of accurate modeling, on the timing of the outcome, which is the moment of the actual, final, and authoritative reception of the complete packet. Thus, for convenience and brevity of description, the address check is logically combined in our model into the *BMP/EMP* events.[14]

In Sect. 8.1.2, we shall see how the events triggered by a packet being received are qualified by the channel model to account for signal attenuation, interference, background noise, and so on. For now, let us complete the model of the hub station with the following *setup* method:

```
void Hub::setup ( ) {
        ALOHADevice::setup ( );
        AB = new unsigned int [NTerminals];
        memset (AB, 0, NTerminals * sizeof (unsigned int));
        AQ.setLimit (NTerminals);
        ABuffer.fill (getId ( ) , 0, ACK_LENGTH + FRAME_LENGTH, ACK_LENGTH);
        create HTransmitter;
        create HReceiver;
}
```

Note that the size of the mailbox representing the acknowledgment queue is limited to *NTerminals* (this global variable tells the number of terminal stations). This may not be extremely important (the size need not be limited in the model), but this little detail brings in one more touch of realism into the model. In real life, the size of the acknowledgment queue would have to be limited (and the cost of storage in the prime days of ALOHA was many orders of magnitude higher than it is today). *NTerminals* is a safe upper bound on the number of acknowledgments that may ever have to be present in the queue at the same time.

---

[14]Of course, it doesn't absolutely have to. If, for some reason, the receiving process wants to model the timing of address recognition in the packet's header, it can delay for the prescribed amount of time and carry out an integrity assessment of the initial fragment of the packet, as received so far (e.g., applying the *dead* method of *Transceiver*, see Sects. 8.2.5 and 8.2.1).

2.4.4 The Terminals

The most natural way to model the behavior of a half-duplex node is to use a single process, which agrees with the way the device was run in real life. Therefore, a terminal station will run a single process carrying out both types of duties, i.e., transmission as well as reception. The terminal station structure extends the base type *ALOHADevice* in this way:

```
station Terminal : ALOHADevice {
        unsigned int AB;
        TIME When;
        void setup ( );
        Boolean getPacket (int st) {
                if (When > Time) {
                        Timer->wait (When – Time, st);
                        return NO;
                }
                if (Buffer.isFull ( ))
                        return YES;
                if (Client->getPacket (Buffer, 0, PACKET_LENGTH,  FRAME_LENGTH)) {
                        Buffer.AB = AB;
                        return YES;
                }
                Client->wait (ARRIVAL, st);
                return NO;
        };
        void backoff ( ) {
                When = Time + etuToItu (dRndUniform (MIN_BACKOFF, MAX_BACKOFF));
        };
        void receive (Packet *p) {
                if (p->TP == NONE) {
                        if (Buffer.isFull ( ) &&
                            ((AckPacket*)ThePacket)->AB == Buffer.AB) {
                                Buffer.release ( );
                                AB = 1 – AB;
                        }
                } else {
                        Client->receive (p, &HTI);
                }
        };
};
```

Attribute *AB* holds the alternating bit value for the next outgoing packet, and *When* is used to implement backoff delays separating packet retransmissions while the node is waiting for an acknowledgment. The role of *getPacket* is like that of its hub station counterpart, i.e., the method is called to try to acquire a data packet for transmission. In contrast to the hub case, there are no acknowledgment packets to transmit.

The method first checks if *When* is greater than *Time*, which will happen when the node is waiting on the backoff timer. In such a case, the variable holds the time when the backoff delay will expire, so *getPacket* issues a *Timer* wait request for the residual amount of waiting time and fails returning *NO*. Note that the terminal variant of *getPacket* doesn't automatically sleep (as was the case with the hub station). This is because the (single) half-duplex process executing the method must attend to other duties, if no packet is available for transmission.

If the node need not delay the transmission, the method checks whether the data packet buffer already contains a packet (the *isFull* method of *Packet*). This covers the case of a retransmission, i.e., a data packet was previously obtained from the *Client*, and is already available in the buffer, but it has not been transmitted yet. If the buffer is empty, *getPacket* tries to get a new packet from the *Client*. If the operation succeeds, the alternating bit of the new packet is set to *AB*. Otherwise, the method issues a wait request to the *Client* for the *ARRIVAL* event and fails.

The *backoff* method is executed by the station's process to set *When* to current time (*Time*) plus a uniformly distributed random delay between *MIN_BACKOFF* and *MAX_BACKOFF*. As the two constants express the delay bounds in *ETUs*, the value produced by *dRndUniform* is converted to *ITUs* before being added to *Time*.

The *receive* method is called after a packet reception. The standard *TP* (type) attribute of the received packet can be used to tell an acknowledgment from a data packet. For the latter, *TP* is the index of the *Client's* traffic pattern responsible for the packet (a number greater than or equal to zero). The default value of *TP* for a packet that doesn't originate in the *Client* (whose contents are set up by the *fill* method of *Packet*, see the *setup* method of the *Hub* station in Sect. 2.4.3) is *NONE* (−1).

When an acknowledgment packet is received, the method checks if the data packet buffer is full, which means that a pending, outgoing packet is present, and if its alternating bit equals the alternating bit of the acknowledgment. In such a case, the packet buffer is *released*, i.e., the *Client* is notified that the packet has been successfully expedited, and the packet buffer is marked as empty. The alternating bit for the next outgoing packet is also flipped. For a data packet arriving from the hub, the method simply executes the *receive* method of the *Client*. No duplicates are possible and packets can be lost. The *AB* attribute of the data packet is not used for the hub-to-terminal traffic.

Here is the single process run by the terminal station:

```
process THalfDuplex (Terminal) {
        Transceiver *Up, *Down;
        states { Loop, GotBMP, GotEMP, Xmit, XDone };
        void setup ( ) {
                Up = &(S->THI);
                Down = &(S->HTI);
        };
        perform {
                state Loop:
                        if (S->getPacket (Loop)) {
                                Timer->delay (SWITCH_DELAY, Xmit);
                                sleep;
                        }
                        Down->wait (BMP, GotBMP);
                        Down->wait (EMP, GotEMP);
                state GotEMP:
                        S->receive (ThePacket);
                transient GotBMP:
                        skipto Loop;
                state Xmit:
                        Up->transmit (S->Buffer, XDone);
                state XDone:
                        Up->stop ( );
                        S->backoff ( );
                        Timer->delay (SWITCH_DELAY, Loop);
        };
};
```

If we ignore the *if* statement at the beginning of state *Loop*, then the first three states from the code method of the above process look exactly like the code method of the hub receiver. The difference amounts to checking, on every firing of the *Loop* state, whether a packet transmission can commence now. If the result of that test is negative (the *if* fails), the process will be waiting for the nearest opportunity to transmit (this is taken care of by *getPacket*), transiting back to *Loop* when the opportunity materializes. That event will preempt the reception procedure.

When the *if* succeeds, i.e., a packet can be transmitted, the process waits for the switchover delay and transits to state *Xmit* where it starts the packet transmission. Then, in state *XDone*, the transmission is completed. The process generates a back-off delay until the retransmission and transits back to *Loop* after another switchover delay. The process will not be able to transmit anything at least for the minimum backoff delay, so it effectively switches to the "receive mode" expecting an acknowledgement.

Note that the current packet is removed from the retransmission cycle by the station's *receive* method upon the acknowledgment reception. The buffer status is then changed to empty, and a new packet acquisition from the *Client* is needed for the *if* statement in state *Loop* to succeed again.

One potentially controversial issue is whether to reset the pending backoff delay when the current outgoing packet has been acknowledged. The backoff timer is armed by setting *When* to the time (ahead of current time) when the backoff delay will expire. For as long as *When* is still ahead of current time (*When > Time*), *getPacket* will fail. If a packet is available for transmission while the condition holds, the transmission will be postponed. Note that *When* is set by the *backoff* method, but never reset. Thus, when the current outgoing packet is acknowledged, any residual backoff delay will apply to the next packet (should it become available before the delay expires).

The problem (if it is a problem) can be remedied by setting *When* to zero upon the receipt of an acknowledgment, e.g., after the packet buffer is *released* (in the *receive* method). One can voice arguments in favor of both solutions. The popular, informal description of the protocol states that a new packet arriving from the *Client* is transmitted right away, only its retransmissions are scheduled after a randomized delay. But if the terminal has several outgoing packets in its queue, then the boundaries between those packets will not be randomized, which is probably not what the "founding fathers" had in mind, because deterministic (and short) intervals between consecutively transmitted packets increase the potential for collisions. The problem was probably irrelevant from the viewpoint of the real ALOHA network, because the terminals had no buffer space to queue a nontrivial number of outgoing messages.

Here is the postponed *setup* method of the *Terminal* station:

```
void Terminal::setup ( ) {
        ALOHADevice::setup ( );
        AB = 0;
        When = TIME_0;
        create THalfDuplex;
}
```

Note that any value of *When* lower than or equal to *Time* (current time) effectively disables the backoff timer.

2.4.5 The Channel Model

Unlike the built-in models of wired links (Sect. 4.2), which are almost never extended by SMURPH programs, the built-in model of a wireless channel occurs in a single and necessarily open-ended variant. This means that the protocol program in SMURPH can/should extend the built-in model, because it does not provide enough function-ality for even a simple, naive operation. This is because there is no such thing as a generic, universal model of a wireless channel (as opposed to, say, a generic, univer-sal model of a piece of wire), and every wireless channel model to speak of requires some attention to idiosyncratic details.

The channel model that we are going to create for our ALOHA project is going to be rather simple and naive, as channel models go. We want to use it to illustrate how channel models are built, focusing on the role and interface of the different components provided by the user, without complicating those components beyond the basic needs of the illustration. SMURPH offers library models of more realistic channels (see APPENDIX A) which can be used without fully understanding how they interface to the simulator. Here we build one completely from scratch.

The definition of a wireless channel model involves declaring a class descending from the built-in type*RFChannel* and equipping it with a few methods. The idea is that the built-in components implement all the tricky dynamics needed to propagate packets through the wireless medium and intercepting all the events that the model might want to perceive, while the user-provided extensions are just formulas, i.e., usually straightforward functions transforming various values into other values and encapsulating the parameterizable parts of the model into easy to comprehend, static expressions. All this will be described in detail in Sects. 4.4.3 and 8.1. For now, we only need a general understanding of the unifying philosophy of all channel models, which can be explained as follows.

Packets and bits propagating through a wireless channel are characterized by properties affecting the way they attenuate, combine, and interfere. One of such properties is the signal strength. A packet is transmitted at some signal strength (transmit power level) which (typically) decreases (attenuates) as the packet moves away from the transmitter. When a packet is being received, its signal competes with the noise caused by other (interfering) packets and also with the ubiquitous background (floor) noise, which is there all the time. The quality of reception, i.e., the likelihood that all bits of the packet will be received correctly, depends on how strong the packet's signal appears at the receiver compared to the combined noise. The signal to noise ratio, denoted SNR (or sometimes SIR, for signal-to-interference ratio) is one important characteristic directly translating into the chances for the packet's successful reception. One problem with high fidelity simulation (which is the kind of simulation that we are into) is that SNR may change many times during the reception time of any given packet. The need to account for this is one example of a (messy) problem that SMURPH solves internally (for all channel models), so the user specification of a channel model need not worry about it.

In radio communication (and in many other areas involving signal processing), ratios and factors are often expressed in decibels (dB). Assuming r is a straightfor-ward, linear ratio, its value expressed in decibels is $10 \times log_{10}r$. The dB notation is more compact and convenient when the ratios involved naturally tend to span many

Fig. 2.7 A sample SIR to
BER function

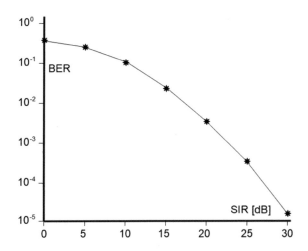

magnitudes. This is the case for SIR which may range, say, from 100 dB (10^{10})
to 0 dB (1) within the same system. Two built-in SMURPH functions, *linTodB* and
dBToLin (Sect. 3.2.10) provide conversions from linear to dB and from dB to linear,
respectively.

So, packets are transmitted at some signal levels, those signals attenuate as the
packets propagate through the wireless medium, and they arrive at their receivers at
some different (much weaker) signal levels with varying degrees of noise contribution
from other packets and from the environment. To close the open ends of the built-in
part of the channel model, the user must provide formulas telling the model:

- how the signals attenuate
- how they combine
- how their relationship to the noise affects the opportunities for packet reception.

Usually, but not necessarily, the last point involves a function known as the bit
error rate (BER) telling, for a given value of SNR, the probability that a random bit
of the packet will be received erroneously. The exact shape of that function depends
on many subtle properties of the RF channel (frequency, modulation, encoding,
transmission rate) and thus cannot be hardwired as a constant, built-in component of
the channel model. In many practical cases, the function is determined experimentally
and described as a collection of discrete points (see Fig. 2.7). In our model, the SIR-
to-BER function will be specified in the input data file to the simulator as a sequence
of pairs of values handled by an object of this class:

```
class SIRtoBER {
        double *Sir, *Ber;
        int NEntries;
        public:
        SIRtoBER ( ) {
                int i;
                readln (NEntries);
                Sir = new double [NEntries];
                Ber = new double [NEntries];
                for (i = 0; i < NEntries; i++) {
                        readln (Sir [i]);
                        readln (Ber [i]);
                }
        };
        double ber (double sir) const {
                int i;
                for (i = 0; i < NEntries; i++)
                        if (Sir [i] <= sir)
                                break;
                if (i == 0)
                        return Ber [0];
                if (i == NEntries)
                        return Ber [NEntries – 1];
                return Ber [i–1] + ((Sir [i–1] – sir) / (Sir [i–1] – Sir [i])) * (Ber [i] – Ber [i–1]);
        };
};
```

The class consists of a pair of arrays, *Sir* and *Ber*, of size *NEntries*. All the requisite values are read (by the class constructor) from the input data file when a class object is created. The first element of a pair ($Sir [i]$) is a signal to noise ratio (in dB), the second element is the corresponding bit error rate interpreted as the probability that a random bit received at the given SIR level will be damaged.

Method *ber* tries to turn the collection of points into a continuous function. Its argument is a SIR value in dB, and the method is expected to return the corresponding error rate. It accomplishes that by interpolation, assuming that the *Sir* values occur in the array in the decreasing order (which implies that the corresponding *Ber* values are increasing). The discrete points become effectively connected with straight line segments, as in Fig. 2.7. If the argument is less than the first entry in *Sir* or greater than the last entry, the method returns the respective boundary values of *Ber*. Note that BER can never be greater than 0.5.

Below is the complete definition of the channel model:

```
rfchannel ALOHARF {
        double BNoise, AThrs;
        int MinPr;
        const SIRtoBER *StB;
        double pathloss_db (double d) {
                if (d < 1.0)
                        d = 1.0;
                return −27.0 * log10 ((4.0*3.14159265358979323846/0.75) * d);
        };
        TIME RFC_xmt (RATE r, Packet *p) {
                return (TIME) r * p->TLength;
        };
        double RFC_att (const SLEntry *xp, double d, Transceiver *src) {
                return xp->Level * dBToLin (pathloss_db (d));
        };
        Boolean RFC_act (double sl, const SLEntry *sn) {
                return sl * sn->Level + BNoise > AThrs;
        };
        Long RFC_erb (RATE r, const SLEntry *sl, const SLEntry *rs, double ir, Long nb) {
                ir = ir * rs->Level + BNoise;
                return (ir == 0.0) ? 0 :
                        IRndBinomial (StB->ber (linTodB (sl->Level * rs->Level) / ir)), nb);
        };
        Boolean RFC_bot (RATE r, const SLEntry *sl, const SLEntry *sn, const IHist *h) {
                return (h->bits (r) >= MinPr) && !error (r, sl, sn, h, −1, MinPr);
        };
        Boolean RFC_eot (RATE r, const SLEntry *sl, const SLEntry *sn, const IHist *h) {
                return TheTransceiver->isFollowed (ThePacket) && !error (r, sl, sn, h);
        };
        void setup (   Long nt,        // The number of transceivers
                       double r,       // Rate in bps
                       double xp,      // Default transmit power (dBm)
                       double no,      // Background noise (dBm)
                       double at,      // Activity threshold (dBm)
                       int pr,         // Minimum required preamble length
                       const SIRtoBER *sb) {
                RFChannel::setup (nt, (RATE)(Etu/r), PREAMBLE_LENGTH, dBToLin (xp),
                        1.0);
                BNoise = dBToLin (no);
                AThrs = dBToLin (at);
                StB = sb;
                MinPr = pr;
        };
};
```

The model is described by a few numerical attributes and methods. The methods named *RFC_*... are referred to as *assessment methods* and declared as *virtual* in the superclass (*RFChannel*), so they are invoked by the built-in parts (internals) of the model. They provide formulas for some dynamic, numerical values needed by the model, or prescriptions for deciding when and whether some events should be triggered. The built-in *setup* method of *RFChannel*, referenced from the *setup* method of *ALOHARF*, takes five arguments:

- the number of transceivers interfaced to the channel
- the transmission rate (or rather its reciprocal expressed in *ITUs* per bit, Sect. 2.2.5)
- the preamble length in bits
- the default transmit power
- the default (antenna) gain of a receiver.

Except for the first argument, all the remaining ones refer to the (default) attributes of individual transceivers, which are settable on a per-transceiver basis, e.g., when the transceivers themselves are created. As the transceivers, in our present case, are created implicitly, as class-type variables of stations (Sect. 2.4.3), their attributes are assigned from the channel defaults, which provides a convenient shortcut for initializing them all in the same way.

The interpretation of transmit power (the fourth argument of the *RFChannel setup* method) is up to the experimenter. The model never assumes any absolute reference for interpreting signal strength (it only deals with ratios), which means that as long as all the relevant values are expressed in the same units, the actual units are not important. To make the simulation data look natural, we often assume that transmission power is specified in milliwatts, or, as it usually happens in RF communication, in dBm, i.e., decibel fractions of 1 milliwatt. The built-in model (*RFChannel*) assumes straightforward, linear interpretation of signal levels, so the transmit power level is converted to linear (assuming that the value passed to the *setup* method of *ALOHARF* is in dBm).

Sometimes, it may difficult to remember where a conversion to or from linear is needed. Generally, all built-in elements of the RF channel model deal with the linear representation of signal levels and ratios, for the ease of computation.[15] The only reason for (occasional) conversion is that we often prefer to provide the numerical parameters for the model in dB/dBm. Then the exact place where to do the conversion is up to the user. For example, the argument passed to the *setup* method of *ALOHARF* could be converted to linear after being read from the input data set, before passing it to the channel. Or we could simply specify it linearly in the input data set, so no conversion would be needed at all. The model assumes that the transmission rate provided as the second argument to the *setup* method of *ALOHARF* is expressed in natural units, i.e., in bits per *ETU* (set to 1 s in the model, see Sect. 2.4.6), and must be converted to the internal representation (*ITUs* per bit) as well.

---

[15]Some people would argue that addition/subtraction (in the logarithmic domain) is simpler than the multiplication needed in the linear domain. Unfortunately, while calculating dynamic interference, SMURPH must frequently add multiple signals in the linear domain, which operation would be messy, if the signals were represented in the logarithmic domain.

Note that the conversion is also applied to two (local) attributes of *ALOHARF*: *BNoise* and *AThrs*. The first of them represents the power of the ambient background noise, i.e., the omnipresent, constant interference signal affecting all receptions. It should be expressed in the same units as the transmit power level. The second attribute is not really taken advantage of by the protocol program. It indicates the minimum received signal level at which the transceiver can tell that some RF activity is present in the neighborhood. In the ALOHA protocol, transceivers never listen to the channel before transmission. Note that the transmitter and the receiver of the same node are formally tuned to different frequencies.

Attribute *StB* is the pointer to an object of class *SIRtoBER* describing the SIR-to-BER function. The interpretation of that function is up to the user-provided component of the model, i.e., it is not passed to *RFChannel*. The latter does not presume any particular method of calculating bit errors. Similarly, the procedure of interpreting boundary events triggered by potentially receivable packets is determined by the user part. In particular, attribute *MinPr* (minimum preamble length) plays a role in the procedure of detecting the beginning of packet events (*BOT* and *BMP*), as explained below. The dynamics of those procedures are taken care of by the SMURPH interiors (so the user does not have to program too much). The user part of the channel model specification only provides some model-specific "static" calculations (functions) parameterizing those procedures. Those functions come as the set of *RFC_...* methods which lie at the heart of the channel model specification (Sect. 4.4.3).

Going from the top, *RFC_xmt* calculates the amount of time needed to transmit (insert into a transceiver) a packet at rate r. In the most straightforward case (as it happens in our model), the result is the direct product of the packet's total length (*Packet* attribute *TLength*) and r (note that the rate is expressed internally in *ITUs* per bit).[16] Sometimes the function must factor in subtle encoding issues, like the fact that groups of logical (information) bits of the packet are encoded into symbols represented by more physical bits (i.e., ones that count to the transmission time, but not to the payload). In some models, different portions of the same packet are transmitted at different rates.

The role of *RFC_att* is to compute the attenuated value for a signal originating at a given source transceiver (*src*) and arriving at the present transceiver (on whose behalf the method is called). The latter is implicitly provided in the (global) variable *TheTransceiver*, which points to the current (destination) transceiver whenever the method is invoked. The method can carry out its calculation using detailed information extracted from the two transceivers (including their locations), and any other data structures, e.g., defined within the channel model. It can also randomize the result before returning the value. The signal level is interpreted linearly. As the most natural parameter for calculating signal attenuation, one being used by many models, is the distance between the two transceivers, the second argument of the method provides its ready value expressed in the distance units (*DUs*) assumed by the model, see Sect. 3.1.2).

---

[16]This also happens to be the behavior of the default copy of *RFC_xmt*, so its inclusion in our model is superfluous.

The first argument of *RFC_att* points to an object of type *SLEntry* (signal level entry) which contains a standard description of what we call the transmit signal level (Sect. 8.1). The description contains the linear numerical value of the signal power augmented with a generic attribute called the signal *Tag*. The latter can be used in advanced models to represent special signal features, like the channel number or code (in CDMA systems [21, 22]) to offer more parameters needed by tricky attenuation functions.

For our simple model, signal attenuation is determined based solely on the distance using this rule-of-thumb formula [12]:

$$A = \alpha \times 10 \, log_{10}\left(\frac{4\pi d}{\lambda}\right)$$

where A is the attenuation in dB, d is the distance, λ is the wavelength (expressed in the units of distance), and α is the loss exponent (typically set to 2 for open space propagation). The formula does not properly account for the gain of a specific antenna, which basically means that the $4\pi d/\lambda$ factor should be treated as a parameterizable constant. If we express the attenuation linearly, the formula will take this simple shape:

$$A_{[lin]} = F \times d^{\alpha}$$

where F is a constant. Basically, the only generic feature of this function is that the signal attenuates as the α power of the distance. Note that d should be interpreted relative to some minimum reference distance (because, e.g., $d = 0$ is meaningless in both versions of the formula).

Our version of *RFC_att* multiplies the source signal level by the linearized value calculated by *pathloss_db*. Note that the value in dB returned by *pathloss_db* is negative, because the attenuation is in fact used to multiply the source signal level (i.e., it is factored in as gain). The exponent α is assumed to be 2.7. The model makes no claims of being realistic, but this is not its purpose. The parameters should be viewed as examples, and the user is encouraged to play with other values.

The reader has probably noticed that, notwithstanding the naive character of our model, it is not implemented in the most efficient of ways. Considering that, in realistic performance studies (long runs), *RFC_att* tends to be invoked many times, one should try to make its calculations efficient. Obviously, *pathloss_db* could be easily converted to return a linear attenuation factor, and the call to *dBToLin* (which basically undoes the logarithm so laboriously and unnecessarily computed by *pathloss_db*) could be eliminated.

One important method from the *RFC_...* pack, which our model does not redefine (because its default version defined by *RFChannel* is quite adequate), is *RFC_add* (Sect. 4.4.3) whose responsibility is to add the levels of multiple signals sensed by a transceiver at any given moment. The default method simply adds those signals together, which is the most natural (and most realistic) thing to do.

The next method brought in by *ALOHARF*, *RFC_act*, is called whenever the model must answer a query to the transceiver asking whether it senses any RF activity at the time. Its implementation usually involves a threshold for the combined signal level at the transceiver. The first argument of *RFC_att* is the combined signal level (as calculated by *RFC_add*), and the second one contains the receiver gain of the current (sensing) transceiver. Our protocol model does not issue such queries (so skipping the method would not affect the program). We make it available, because one tempting exercise is to add to the protocol a "carrier sense" (or "listen-before-transmit") feature to see how it improves the performance of the original scheme. Note that formally that would require outfitting the nodes with extra hardware, because their (normal) receivers and transmitters are tuned to different RF channels.

The most interesting part of the model is the procedure of generating bit errors based on signal attenuation and interference and transforming them into the packet-reception events perceived by a receiver process. This is clearly the part whose quality (understood as closeness to reality) determines the quality of the entire model. The components we have discussed so far are merely prerequisites to this operation.

The main problem with the procedure is the variability of the interference (and thus the signal to noise ratio) experienced by a packet during an attempt at its reception. This happens, because the packet may overlap with other packets, coming from multiple sources within the receiver's neighborhood, whose transmissions may start and terminate at various times. In consequence, the history of the packet's signal, as seen by the receiver during its reception, may consist of segments perceived with different SIR. Consequently, the bit error rate is different for the different segments. One of the responsibilities of the built-in part of the model is to keep track of those segments and to properly turn their varying bit error rates into a realistic bit error rate generator for the entire packet (see Sect. 8.2.4). Luckily, the user part of the model does not have to deal with the complexity of that process.

Formally, from the viewpoint of the (user) process (or processes) monitoring a transceiver and trying to receive packets on it, the only thing that matters is the detection of events marking the ends of the received packets, i.e., *EOT* and *EMP*. Following the end of a packet's transmission at the source (the *stop* operation), those events will be eligible to occur at a destination transceiver after the propagation delay separating the source from the destination (in the case of *EMP* the packet must also be addressed to the destination station). Whether the eligibility translates into an actuality is determined by *RFC_eot*. The method decides whether a packet formally eligible to trigger *EOT* or *EMP* is actually going to trigger the respective event.

In a realistic channel model, the decision must involve models of the physical stages of the packet's reception, as they happen in real life. To arrive at the end of the packet, the receiving process must first perceive the packet's beginning (events *BOT*, *BMP*), for which some specific criteria have to be met. To that end, method *RFC_bot* is invoked at the time when *BOT* or *BMP* can be potentially triggered to decide whether they in fact should.

The two assessment methods, *RFC_bot* and *RFC_eot*, return Boolean values representing yes/no decisions. Those decisions can be based on whatever criteria the methods find useful, but it is natural to apply some idea of bit errors in the received packets. The built-in part of the model offers shortcuts for bit-error-based assessment (Sect. 8.2.4). To make sure that no elements of the history of the packet's "passage" through the transceiver are excluded from the view of the assessment methods, they receive as arguments essentially the complete history, expressed as the series of the different interference levels experienced by the packet along with their durations. The data structure holding such a history, called the *interference histogram*, is represented by class *IHist* (see Sect. 8.1.5). Owing to the built-in shortcuts mentioned above, interference histograms are seldom handled directly by the assessment methods.

The argument lists of the two assessment methods look identical and consist of the transmission rate, the attenuated signal level at which the packet is heard (together with the *Tag* attribute of the sender's transceiver), the receiving transceiver gain (and its *Tag*), and the interference histogram of the packet, as collected at the receiving transceiver. For *RFC_bot*, the interference histogram refers to the portion of the packet received prior to its beginning, i.e., the preamble. The Boolean condition evaluated by our instance of the method means this: "the histogram contains at least *MinPr* bits and the last *MinPr* of those bits have been recognized without a single bit error." This is a typical condition checked by a real-life receiver when tuning to an incoming packet.

The first step in unraveling the condition is to note that the *bits* method of *IHist* returns the number of bits covered by the interference histogram. The segments in the histogram are delimited by time intervals, so the rate argument is needed to interpret the histogram as a sequence of bits. The second part involves an invocation of *errors* which is a method of *RFChannel*. The method returns the number of erroneous bits within the bit sequence represented by an interference histogram, more specifically, within the selected fragment of that sequence. The method has a few variants for identifying different interesting portions of the histogram. For the case at hand, the last two arguments specify the section of the histogram corresponding to the last *MinPr* bits (see Sect. 8.2.4 for details).

To accomplish its task of generating the number of erroneous bits in a histogram potentially consisting of many segments representing different levels of interference, the *error* method resorts to one of the assessment methods provided by the model. That assessment method, *RFC_erb*, solves the simpler problem of calculating the number of errors in a sequence of bits received at a steady interference level. It is usually simple, because it only needs to directly apply the SIR-to-BER function. Arguments *sl* and *rs* describe the attenuated, received signal level and the receiver's gain, *ir* is the (steady) interfering signal level at the receiver, and *nb* is the total number of bits. Note that *RFC_erb* mentions no interference histograms and deals solely with bits, signal levels, and (in more advanced cases) signal attributes encoded in *Tags*. Thus, it captures the essence of the model in a way that should be reasonably easy to specify by its designer. The internal *error* method builds on this specification and does the tricky job of extrapolating the bit errors onto packets received at varying interference levels whose boundaries may fall on fractional bits.

RFC_erb starts by calculating the total, perceived interference level, by multiplying the interference by the receiver gain and adding the background noise. Then it generates a randomized number of "successes" in a Bernoulli trial (function *lRnd-Binomial*) involving *nb* tosses of a biased "coin," where the probability of success is equal to the bit error rate. The latter is obtained by invoking the SIR-to-BER function. Note that the function expects the SIR in dB, so the linear value must be converted before being passed as the argument.

Looking at the packet reception code in Sects. 2.4.3 and 2.4.4, we see no good reason for awaiting the *BMP* event, because *EMP* seems to be the only event that matters: following that event, the packet is formally received and, apparently, the interception of *BMP* serves no useful purpose. In fact, there exists a connection between the two events. The interception of *BMP* marks the packet triggering the event as being "followed" (Sect. 8.2.1), which effectively associates a flag with the packet indicating that its beginning has been recognized by the transceiver. This flag is checked by the second assessment method, i.e., *RFC_eot* (via *isFollowed*), as a prerequisite for the affirmative decision. This way, the assessment by *RFC_eot* can be thought of as a continuation of the packet's reception procedure initiated by *RFC_bot*. Only a packet that has been positively assessed by *RFC_bot* stands a chance of triggering *EOT* (or *EMP*). The second part of the condition applied by *RFC_eot* involves the built-in *error* method asserting that the packet's entire interference histogram (reception history) contains no bit errors.

Our network model uses two separate RF channels created as two independent instances of *ALOHARF* (see Sect. 2.4.6). This way the two channels are isolated in the model and, in particular, they do not interfere. The real-life frequencies of the two channels are relatively close (407.350 and 413.475 MHz), thus, some crosstalk is bound to occur in real life, especially at the fully duplex hub station, when it transmits and receives at the same time. We do not want to split hairs for this simple exercise, so we shall ignore this issue for now. A more realistic way to model the radio setup would be to have a single channel (or medium) for both ends of communication, with different *channels*[17] modeled via transceiver *Tags*. That approach would make it possible to model the channel crosstalk interference, as it is done in the library model discussed in Sect. A.1.2.

---

[17]The terminology is a bit unfortunate and potentially confusing. Yielding to the overwhelming tradition, we have been using the word "channel" (as in "wireless channel model") to denote an object representing an RF propagation medium (band) and modeling RF signal passing in that medium. Now, we can think of having multiple *channels* (say frequencies or codes) within such a model, with definite crosstalk parameters affecting interference levels among those channels.

2.4.6   *Wrapping It Up*

To complete the model, we must build the network and define the traffic patterns in it; then, when the experiment is over, we want to see some performance results. This is the rather routine *Root* process to take care of these tasks:

```
process Root {
        void buildNetwork ( );
        void initTraffic ( );
        void setLimits ( );
        void printResults ( );
        states { Start, Stop };
        perform {
                state Start:
                        buildNetwork ( );
                        initTraffic ( );
                        setLimits ( );
                        Kernel->wait (DEATH, Stop);
                state Stop:
                        printResults ( );
                        terminate;
        };
};
```

and here is how the network is built:

```
Long NTerminals;
ALOHARF *HTChannel, *THChannel;

void Root::buildNetwork ( ) {
        double rate, xpower, bnoise, athrs;
        int minpreamble, nb, i;
        SIRtoBER *sib;

        setEtu (299792458.0);
        setDu (1.0);

        readln (xpower);
        readln (bnoise);
        readln (athrs);
        readln (minpreamble);
        sib = new SIRtoBER ( );
        readln (NTerminals);
        HTChannel = create ALOHARF (NTerminals + 1,
                        DATA_RATE,
                        xpower,
                        bnoise,
                        athrs,
                        minpreamble,
                        sib);
        THChannel = create ALOHARF (NTerminals + 1,
                        DATA_RATE,
                        xpower,
                        bnoise,
                        athrs,
                        minpreamble,
                        sib);
        for (i = 0; i < NTerminals; i++)
                create Terminal;
        create Hub;
}
```

The parameter of *setEtu* is equal to the speed of light in vacuum expressed in meters per second. This way, if we assume that 1 *ETU* corresponds to 1 s of real time, then 1 *ITU* will represent the amount of time needed for a radio signal to cross the distance of 1 m. This is the granularity of time in our model.

With *setDu*, we set the distance unit, i.e., declare the amount of time (in *ITUs*) corresponding to a natural basic measure of propagation distance. The argument is 1 meaning that distance will be expressed in meters. This will apply, e.g., to node coordinates and to the second argument of *RFC_att* (in Sect. 2.4.5).

Next, *buildNetwork* reads a few numbers and creates the two RF channels, as explained in the previous section. The two channels use identical parameters. Finally, the terminals and the hub are created. Recall that the protocol processes are started from the *setup* methods of their owning stations.

The way the network stations are created by *buildNetwork*, they are assigned internal identifiers (SMURPH indexes) from 0 to *NTerminals*, where the hub receives the index *NTerminals*, while the terminals are numbered 0 through *NTerminals–1*. Occasionally, one may want to access a station object from a process that does not belong to the referenced station (if it does, the station is naturally accessible through the *S* pointer, Sect. 5.1.2). That may happen at the initialization or termination, e.g., when calculating and printing performance results. The global function *idToStation* transforms a station index into a pointer to the generic type *Station* (see Sect. 4.1.1). In our case, the following two macros will wrap it into friendlier forms by casting the pointers to the right types:

#define TheHub   ((Hub*)idToStation (NTerminals))
#define TheTerminal(i)   ((Terminal*)idToStation (i))

Thus, *TheHub* produces the pointer to the hub node, and *TheTerminal (i)* returns the pointer to the terminal node number *i*.

The model runs two independent traffic patterns described by objects of the same type:

traffic ALOHATraffic (Message, DataPacket) {};

and represented by two pointers:

ALOHATraffic *HTTrf, *THTrf;

The two traffic patterns model the traffic in two separate directions: hub to terminal and terminal to hub, implemented as two independent streams of data. The lack of correlation is not very realistic. In the real network, most of the hub-to-terminal traffic was probably directly caused by packets arriving from the terminals. Our traffic patterns are created as follows:

```
void Root::initTraffic ( ) {
        double hmit, tmit;
        Long sn;
        readln (hmit);
        readln (tmit);
        HTTrf = create ALOHATraffic (MIT_exp + MLE_fix, hmit,
                (double) PACKET_LENGTH);
        THTrf = create ALOHATraffic (MIT_exp + MLE_fix, tmit,
                (double) PACKET_LENGTH);
        for (sn = 0; sn < NTerminals; sn++) {
                HTTrf->addReceiver (sn);
                THTrf->addSender (sn);
        }
        HTTrf->addSender (NTerminals);
        THTrf->addReceiver (NTerminals);

}
```

The message arrival processes for both traffic patterns are Poisson (*MIT_exp*) with fixed message length (*MIT_fix*) exactly matching the maximum packet length (80 bytes). The *for* loop dynamically constructs the sets of receivers (for the hub-to-terminal traffic) and senders (for the terminal-to-hub traffic). The set of senders for the hub-to-terminal traffic consists of a single station, i.e., the hub, which is the same as the set of receivers for the terminal-to-hub traffic. For defining those sets, the stations are represented by their SMURPH indexes (see above). Recall that in Sect. 2.2.5, while defining the simple traffic pattern for the Alternating Bit model, we were referencing the stations by their object pointers. Both ways are OK.

In a simple traffic pattern, like the ones defined above, a set of stations provides a choice. For example, when a message belonging to the hub-to-terminal traffic arrives from the *Client* for transmission, it is always queued at the hub (because the hub is the only member of the senders set), but its recipient is generated at random from the set of receivers, i.e., terminals. Each terminal can be selected with the same probability. Conversely, for a terminal-to-hub message, the receiver is always known in advance, while the sender (the station at which the message is queued for transmission) is selected at random from the set of senders (terminals).

The last step of the startup sequence defines the termination conditions for the experiment:

```
void Root::setLimits ( ) {
        Long maxm;
        double tlim;
        readIn (maxm);
        readIn (tlim);
        setLimit (maxm, tlim);
}
```

The two-argument variant of *setLimit* (Sect. 10.5) defines two simultaneous limits: the maximum number of messages received at their destinations, and the simulated time limit in *ETUs* (i.e., seconds in our present case). The received messages are counted over all (i.e., two in our case) traffic patterns.

This *Root* method will be invoked when the experiment terminates:

```
void Root::printResults ( ) {
        HTTrf->printPfm ( );
        THTrf->printPfm ( );
        Client->printPfm ( );
        HTChannel->printPfm ( );
        THChannel->printPfm ( );
}
```

It will write to the output file the standard set of performance measures for each of the two traffic patterns plus the combined measures for both traffic patterns, i.e., the set of standard performance measures of the *Client* (Sect. 10.2).

2.4.7   Experimenting with the Model

To play with the model, we will build a network consisting of six terminals and a hub shown in Fig. 2.8. This is not the real ALOHA network (or even *a* real ALOHA network). I could find no information regarding the actual distribution of nodes, except for the mention that the hub was located on the Oahu Island. Of course, the network would evolve and the configuration of nodes would change over the years of its deployment. Apparently, not all nodes could be reached directly from the hub, hence the need for repeaters [15] which we have conveniently skipped in our model. Thus, our channel model makes no pretense of accurately reflecting the real RF characteristics of ALOHA. We can easily play with the channel to try to make it

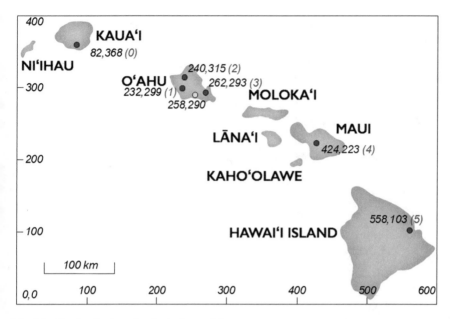

Fig. 2.8 The distribution of nodes in the model

closer to reality, if the parameters of the real-life ALOHA channel become available, but this is not the point. We must draw a line somewhere. There is no need to come close to the real ALOHA network with the model, because the real network is just a picturesque story, hopefully adding some spice to our otherwise dull (and abstract) exercise.

The map in Fig. 2.8 shows the geographical distribution of nodes with their planar coordinates (in kilometers), assuming that the left bottom corner of the map is the (arbitrary) origin. The yellow node at (258,290) is the hub. Below is a sample data set describing a network with the above layout. The numbers in parentheses, following the second coordinate in Fig. 2.8, are node indexes in the data file (and in the model). Thus, for example, the Maui terminal, with coordinates (424,223) has the index 4 (and its location is described by the 4th entry, counting from zero, in the "terminals" section of the data set).

Power	40.0	dBm
Floor noise	-166.0	dBm
Activity threshold	-140.0	dBm
Minimum preamble	32	bits
SIR/BER	7	entries

30.0	0.00001
25.0	0.0005
20.0	0.003
15.0	0.01
10.0	0.1
5.0	0.4
0.0	0.5

Number of terminals   6

82000.0, 368000.0
232000.0, 299000.0
240000.0, 315000.0
262000.0, 293000.0
424000.0, 223000.0
558000.0, 103000.0
258000.0, 290000.0

Message interarrival times:

2.0	hub
1.0	terminal

Message number limit	200000	
Time limit	36000.0	ten hours

The above sequence of numbers matches the sequence of calls to *readIn* issued from the code initializing and starting the model (Sect. 2.4.6). The transmission power (the same for all nodes) is set to 40 dBm (10 W), which figure is mentioned in [15]. Activity threshold is not used by the protocol model, although the channel model includes it as a parameter (thus making the related inquiries available to the protocol).

The collection of points describing the SIR-to-BER function roughly matches Fig. 2.7. The *ber* method of the *SIRtoBER* class (Sect. 2.4.5) assumes that the points are entered in the decreasing order of SIR. There is no real-life backing for those numbers (in other words, I have pulled them out of thin air), so they should be taken with a big grain of salt. Note that the minimum bit error rate is 10^{-5}, assumed for the SIR of 30 dB and more. Packets become practically unreceivable at the SIR of 10 dB.

The node coordinates are taken directly from Fig. 2.8 (the last pair refers to the hub) and are expressed in meters. Note that 1 m is the official distance unit of the model. The next two numbers provide the (independent) mean inter-arrival times of the two traffic patterns expressed in seconds. We will be changing them to see how the network fares under diverse loads. The model will run for ten hours of virtual time, or until the network has delivered 200,000 messages, whichever happens first.

When we compile the model (by running *mks* in its directory, Sect. B.5) and then execute it on the above data set (by running *./side data.txt output.txt*), we get in the output file (within a few seconds) the familiar (from Sect. 2.2.6) collection of *Client* performance data describing the observed statistics of the traffic patterns. This time we receive three versions of such data: one for each of the two traffic patterns, plus one global for the entire *Client*, i.e., for the two traffic patterns combined. The first of the three sets refers to the hub-to-terminal traffic (see the *printResults* method of *Root* in Sect. 2.4.6). The most interesting part of that data is probably in the second set (for the terminal-to-hub traffic); it relates to the essence of ALOHA, which (in numerous adaptations and extensions) has been subjected to extensive performance studies since its inception. The *Client* data is followed by the sets of counters for the two wireless channels, which look similar to those we saw in Sect. 2.2.6 for wired links.

The elapsed (virtual) time at the end of simulation is just slightly above 36000 s, which means that the simulator has hit the time limit, rather than the limit on the number of received messages. We could guess that that would be the case by looking at the inter-arrival times for the traffic patterns. The combined mean inter-arrival time is about 0.67 s (the harmonic average of the two), which means that processing 200,000 messages would take about 37 h of virtual time. The fact that the inter-arrival times of the two traffic streams are round numbers allows us to easily verify the basic parameters of the model. The number of *Client* messages (or packets, which in our model correspond 1–1 with messages) received by the hub is 36,090 (which happens to be close enough to 36,000). The number of generated messages is equal to the number of received messages +1. One message remains queued at a terminal when the simulation stops, which is quite normal: the load is definitely non-saturating, and all terminal messages are (eventually) received by the hub. For the hub-to-terminal traffic, the number of transmitted messages/packets is 18,145 (which well matches the inter-arrival time of 2 s), but the number of received messages is only 17,302, i.e., slightly over 95%. At first sight, the loss may be attributable to the flaky nature of the radio channel, but the issue is not that simple. As the terminals operate in the half-duplex mode, and occasionally have to transmit packets, some loss should be attributed to those circumstances when the terminal happens to be transmitting just at the moment when a packet addressed to it arrives from the hub. The amount of time needed by the terminal to expedite a packet and switch (twice) between the receive and transmit modes is about 0.1 s. Now, according to our reception model (Sect. 2.4.4), a received packet is lost, if its reception time overlaps that interval in any way, which means that it starts at most one packet time (0.083 s) before the beginning, and at most at the end of the interval,[18] yielding a deaf window of ca. 0.18 s. How often is that window deployed? By looking at the counters reported for the terminal-to-hub channel, we see that the number of packets transmitted by all terminals was 41,931 (this covers the original packets arriving from the *Client* at the average rate of one packet per second, as well as their retransmissions). Thus, the combined, effective transmission rate is 41,931/36,000 ≈ 1.16 packets per seconds. Consequently, 1.16 × 0.18/6 = 0.0348 (about 3.5%) yields the fraction of the packets randomly arriving at a terminal that will be lost on the deaf window caused by terminal transmissions.

Of course, the above figure can be verified experimentally by our model (see below), but simple calculations like this are useful, because they help us understand the model and diagnose its departures from realism (which is a euphemism for bugs). The first step in a model's verification is to check whether the obvious aspects of the system's behavior are captured correctly.

The throughput measure is only printed for the *Client*, i.e., for the two traffic patterns combined, and it yields about 949 bits/s. Note that this time, in contrast to Sect. 2.2.6, we measure time in seconds (not in bits), so the throughput figure is not normalized to 1 per channel. Theoretically, with the two channels operating at 9600 bps, the upper bound for throughput is 19,200 bps. More realistically, as the *Client* throughput reported by the simulator is *effective* (i.e., it only covers packet payloads) we should subtract from the bounds the useless margins, which amount to 154 bits (i.e., *FRAME_LENGTH* + *PREAMBLE_LENGTH* in Sect. 2.4.3) per every packet, whose useful (payload) length is 80 × 8 = 640 bits. Thus, the upper bound per channel is 9600 × 640/(640 + 154) ≈ 7738, i.e., about 80% of the nominal capacity.

When we look at the last set of counters in the output file, for the terminal-to-hub channel, we see that the total number of transmissions in the channel was 41,931, while the number of received packets was 36,090 (the same as the number reported in the traffic pattern's data). So, the loss fraction is about 14%, which is significantly more than for the other channel. Of course, we know the reason: the hub-to-terminal channel is interference-free, so packets only get lost because of a) random errors caused by background noise, b) momentary lockouts of the terminal receiver when the node is transmitting. On the terminal-to-hub-channel, packets are also destroyed by collisions.

For a more systematic assessment of the network's performance, we should run the network under cleaner traffic conditions. To eliminate the impact of receiver lockouts at terminals, let us make sure that it never happens, so the terminals never transmit. For that, let us set the inter-arrival time of the terminal-to-hub traffic to something huge, e.g., 1,000,000.0 (leaving the remaining data unchanged). Looking at the hub-to-terminal traffic, we now see 18,023 transmitted messages and 17,789

---

[18]Note that transmission in our model has absolute priority over reception, and, in particular, it preempts a reception in progress. It is possible to change the model in a way that would reduce the window slightly by locking the receiver upon the recognition of a packet preamble. The reader may want to implement this enhancement as an exercise.

received ones, i.e., over 98.7% of the packets have been delivered.[19] This tallies up with our "analytical" estimation above. When we subtract the losses on deaf windows caused by terminal transmissions (about 3.5%), we get the 95%+ packet delivery fraction from the original output file. The present 98.7% figure represents the "pure" PDF on the hub-to-terminal channel. The 1.3% loss results solely from incidental bit errors caused by background noise.

The most interesting aspect of the ALOHA network is the performance of the shared terminal-to-hub channel. One intriguing number to be determined is the maximum sustainable throughput of that channel. To see it clearly, let us switch off the hub-to-terminal traffic, in the same way as before, and saturate the terminals with outgoing packets by making sure that they arrive faster than they can ever be sent. As it takes at least 83 ms to transmit a packet (not to mention receiving an acknowledgment for it), setting the inter-arrival time to 0.05 will surely do. When we run the model and look at the counters for the terminal-to-hub traffic pattern, we see that only 117,722 of the 720,012 messages arrived from the *Client* have been transmitted, so the network is oversaturated about 6 times. The throughput reads 2093 (payload) bits per second. Comparing it to the (effective) absolute maximum (7738 bps), we get about 27%. This is good news, because the theoretical, asymptotic maximum for a pure ALOHA channel is about 18.4% [14]. This is not unexpected. The formula for the asymptotic performance of ALOHA has been arrived at under these two assumptions that do not quite hold in our case:

1. the population of transmitting nodes is infinite, and all transmissions (and retransmissions) are independent events
2. collisions are always destructive for all the packets involved

Our network consists of a finite and small number of terminals. Clearly, if there were a single terminal, there would be no collisions at all, so the theoretical model wouldn't apply. With two terminals, depending on the distribution of the backoff delay,[20] the likelihood of an interference from the other node would be smaller than in a situation when the number of nodes is large (and the asymptotic statistics kick in). The second assumption is probably even more relevant. Owing to the disparate distances in our network (some terminals basically touch the hub, while some others are hundreds of kilometers away), when a close terminal transmits simultaneously with a distant one, the transmission from the closer terminal is likely to succeed despite the collision. This kind of "capture effect" is one of the aspects of reality that our model reflects quite adequately, or it could in principle, if we had access to the parameters of the RF channel from the original network.

---

[19]The number of transmitted messages is slightly different than before, because the simulator runs on modified data, so the sequence of (pseudo) random numbers generated during the second run is different.

[20]The backoff procedure is another important factor contributing to traffic correlation and thus strongly influencing the network performance in real-life scenarios. The analytical model of ALOHA ignores that procedure assuming that retransmissions are uncorrelated effectively being part of the same Poisson arrival process as the *Client* traffic arrival.

We will shortly return to the problem of the bias in service received by different terminals. Before we finish with the maximum global throughput, we should ask ourselves this question: is the throughput figure that we have obtained under saturation truly the maximum achievable figure? Note that it doesn't have to be, and it obviously wouldn't be in a network consisting of a large number of terminals. This is because, when the population of terminals is large and they are all saturated, any transmission (or retransmission) attempt is virtually guaranteed to collide with another attempt, so the throughput drops to zero, as in the analytical model of ALOHA [14].

To answer the question, we should reduce the network load to below the saturation level from the previous experiment, slightly above one that would yield exactly the observed throughput figure, to see if the throughput cannot be pushed any further by making the load less intimidating. Dividing 2093 by 640 (the payload length in bits) we get about 3.27 (the number of packets/messages per second) and the inverse of this number, 0.306, gives us the mean message inter-arrival time that will produce exactly the observed throughput without saturating the network (running it just at the threshold). Clearly, by increasing the inter-arrival time above this value, we will reduce the throughput. So, we should try decreasing it a bit. When we do that, we see that it never helps, i.e., the throughput stays basically the same. This is because the population of terminals is small, so the terminals throttle their combined, offered load to the network by saturating themselves rather than the network's medium.

Now we can get back to the issue of the symmetry of service with respect to the different terminals, or, as we should properly call it, the issue of *fairness*. Different terminals are located at different distances from the hub, which affects signal attenuation, the quality of reception, the bit error rate, and thus the packet delivery fraction. It also affects the opportunities for a successful packet capture in case of collision. For the hub-to-terminal traffic, there are no collisions, no acknowledgments, and every message/packet arriving from the *Client* undergoes a single transmission attempt. Thus, the packet delivery fraction is the only performance measure of interest. For the terminal-to-hub traffic, all packets eventually get through, so we should ask about delays, or, what is essentially the same thing, about the (average) number of retransmissions per packet. Note that knowing that number, plus the average backoff time, we can easily compute the average delay experienced by a packet.

Let us note that the global average number of retransmissions per packet, for the terminal-to-hub traffic, is already there in the output data. Looking at the saturated network, the list of counters associated with the terminal-to-hub channel (at the very end of the output file) shows 229,038 packet transmissions and 117,722 receptions. Recall that a transmission is counted whenever a node executes *transmit* (Sect. 2.4.4) on the channel's transceiver, while a reception means that the packet has been in fact received and passed to the *Client*. Thus, the ratio of those numbers, 1.95, gives the average number of transmission attempts per packet. Note that in pure ALOHA, that number (at the maximum throughput) is about 2.72 (the number e), so our network fares better.

The list of standard performance measures associated with traffic patterns includes global statistics of *packet access time* (PAT) and *absolute packet delay* (APD) which, in the case of our model, relate to the number of retransmissions. Both measures (see

Sect. 10.2.1) refer to the amount of time elapsing from the moment a packet becomes ready for transmission at the sender, until it has been processed. For PAT that means *releasing* the packet at the sender, while for APD, the fulfilling event is the packet's reception (*receive*) at the destination.

The average values of PAT and APD (for the saturated terminal-to-hub traffic) are 1.83 and 1.77 (s), respectively. Note that PAT is larger, because a packet must be received before it can be acknowledged and ultimately released (which action takes extra time). Considering that the average backoff time is 0.85 s and the average number of transmissions per packet is 1.95, the delay can be (roughly) calculated as $(0.85 + 0.083) \times 1.95 = 1.82$, where 0.083 s covers the packet's transmission time.

We still do not see how the global performance measures spread over the different terminals. To see that, we need to collect additional data. We will look at both types of traffic. Starting from the hub-to-terminal end, we add two arrays of counters:

Long *HXmt, *TRcv;

to collect, for every terminal, the number of hub packets transmitted to it, as well as those actually delivered. The arrays should be initialized by *Root* (anywhere before the protocol is started):

HXmt = new Long [NTerminals];
TRcv = new Long [NTerminals];
bzero (HXmt, sizeof (Long) * NTerminals);
bzero (TRcv, sizeof (Long) * NTerminals);

The first counter is updated whenever a packet is transmitted by the hub, in the code method of *HTransmitter* (see Sect. 2.4.3):

...

```
state PDone:
        Up->stop ( );
        if (Pkt == &(S->Buffer)) {
                HXmt [Pkt->Receiver]++;
                S->Buffer.release ( );
        }
        proceed NPacket;
```

...

and the second one in the *receive* method of *Terminal* (Sect. 2.4.4):

```
void receive (Packet *p) {
        if (p->TP == NONE) {
                if (Buffer.isFull ( ) && ((AckPacket*)ThePacket)->AB == Buffer.AB) {
                        Buffer.release ( );
                        AB = 1 – AB;
                }
        } else {
                Client->receive (p, &HTI);
                TRcv [p->Receiver] ++;
        }
};
```

To measure the average number of retransmissions for the terminal-to-hub traffic, we will use a random variable, i.e., an object of type *RVariable* (Sect. 10.1), which is the standard way to track samples in SMURPH and calculate their distribution parameters. We begin by adding these declarations to the set of attributes of a terminal station (Sect. 2.4.4):

```
station Terminal : ALOHADevice {
        unsigned int AB;
        TIME When;
        RVariable *RC;
        int rc;
        ...
```

The random variable is created in the *setup* method of *Terminal* with this statement:

RC = create RVariable;

which can be put anywhere inside the method. Then, whenever the node acquires a new packet for transmission from the *Client*, it will zero out *rc* (see the *getPacket* method of *Terminal* in Sect. 2.4.4):

```
        ...
if (Client->getPacket (Buffer, 0, PACKET_LENGTH,  FRAME_LENGTH)) {
        Buffer.AB = AB;
        rc = 0;
        return YES;
}
        ...
```

The variable will count the number of transmission attempts for the current outgoing packet. For that, we need to add one statement to the *Xmit* state of the *THalfDuplex* process:

...

state Xmit:

```
S->rc++;
```

Up->transmit (S->Buffer, XDone);

...

We need just one more addition to the *receive* method of *Terminal*, to update the random variable when the packet is eventually acknowledged:

```
void receive (Packet *p) {
        if (p->TP == NONE) {
                if (Buffer.isFull ( ) && ((AckPacket*)ThePacket)->AB == Buffer.AB) {
                        RC->update (rc);
                        Buffer.release ( );
                        AB = 1 − AB;
                }
        } else {
                Client->receive (p, &HTI);
                TRcv [p->Receiver] ++;
        }
};
```

The update method of *RVariable* adds the new sample (the number of transmission attempts for the current packet) to the history of the random variable, automatically calculating the mean (and possibly other statistics, see Sect. 10.1.1). To write the new performance measures to the output file at the end of the simulation experiment, we add this method to the *Terminal* class:

```
void printPfm ( ) {
        LONG sc;
        Long id = getId ( );
        double dist, min, max, ave [2];
        dist = THI.distTo (&(TheHub->THI));
        Ouf << getOName ( ) << i, distance i << form (i%3.0f:\nî, dist / 1000.0);
        Ouf << " From hub ---> " <<
                form (" Sent: %8d    Rcvd: %8d    PDF: %5.3f\n",
                HXmt [id], TRcv [id],
                HXmt [id] ? (double) TRcv [id]/HXmt [id] : 0.0);
        RC->calculate (min, max, ave, sc);
        Ouf << " To hub <----- " <<
                form (" Sent: %8d    Retr: %8.6f\n\n", (Long) sc, ave [0]);
};
```

With some creative formatting, the method transforms the per-terminal performance data into presentable information. *Ouf* is a global variable pointing to the

Table 2.1 PDF for the
hub-to-terminal traffic

Terminal	Distance (km)	Sent	Received	PDF
0	193	2990	2973	0.994
1	28	3085	3085	1.000
2	31	3045	3041	0.999
3	5	2963	2963	1.000
4	179	3005	2997	0.997
5	354	2935	2730	0.930

output file (Sect. 3.2.2), if there is a need to write something to it directly, behind the scenes of the standard output functions offered by SMURPH. The string returned by *getOName* (invoked here as a *Station* method) identifies the terminal (see Sect. 3.3.2). Following the terminal identifier, we show its distance to the hub rounded to full kilometers. The *distTo* method of *Transceiver* returns the transceiver's distance to another transceiver belonging to the same radio channel. The next line shows the number of packets transmitted by the hub and intended for this terminal, the number of received hub packets, and their ratio. Then, the *calculate* method of *RVariable* extracts four standard parameters of the distribution: the minimum, the maximum, the list of moments, and the number of samples (see Sect. 10.1.2). We only care about the number of samples and the first value returned in the array *ave* (which happens to be the average), which are written to the output file with the last statement.

Finally, a call to *printPfm* must be included in the *printResults* method of *Root* (Sect. 2.4.6), e.g., like this:

```
void Root::printResults ( ) {
    Long i;
    HTTrf->printPfm ( );
    THTrf->printPfm ( );
    Client->printPfm ( );
    HTChannel->printPfm ( );
    THChannel->printPfm ( );
    for (i = 0; i < NTerminals; i++)
        TheTerminal (i)->printPfm ( );
}
```

There are many experiments that one may want to conduct with the model, especially that it runs fast: a typical run takes just a few seconds. It is not the purpose of this section to discuss them all, but rather to bring in a few useful tools and suggest ways of playing with the model. For example, to clearly see how the PDF depends on station location, we may now redo the experiment with the terminal-to-hub traffic switched off. Table 2.1 shows the per-terminal packet delivery statistics extracted from the new performance data that now appears at the end of the output file.

Table 2.2 Average number of transmissions per packet

Terminal	Distance (km)	Sent	Average number of transmission
0	193	14,969	2.543
1	28	22,312	1.712
2	31	21,386	1.785
3	5	31,386	1.229
4	179	31,088	2.525
5	354	12,839	2.975

Our channel model may not be very realistic (the original network would not transmit over 300+ km without a repeater), but it does show that the packet delivery fraction drops with distance. Table 2.2 illustrates the location dependence of the average number of retransmissions for the (pure) terminal-to-hub traffic. The bias in the quality of service perceived by the different terminals is amplified by two new factors: the acknowledgments, which also experience a distance-dependent rate of damage, and the contention to the channel. The most distant terminal is able to expedite less than half of the packets that a more fortunately situated terminal can successfully transmit within the same time.

To wrap up this section, let us see how the network fares when saturated by both traffic patterns. For that, we set both mean inter-arrival times to 0.05. This time the run terminates by reaching the 200,000 limit on the number of received messages after about 16,137 s of virtual time. The global effective throughput of the network (reported in the combined *Client* section) is 7931 bps, which is slightly more than ½ of the maximum nominal capacity of the two channels combined (2 × 7738 bps). To calculate the contribution of each traffic pattern, we should divide the respective numbers of received bits by the simulated time. For the hub-to-terminal traffic we get 5842 bps. The balance of the bandwidth on the hub-to-terminal channel (7738 − 5842 = 1896 bps) is taken by acknowledgments (which take precedence over data packets), bit-error losses, and switchover losses at the (half-duplex) terminals.

The global packet delivery fraction for the hub-to-terminal traffic is about 0.8. The drop is caused by the fact that the (half-duplex) terminals transmit a lot, thus remaining deaf to the hub traffic for relatively long periods of time. The by-terminal numbers are shown in Table 2.3.

Table 2.3 Per-terminal data for the two traffic patterns combined

Terminal	Distance (km)	Sent T-H	Avg Xmt	Sent H-T	Rcvd H-T	PDF
0	193	6685	2.561	30846	24840	0.805
1	28	9988	1.716	30576	24711	0.808
2	31	9622	1.789	30730	24715	0.804
3	5	13947	1.233	31119	25140	0.808
4	179	6704	2.555	30845	24920	0.808
5	354	5751	2.979	30685	22976	0.749

References

1. B. Stroustrup, *The C++ Programming Language* (Pearson Education India, 2000)
2. J. Alger, *C++ for Real Programmers* (Academic Press, Inc., 2000)
3. S.M. Ross, *Introduction to Probability and Statistics for Engineers and Scientists* (Elsevier Science, 2014)
4. K.A. Bartlett, R.A. Scantlebury, P.T. Wilkinson, A note on reliable full-duplex transmission over half-duplex lines. Commun. ACM **12**, 260–265 (1969)
5. W.C. Lynch, Reliable full-duplex transmission over half-duplex telephone lines. Commun. ACM **11**, 407–410 (1968)
6. A. O'Dwyer, *Handbook of PI and PID Controller Tuning Rules* (World Scientific, 2009)
7. D.E. Rivera, M. Morari, S. Skogestad, Internal model control: PID controller design. Industrial & Engineering Chemistry Process Design and Development **25**, 252–265 (1986)
8. W.R. Stevens, B. Fenner, A.M. Rudoff, *UNIX Network Programming*, vol. 1 (Addison-Wesley Professional, 2004)
9. W.W. Gay, *Linux Socket Programming: By Example* (Que Corp., 2000)
10. Åström, T. Hägglund, *PID Controllers: Theory, Design, and Tuning*, vol. 2 (Isa Research Triangle Park, NC, 1995)
11. T.S. Rappaport and others, *Wireless Communications: Principles and Practice*, vol. 2 (Prentice hall PTR New Jersey, 1996)
12. M.A. Ingram, Six time-and frequency-selective empirical channel models for vehicular wireless LANs. IEEE Veh. Technol. Mag. **2**, 4–11 (2007)
13. A. Aguiar, J. Gross, *Wireless Channel Models* (2003)
14. N. Abramson, The aloha system—another alternative for computer communications, in *Fall Joint Computer Conference, AFIPS Conference Proceedings* (1970)
15. R. Binder, N. Abramson, F. Kuo, A. Okinaka, D. Wax, ALOHA packet broadcasting: a retrospect, in *Proceedings of the May 19–22, 1975, National Computer Conference and Exposition* (1975)
16. N. Abramson, Development of the ALOHANET, in *IEEE Transactions on Information Theory*, vol. IT-31 (1985), pp. 119–123
17. R.M. Metcalfe, D.R. Boggs, Ethernet: distributed packet switching for local computer networks. Commun. ACM **19**, 395–404 (1976)
18. P. Gburzyński, P. Rudnicki, Modeling low-level communication protocols: a modular approach, in *Proceedings of the 4th International Conference on Modeling Techniques and Tools for Computer Performance Evaluation* (Palma de Mallorca, Spain, 1988)
19. P. Gburzyński, *Protocol design for local and metropolitan area networks* (Prentice-Hall, 1996)
20. D.R. Boggs, J.C. Mogul, C.A. Kent, *Measured Capacity of an Ethernet: Myths and Reality* (Digital Equipment Corporation, Western Research Laboratory, Palo Alto, California, 1988)

21. J.S. Lee, L.E. Miller, *CDMA Systems Engineering Handbook* (Artech House, Inc., 1998)
22. M.R. Souryal, B.R. Vojcic, R.L. Pickholtz, Adaptive Modulation in Ad Hoc DS/CDMA Packet Radio Networks. IEEE Trans. Commun. **54**, 714–725 (2006)

Chapter 3
Basic Operations and Types

In this chapter, we discuss the add-ons to the standard collection of types defined by C++ that create what we would call the look and feel of SMURPH programs, and introduce a few basic operations laying out the foundations for developing models in SMURPH. Note that SMURPH does not hide any aspects of C++; the specific syntax of some SMURPH constructs is handled by a specialized preprocessor (Sect. 1.3.2), which only affects the parts it understands, or by the standard C preprocessor (as straightforward macros). All C++ mechanisms are naturally available to SMURPH programs.

3.1 Numbers

Having been built on top of C++, SMURPH naturally inherits all the arithmetic machinery of that language. Most of the SMURPH extensions in this area result from the need for a high precisionrepresentation of the modeled discrete time. This issue is usually less serious in control programs (where the time resolution of order microsecond is typically sufficient) than in simulators (where the ultra-fine granularity of time helps identify race conditions and other timing problems).

3.1.1 Simple Integer Types

Different machine architectures may represent types *int* and *long* using different numbers of bits. Moreover, some machines/compilers may offer multiple non-standard representations for integer numbers, e.g., type *long long*. Even though SMURPH mostly runs with GNU C++, which introduces a standard, consistent collection of names for numerical types of definite sizes, it insists on its own naming scheme for

© Springer Nature Switzerland AG 2019
P. Gburzyński, *Modeling Communication Networks and Protocols*,
Lecture Notes in Networks and Systems 61,
https://doi.org/10.1007/978-3-030-15391-5_3

the integer numerical types, primarily for historical reasons.[1] Specifically, SMURPH defines the following types:

Long	This type is intended to represent numbers with a sufficient precision to cover the range of object identifiers (*Id's*, see Sect. 3.3.2). Generally, type *Long* should be used for integer numbers (possibly signed) which require more than 16 but no more than 32 bits of precision. The postulated, minimum precision of type *Long* is 24 bits (in practice, it is always at least 32 bits)
LONG	This type represents the maximum-range signed integer numbers available on the system. Type *LONG* is used as the base type for creating numbers of type *BIG* (Sect. 3.1.3)
IPointer	This is the smallest signed integer type capable of holding a pointer, i.e., an address, on the given system
LPointer	This is the smallest type capable of holding both a pointer and a *LONG* integer number

The distinction between the above integer types is meaningful when the same program must consistently execute on systems with diverse sizes for types *int*, *long*, and so on. For example, on a 64-bit system, type *int* may be represented with 32 bits, while both *long* integers and pointers are 64-bit long. On such a machine, *Long* will be equivalenced to *int* and the remaining three types (i.e., *LONG*, *IPointer*, and *LPointer*) will be all defined as *long int*.

On a legacy 32-bit system, where types *int* and *long*, as well as pointer types, take 32 bits, types *Long*, *LONG*, and *LPointer* are all set to *long*, and type *IPointer* is declared as *int*. If the C++ compiler offers the type *long long*, the user may select this type as the base type for creating type *BIG* (see Sect. 3.1.3). In such a case, types *LONG* and *LPointer* will be declared as *long long*, type *Long* will be equivalenced to *long*, and type *IPointer* will be defined as *int* (unless pointers are wider than *int*). Other combinations are conceivable, depending on the subtleties of the machine architecture and the assortment of integer types offered by the compiler.

In most cases, the interpretation of the SMURPH-specific integer types happens inside SMURPH, so the user seldom should worry about the details. However, those types are used to declare some arguments of standard functions, or some attributes of standard class types. While natural conversions often take place automatically, there exist cases where the user should be aware of the differences (or at least the rationale) behind the different integer types.

---

[1]In the days when SMURPH was programmed, C++ was a much less established tool than it is today.

3.1.2 Time in SMURPH

As all simulators, SMURPH provides tools for modeling the flow of time in a class of physical systems (Sect. 1.2.3). Time in SMURPH is discrete, which means that integer numbers are used to represent time instants and intervals.

The user is responsible for choosing the granularity of the modeled time (which can be practically arbitrarily fine). This granularity is determined by the physical interpretation of the indivisible time unit (*ITU*, Sect. 2.1.4), i.e., the shortest possible interval between different, consecutive time instants. In the virtual (simulation) mode, the correspondence between the *ITU* and an interval of real time is not defined explicitly. All time-dependent quantities are assumed to be relative to the *ITU*, and it is up to the interpreter of the input data and output results to assign a meaning to it. In the real mode, by default, one *ITU* is mapped to one microsecond of real time. This binding can be redefined by the user. In the visualization mode, the user must set up the correspondence between the internal time unit and an interval of real time. This correspondence must be explicitly established when the visualization mode is entered (Sect. 5.3.4).

If two events modeled by SMURPH occur within the same *ITU*, there is generally no way to order them with respect to their *actual* succession in time. In many cases (especially in any of the two simulation modes) the order in which such events are presented (trigger actions in the user program) is nondeterministic (Sect. 1.2.4). It is possible, however, to assign priorities to awaited events (Sect. 5.1.4). Then if multiple events occur at the same time, they will be presented in the order of their priorities. It is also possible (see Sect. B.5) to switch off the nondeterminism in SMURPH (which is generally much less desirable in control than in simulation). Of course, events triggered by actual physical components of a controlled system always occur at some definite real-time moments.

Besides the *ITU*, there exist two other time-related units, whose role is to hide the internal time unit and simplify the interpretation of the input and output data. One of them is *ETU*, i.e., the experimenter time unit (Sect. 2.1.4), which determines the primary unit of time being of relevance to the user. The user declares the correspondence between the *ITU* and the *ETU* by calling one of the following two functions:

void setEtu (double e);
void setItu (double i);

The argument of *setEtu* specifies how many *ITU*s are in one *ETU*; the argument of *setItu* indicates the number of *ETU*s in one *ITU*; The call:

setEtu (a);

is equivalent to the call:

setItu (1/a);

One of the two functions (any of them) should be called only once, at the beginning of the network creation stage, before the first object of the network has been created

(Sect. 5.1.8). Two global, read-only variables *Itu* and *Etu* of type *double* store the relationship between the *ITU* and *ETU*. *Itu* contains the number of *ETUs* in one *ITU* and *Etu*, the reciprocal of *Itu*, contains the number of *ITUs* in one *ETU*.

The simulator's clock, i.e., the amount of virtual time elapsed from the beginning of the model's execution, is available in two global variables declared as:

TIME Time;
double ETime;

Type *TIME* is formally introduced in Sect. 3.1.3. The first variable tells the elapsed, simulated time in *ITUs*, while the second one does it in *ETUs*.[2]

The third fundamental unit is called *DU*, which stands for *distance unit*. Its purpose is to establish a convenient measure of distance (typically propagation distance between network nodes), which internally translates into a specific number of *ITUs* required by the signal to travel a given distance. The function:

void setDu (double d);

sets the *DU* to the specified number of *ITUs*. The corresponding global variable (of type *double* is called *Du* and contains the number of *ITUs* in one standard unit of distance.

For illustration, suppose that we want to model a mobile wireless network operating at 100 Mb/s, with the mobility grain of (at least) 10 cm (this is the minimum discernible distance that we care about). Assuming the signal propagation speed of 3×10^8 m/s, the amount of time needed by the signal to cross the distance of 10 cm is $1/(3 \times 10^9) \approx 0.33 \times 10^{-9}$ s. We want the *ITU* to be at least as small as this interval. Also, we must keep in mind that the transmission rate of a transceiver (or port) should be expressible as an entire number of *ITUs* per bit with a negligible error. At 100 Mb/s, one bit translates into 10^{-8} s, thus, when we assume that 1 *ITU* corresponds to 0.33×10^{-9} s, the insertion time of a single bit becomes about 30.3 ITUs. As the actual rate must be an integer number, we will end up with 30 *ITUs* per bit, which will incur an error of about 1%. If this is acceptable, we can set the *DU* to 10.0, which will simply mean that distances in the model are expressed in meters. These two statements do the trick:

setEtu (3e9);
setDu (10.0);

The first statement defines the *ETU* relative to the *ITU*; we can easily see that it corresponds to one second. If we don't like the error in the rate representation, then we can reduce the *ITU*, e.g.,

setEtu (3e10);
setDu (100.0);

---

[2]This "variable" is in fact a macro referencing *Time*, so current time is stored in one place.

(by the factor of ten) which will also reduce the error tenfold without changing the interpretation of the user-preferred, high-level units.

Alternatively, what may be more natural in a network with a homogeneous transmission rate, we can start from the other end, e.g., assuming that the *ITU* is equal to 1/100 of the bit insertion time, i.e., 10^{-10} s. This amount of time translates into 3.3 cm, so setting things this way:

setEtu (1e10);
setDu (30.3);

will again make (a) 1 m the unit of distance, (b) 1 s the unit of time, (c) the grain of distance fine enough, (d) the rate expressible as an integer number of *ITUs* with no error. The distance grain does not divide the distance unit exactly, but this kind of problem is usually easier to live with than a serious discrepancy in the transmission rate. And, of course we can always reduce the *ITU* to make the error as small as we want. There is little penalty for that. Note that the *ITU* is never defined explicitly: its meaning is determined by the values of the remaining two units, as well as by the transmission rate assigned to ports and/or transceivers.

In most models, the three representations of time, in *ITUs*, in *ETUs*, and in *DUs*, yield three different numerical values for the same interval, unless that interval happens to be zero. Conversions are simple and involve the factors *Itu,Etu,*and *Du*. SMURPH provides the following functions (macros) whose consistent usage may help the program to avoid confusion or ambiguity:

double ituToEtu (TIME);
TIME etuToItu (double);
double ituToDu (TIME);
TIME duToItu (double);

The first function converts an interval from *ITUs* to *ETUs* (multiplying it by *Etu*). The multiplications carried out by the remaining functions are easy to guess.

The issues of precision brought in by conversions between *TIME* and *double* are usually irrelevant, if the *ITU* is fine enough and the *double* values are used to store intervals between "typical" events rather than absolute clock indications. The distance between a pair of nodes is usually a good representative of the former: even with an extremely fine precision of *TIME* its (properly scaled) representation by *double* is going to very accurate. On the other hand, as we argued in Sect. 2.1.4, the virtual clock may reach humongous values, so the value of *Time* should be considered authoritative (and precise), while *ETime* should be viewed as its presentable and possibly more "natural," but approximate, representation.

In the real mode, the *DU* is not used (the concept of propagation distance only applies to the models of communication channels), and the *ETU* immutably stands for one real second. There exists an alias for *Etu*, namely *Second*; as well as an alias for *setEtu*, namely *setSecond*. These functions make it possible to define the internal granularity of time. By default, *Second (Etu)* is set to 1, 000, 000 (and *Itu* contains 0.000001), which means that 1 *ITU* maps to one microsecond.

The visualization mode brings no restrictions compared to the virtual mode; however, it still makes sense to agree to interpret the *ETU* as one second, as it probably does in most models, regardless of the mode. This is because the operation of entering the visualization mode (Sect. 5.3.4) specifies the number of real milliseconds to be applied as the synchronization grain (at those intervals the simulator will be synchronizing its virtual time to real time) and associates with it a given fraction of the *ETU*. Things are easier to comprehend, if the *ETU* corresponds to a nice interval, and one second looks like a safe and sane choice.

To continue with the above example, suppose that we would like to visualize the behavior of our network in slow motion, with the sync granularity of $1/4$ s, assuming that one virtual second spreads over ten seconds of real time visualization. Here is the operation to enter the visualization mode:

setResync (250, 0.025);

The first argument specifies 250 ms as the "resync granularity," and the second argument maps these 250 ms (of real time) to 0.025 *ETUs*. Thus, assuming that 1 *ETU* stands for 1 s of virtual time, one second of real time will correspond to $1/10$ of a second in the model. This means that the visualization will be carried out with the slow-motion factor of 10.

The functions introduced above are not available when the simulator (or rather the control program) has been compiled with $-F$ (Sect. B.5). In that case, the *ETU* is "hardwired" at 1,000,000 *ITUs* (*Etu*, *Itu*, and *Second* are not available to the program).

In both simulation modes, *Itu*, *Etu*, and *Du* all start with the same default value of 1.0. If the same program is expected to run in the simulation mode, as well as in the control mode, it usually makes sense to put the statement:

setEtu (1000000.0);

somewhere in the initialization sequence (Sect. 5.1.8). While operating in the real or visualization mode, a SMURPH program maps intervals of virtual time to the corresponding intervals of real time. The actual knowledge of real time (the precise date and time of the day) is usually less relevant. One place where real-time stamps are used is journaling (Sect. 9.5), which is available in the real and visualization modes. Such time stamps are represented as the so-called UNIX-epoch time, i.e., the number of seconds and microseconds that have elapsed since 00:00:00 on January 1, 1970. The following function:

*Long getEffectiveTimeOfDay (Long *usec = NULL);*

returns the number of seconds in UNIX time. The optional argument is a pointer to a *Long* number which, if the pointer is not *NULL*, will contain the number of microseconds after the last second. Function *tDate* (Sect. 3.2.9) converts the value returned by *getEffectiveTimeOfDay* (seconds only) into a textual representation of the date and time.

It is possible to invoke a SMURPH program operating in the real or visualization mode in such a way that its real time is shifted to the past or to the future. This may be useful together with journaling (Sect. 9.5), e.g., to synchronize the program to a past journal file. The $-D$ call argument (Sect. B.6) can be used for this purpose. If the real time has been shifted, then both *getEffectiveTimeOfDay* and *tDate* return the shifted time.

3.1.3 Type BIG and Its Range

With finely-grained, detailed modeling, it may happen that the standard range of integer numbers available on the system is insufficient to represent time. From the very beginning, the SMURPH models, intended to study the performance and correctness of low-level (medium access) protocols, were designed and programmed with obsessive attention to real-life phenomena related to time, including race conditions that might involve extremely short intervals [1–3]. Ultra-fine granularity of time is thus encouraged in SMURPH. Whenever there is even a tiny trace of suspicion that the *ITU* may be too coarse, it is a natural and easy solution to make it finer, even by a few orders of magnitude, just to play it safe. This means that time intervals may easily become huge numbers exceeding the capacity of the machine representation of integers.

Therefore, SMURPH comes with its own type for representing potentially huge nonnegative integer numbers and performing standard arithmetic operations on them. Time instants and other potentially huge nonnegative integers should be declared as objects of type *BIG* or *TIME*. These two types are equivalent. For clarity, *TIME* should be used in declarations of variables representing time instants or intervals, and type *BIG* can be used for other potentially humongous integers, not necessarily related to time.

When a SMURPH program is created by *mks* (see Sect. B.5), the user may specify the precision of type *BIG* with the $-b$ option. The value of this option (a single decimal digit) is passed to SMURPH (and also to the user program) in the symbolic constant *BIG_precision*. This value selects the implementation of type *BIG* by indicating the minimum required range of *BIG* numbers as *BIG_precision* $\times$ 31 bits. If *BIG_precision* is 1 or 0, type *BIG* becomes equivalent to the maximum-length integer type available on the system (type *LONG*, see Sect. 3.1.1), and all operations on objects of this type are performed directly as operations on integers of type *LONG*. Otherwise, depending on the size of type *LONG*, several objects of this type may be put together to form a single *BIG* number. For example, if the size of *LONG* is 32 bits, a single *BIG* number is represented as *BIG_precision LONG* objects. If the *LONG* size is 64 bits, the number of *LONG* objects needed to form a single *BIG* number is equal to[3]:

---

[3]Thus, for *BIG_precision* equals 1 and 2, *BIG* numbers will be represented in the same way, using the standard integer type *long*.

$$\left\lfloor \frac{BIG\_precision + 1}{2} \right\rfloor$$

The tacit limit on the precision of type *BIG* imposed by *mks* (Sect. B.5) is 9 which translates into 279 significant bits. Using this precision, one could simulate the life of the visible universe, more than 10^{20} times over, with the granularity of Planck time.

The default precision of type *BIG* is 2 (in both modes). For the real mode, assuming that one *ITU* is equal to one microsecond, the amount of real time covered by the range of type *BIG* (and *TIME*) is almost 150,000 years (but it is only slightly more that 30 min for precision 1).

3.1.4 Arithmetic Operations

The implementation of type *BIG* is completely transparent to the user. When the precision of *BIG* numbers is larger than 1, the standard arithmetic operators $+$, $-$, $*$, $/$, $\%$, $++$, $--$, $==$, $!=$, $<$, $>$, $<=$, $>=$, $+=$, $-=$, $*=$, $/=$, $\%=$ are automatically overloaded to operate on objects of type *BIG*. Combinations of *BIG* operands with types *int*, *long*, and *double* are legal. If *BIG* is equivalent to *long*, the conversion rules for operations involving mixed operand types are defined by C++. Otherwise, if an operation involves a *BIG* operand mixed with a numerical operand of another type, the following rules are applied:

- If the type of the other operand is *char*, *int*, or *long*, the operand is converted to *BIG* and the result type is *BIG*. One exception is the modulo ($\%$) operator with the second operand being of one of the above types. In such a case, the second operand is converted to *long* and the result type is *long*.
- If the type of the other operand is *float* or *double*, the *BIG* operand is converted to *double* and the result type is *double*.
- An assignment to/from *BIG* from/to other numerical type is legal and involves the proper conversion. A *float* or *double* number assigned to *BIG* is truncated to its integer part.

Overflow and error conditions for operations involving *BIG* operands are only checked for precisions 2–9.[4] Those checks can be switched off (e.g., for efficiency) with the $-m$ option of *mks* (see Sect. B.5). Note that it is formally illegal to assign a negative value to a *BIG* variable. However, for precisions 0 and 1 such operation passes unnoticed (and the result may not make sense).

Explicit operations exist for converting a character string (an unsigned sequence of decimal digits) into a *BIG* number and vice versa. The following function turns a *BIG* number into a sequence of characters:

---

[4] If the size of type *LONG* is 64 bits, the overflow and error conditions are only checked for precisions higher than 2.

char *btoa (BIG a, char *s = NULL, int nc = 15);

where *a* is the number to be converted to characters, *s* points to an optional character buffer to contain the result, and *nc* specifies the length of the resulting string. If *s* is not specified (or *NULL*) an internal (static) buffer is used. In any case, the pointer to the encoded string is returned as the function value. If the number size exceeds the specified string capacity (15 is the default), the number is encoded in the format: *dd...ddEdd*, where *d* stands for a decimal digit and the part starting with *E* is a decimal exponent.

The following function can be used to convert a string of characters into a *BIG* number:

BIG atob (const char *s);

The character string pointed to by *s* is processed until the first character not being a decimal digit. Initial spaces and the optional plus sign are ignored.

3.1.5 Constants

One way to create a constant of type *BIG* is to convert a character string into a *BIG* object. If the constant is not too large, one can apply a conversion from *long int*, e.g., $b = 22987$ (we assume that *b* is of type *BIG*) or from *double*. This works also in declarations; thus, the following examples are fine:

BIG b = 12;
TIME tc = 10e9;
const TIME cnst = 20000000000.5;

The fractional part is ignored in the last case.

Three *BIG* constants: *BIG_0*, *BIG_1*, and *BIG_inf* are available directly. The first two of them represent 0 and 1 of type *BIG*, the last constant stands for an infinite or undefined value and is numerically equal to the maximum *BIG* number representable with the given precision. This value is considered special and should not be used as a regular *BIG* number. For completeness, there exist constants *TIME_0*, *TIME_1*, and *TIME_inf* which are exactly equivalent to the three *BIG* constants.

These two predicates tell whether a *BIG* number is defined (or finite):

int def (BIG a);

returning 1 when *a* is defined (or finite) and 0 otherwise, and

int undef (BIG a);

which is a straightforward negation of *def*.

The duality of the representation of time (*ETU* versus *ITU*) motivates three *double* constants: *Time_0*, *Time_1*, and *Time_inf*, intended as the *ETU* counterparts of the respective constants of type *TIME*. *Time_0* is defined as 0.0 and *Time_1* as *Itu*

(Sect. 3.1.2), so it corresponds to 1 *ITU*, not 1 *ETU* (being just an alias for the global variable *Itu*). We have: *ituToEtu (TIME_x) == Time_x*, with x representing the same suffix (*_0, _1, _inf*) on both sides. The most useful of the three constants is *Time_inf* representing (accurately, in the sense of the above equality) the undefined time value in *ETUs*.

By the same token, there exist three constants named *Distance_0*, *Distance_1*, and *Distance_inf*, intended to correspond to the respective *TIME* constants and matching their values in the "distance domain." Thus, *Distance_0* is defined as 0.0, *Distance_1* as 1/*Du*, and *Distance_inf* represents the undefined distance-time, in the sense that *ituToDu (TIME_inf) == Distance_inf*.

The following arithmetic-related symbolic constants are also available to the protocol program:

TYPE_long	Equal to 0. This is the *LONG* type indicator (see Sects. 3.2.2 and 10.1.1)
TYPE_short	Equal to 1. This is an unused type indicator existing for completeness
TYPE_BIG	Equal to 2. This is the *BIG* type indicator
BIG_precision	Telling the precision of BIG numbers (Sect. 3.1.3)
MAX_long	Equal to the maximum positive number representable with type *long int*. For 32-bit 2's complement arithmetic this number is 2147483647
MIN_long	Equal to the minimum negative number representable with type *long int*. For 32-bit 2's complement arithmetic this number is −2147483648
MAX_short	Equal to the maximum positive number representable with type *short int*. For 16-bit 2's complement arithmetic this number is 32767
MIN_short	Equal to the minimum negative number representable with type *short int*. For *16*-bit 2's complement arithmetic this number is −32768
MAX_int	Equal to the maximum positive number representable with type *int*. Usually this number is equal to *MAX_long* or *MAX_short*
MIN_int	Equal to the minimum negative number representable with type *int*. In most cases this number is equal to *MIN_long* or *MIN_short*

There also exist the following constants: *MAX_Long*, *MIN_Long*, *MAX_LONG*, and *MIN_LONG*, which represent the boundary values of types *Long* and *LONG*, according to the way these types have been aliased to the corresponding basic integer types (see Sect. 3.1.1).

3.1.6 Other Non-standard Numerical Types

Some numbers, not necessarily ones representing time instants or intervals, may also happen to be big, at least in some experiments. For example, a counter accumulating the number of bits transmitted in an entire large network may reach a huge value, especially when the messages/packets are (formally) large in the model. Note that the real (CPU) processing time of a packet in the simulator is typically unrelated to the formal (virtual) size of that packet. Thus, in a short time, the simulator may add a huge number of large values to a counter, possibly overtaxing its range. Generally, how fast a counter grows depends on two things: how often it grows, and what it grows by on each occasion. Both parts can be large in a simulation model, and the size of the second part comes for free and costs no processing time.

Using type *BIG* to represent all numbers that could exceed the capacity of the standard types *int* or *long* would be too costly in many situations, when the experimenter knows in advance, and for a fact, that they won't. Therefore, there exist three flexible non-standard numerical types that can be used to represent nonnegative integer variables. By default, all three types are equivalenced (via *typedef*) to type *LONG*, but those defaults can be changed with compilation options to *BIG* or *Long* (see Sect. B.5), depending on whether they are too tight or too wasteful. The three types are:

DISTANCE	This type is used to represent (internally) propagation distances between ports or transceivers (see Sects. 4.3.3 and 4.5.2). The internal representation of a propagation distance refers to the time (the number of *ITUs*) needed for a signal to cross the distance (Sect. 3.1.2). Although time intervals expressed in *ITUs* can be generally huge, an interval corresponding to the longest propagation distance in a link or rfchannel can still be relatively small, and may be representable using an integer type with a size and overhead lesser than those of type *BIG*
BITCOUNT	This type is used to declare variables counting bits, e.g., transmitted or received globally or at a specific station. Numerous such counters are used internally by SMURPH for calculating various performance measures (Sect. 10.2.2)
RATE	This type is used to represent port/transceiver transmission rates (Sects. 4.3.1, 4.5). Formally, the transmission rate of a port is a time interval expressed internally in *ITUs*, which in most cases is relatively small

Five macros are associated with each of the above three types. For example, the following macros are related to type *DISTANCE*:

TYPE_DISTANCE	This macro evaluates to the type indicator of type *DISTANCE* (Sect. 3.1.5). For example, if *DISTANCE* is equivalenced to *LONG*, *TYPE_DISTANCE* is equal to *TYPE_long*. If *DISTANCE* is equivalenced to *BIG*, *TYPE_DISTANCE* is equal to *TYPE_BIG*
MAX_DISTANCE	This macro is defined as *BIG_inf*, *MAX_LONG*, or *MAX_Long*, depending on whether type *DISTANCE* is equivalent to *BIG*, *LONG*, or *Long*
DISTANCE_inf	The same as *MAX_DISTANCE*
DISTANCE_0	This macro is defined as 0, if type *DISTANCE* is equivalent to *LONG* or *Long*, or as *BIG_0* otherwise
DISTANCE_1	This macro is defined as 1, if type *DISTANCE* is equivalent to *LONG* or *Long*, or as *BIG_1* otherwise

To obtain the corresponding macros for the other two types, replace the word *DISTANCE* with *BITCOUNT* or *RATE*.

One should use their judgment, based on the interpretation of *ITU* and the expected range of values accumulated in bit counters (like the total length of all messages received by a node during a simulation run) to decide whether the default definitions are fine. For example, if the *ITU* corresponds to 1 s (femto-second, i.e., 10^{-15} s), the maximum propagation distance covered by a 32-bit *LONG* integer is slightly over 600 m (assuming $c = 3 \times 10^8$ m/s). Thus, if longer distances are not unlikely to show up in the model (and *LONG* happens to be 32 bits long), *DISTANCE* should be declared as *BIG*. Type *RATE* is generally less prone to this kind of problems. As a rate determines the number of *ITUs* needed to expedite a single bit, the lower the real-life transmission rate, the higher the value that must be stored in a *RATE* variable. For illustration, with 1 *ITU* corresponding to the same 1 fs, the maximum interval stored in a 32-bit number is ca. 2^{-6} s, which translates into the minimum representable rate of 500 kb/s.

The proliferation of the different aliases for integer types in SMURPH is mostly the relic of 32-bit systems, where the basic integer type was sometimes quite clearly insufficient, or just too tight for comfort. For example, when counting the total number of bits received in the network, e.g., to calculate the throughput at the end of a simulation run, the 4G limit of a 32-bit counter would not appear impressive, especially if the counter was global for the entire (possibly sizable) network. As 64-bit systems have become a de facto standard, the issues have ceased to be relevant for most models.

3.2 Auxiliary Operation and Functions

In this section, we mention those functions provided by SMURPH that are only superficially related to its operation as a network simulator or a system controller. They cover random number generation, non-standard I/O, and a few other items that are not extremely important, as the user could easily substitute for them some standard tools from the C++ library. In those early days when the first (C++) version of SMURPH was being developed, there was little consensus regarding the exact shape of the C++ I/O library and other common functions, which tended to differ quite drastically across platforms. Consequently, we wanted to cover them as much as possible by our own tools, so the same SMURPH program could compile and run without modifications on all the then popular platforms. These tools are still present in the most recent version of our package, even though the standard C++ libraries are now much more consistent than they were when our project started.

3.2.1 Random Number Generators

All random number generators offered by SMURPH are derived from the following function:

double rnd (int seed);

which returns a pseudo-random number of type *double* uniformly distributed in
[0,1). By default, this function uses its private congruential algorithm. The user can
select the standard random number generator (the *drand48* family) by compiling the
program with the −8 option (see Sect. B.5).

The argument of *rnd* must be one of these values: 0 (represented by the constant
SEED_traffic), 1 (available as *SEED_delay*), or 2 (*SEED_toss*). It identifies one of
three seeds, each of them representing a separate series of pseudo-random numbers.
The initial values of the seeds (which are of type *long*) can be specified when the
SMURPH program is called (see Sect. B.6); otherwise, some default (always the
same) values are assumed. This way simulation runs are replicable, which may be
important from the viewpoint of debugging.

Intentionally, each of the three seeds represents one category of objects to be
randomized, in the following way:

SEED_traffic	Traffic related randomized values, e.g., intervals between message arrivals, message length, selection of the transmitter and the receiver
SEED_delay	Values representing various randomized delays not related to traffic generation. For example, the randomization of *Timer* delays for modeling inaccurate clocks (Sect. 5.3.3) is driven by this seed
SEED_toss	Tossing coins or dice in situations when SMURPH must decide which one of several equally probable possibilities to follow, e.g., selecting one of two or more waking events scheduled at the same *ITU* (see Sects. 1.2.4 and 5.1.4)

All random numbers needed by SMURPH internally are generated according to
the above rules, but the user program is not obliged to obey them. In most cases, the
program need not use *rnd* directly.

When the same SMURPH program is run twice with the same value of a given
seed, all the randomized objects belonging to the category represented by this seed
will be generated in exactly the same way as before. In particular, two simulation
runs with identical values of all three seeds, and the same input data, will produce
identical results, unless the declared CPU time limit is exceeded (Sect. 10.5.3). In the
latter case, the exact moment when the simulator is stopped depends on the operating
system and is generally somewhat non-deterministic.

The following functions can be used to generate exponentially distributed pseudo-
random numbers:

TIME tRndPoisson (double mean);
LONG lRndPoisson (double mean);
double dRndPoisson (double mean);

When the precision of *TIME* is 1 (Sect. 3.1.3), the first two functions are identical.
The first function should be used for generating objects of type *TIME*, while the
second one generates (*LONG*) integer values. The parameter specifies the mean

value of the distribution whose type is *double* in all three cases. The result is a randomized interval separating two customer arrivals in a Poisson process with the specified mean inter-arrival time. The probability density function is:

$$f(t) = \lambda e^{-\lambda t}$$

where λ is the mean. The interval is generated by transforming a uniformly-distributed pseudo-random value obtained by a call to *rnd* with seed 0 (*SEED_traffic*). The first stage result of this transformation is a *double* floating point number which, for the first two functions, is subsequently rounded to an object of type *TIME* or *LONG*. The last function returns the full-precision *double* result.

Below we list five functions generating uniformly distributed random numbers.

TIME tRndUniform (TIME min, TIME max);
TIME tRndUniform (double min, double max);
LONG lRndUniform (LONG min, LONG max);
LONG lRndUniform (double min, double max);
double dRndUniform (double min, double max);

In all cases the result is between *min* and *max*, inclusively. The functions call *rnd* with *SEED_traffic*.

Sometimes we would like to randomize a certain, apparently constant, parameter, so that its actual value is taken with some tolerance. The following functions serve this end:

TIME tRndTolerance (TIME min, TIME max, int q);
TIME tRndTolerance (double min, double max, int q);
LONG lRndTolerance (LONG min, LONG max, int q);
LONG lRndTolerance (double min, double max, int q);
double dRndTolerance (double min, double max, int q);

They generate random numbers according to a variant of the β distribution which is believed to describe technical parameters whose values are bounded and may vary within some tolerance. The idea, in a nutshell, is to have a bounded and biased distribution looking somewhat like normal, thus being "better informed" than the uniform distribution [4]. The above functions call *rnd* with seed 1 (*SEED_delay*) and transform the uniform distribution into $\beta(q, q)$ which is scaled appropriately, such that the resultant random number is between *min* and *max* inclusively. The parameter q, which must be greater than 0, can be viewed as the "quality" of the distribution. For higher values of q the generated numbers tend to be closer to (*min* + *max*)/2. More formally, the general β distribution has two parameters: $a > 0$ and $b > 0$, and its density function is:

$$f(x) = \frac{x^{a-1}(1 - x)^{b-1}}{B(a, b)}$$

where the denominator only depends on a and b (and is there to make sure that the function integrates to 1). As defined above, the function covers the interval [0, 1]. The SMURPH generator assumes that $a = b = q - 1$ and stretches the function to [min,max]. Thus, for $q = 1$, the distribution is uniform, for $q = 2$, it becomes triangular, and, as q increases, it gets more and more focused around the center of the interval. Note that q is integer, even though it doesn't have to be. Restricting q to integer values simplifies the calculations. Reasonable values of q are between 1 and 10. Slightly more time is needed to generate random numbers for higher values of q.

The following function generates a normally distributed (Gaussian) random number:

double dRndGauss (double mean, double sigma);

with the specified mean and standard deviation. It calls *rnd* with seed 2 (*SEED_toss*). Note that the distribution is symmetric around the mean and, in principle, unbounded.

Here is a function to generate a randomized number of successes in a Bernoulli experiment involving tossing a possibly biased coin:

Long lRndBinomial (double p, Long n);

where p is the probability of success in a single experiment, and n is the total number of trials.

Various instances of practical, biased, Zipf-style distributions [5] can be produced with this function:

Long lRndZipf (double q, Long max = MAX_Long, Long v = 1);

which generates integer random values between 0 and $max - 1$, inclusively, in such a way that the probability of value k being selected is proportional to:

$$F_k = \frac{1}{(v + k)^q}$$

where $k = 0, \ldots, max - 1$, $v > 0$, and $q > 1$. The actual probability P_k is obtained by normalizing the factors, i.e., dividing each F_k by the sum of all factors.

The following function simulates tossing a multi-sided coin or a dice:

Long toss (Long n);

It generates an integer number between 0 and $n - 1$ inclusively. Each number from this range occurs with probability $1/n$. The function calls *rnd* with seed 2 (*SEED_toss*).

There is an abbreviation for the most frequently used variant of *toss*; the following function:

int flip ();

returns 0 or 1, each value with probability 1/2. The function is slightly more efficient than *toss (1)*.

3.2.2 Input/Output

Two standard files, dubbed the (input) *data file* and the (output) *results file*, are automatically opened by every SMURPH program at the beginning of its execution. For a simulation experiment, the results file is intended to contain the final performance results at the end of run. In a control session, the results file may not be needed, although it may still make sense to produce some "permanent" output, e.g., a log or some instrumentation data.

The data file, represented by the global variable *Inf* of the standard type *istream*, is opened for reading and it may contain data parameterizing the modeled system's configuration (hardware) and the protocols, e.g., the number of stations and their positions, the lengths of links, the transmission rates, the TCP/IP ports of the daemons, the mapping of virtual sensors/actuators to their physical counterparts, and so on.

The data file is only accessible during the initialization stage, i.e., while the *Root* process is in its first state (see Sect. 5.1.8). The file is closed before the protocol execution commences. If the program requires a continuous stream of data during its execution, it should open a non-standard input file with that data, or, perhaps, read the data from a bound mailbox, if it executes in the real or visualization mode (Sect. 5.3.4). The same applies to non-standard output files, which, needless to say, a program may open as many as it needs.

All standard functions and methods of C++ related to input/output are available from SMURPH. There exist, however, additional functions introduced specifically by SMURPH, to handle input from the standard data file and output to the standard results file. The most natural way to write structured information to the results file is to *expose* objects (Sect. 12.1).

The standard C++ operators << and >>, with a stream object as the first argument have been overloaded to handle *BIG* numbers. Thus, it is legal to write:

sp >> b;

or

sp << b;

where *sp* is a stream object (it should be an *istream* in the first case and an *ostream* in the second case) and *b* is a number of type *BIG* (or *TIME*). In the first case, a *BIG* number is read from the stream and stored in *b*. The expected syntax of that number is the same as for *atob* (see Sect. 3.1.4), i.e., a sequence of decimal digits optionally preceded by white spaces. A plus sign (+) immediately preceding the first digit is also OK.[5]

The << operation encodes and writes a *BIG* number to the stream. The number is encoded by a call to *btoa* (see Sect. 3.1.4 with the third argument (digit count) equal to 15. If the number has less than 15 digits, the initial spaces are stripped off.

---

[5]Recall that *BIG* numbers cannot be negative.

The following functions read numbers from the standard data file:

void readIn (int&);
void readIn (Long&);
void readIn (LONG&);
void readIn (BIG&);
void readIn (float&);
void readIn (double&);

They all ignore in the data file (and skip over) everything that cannot be interpreted as the beginning of a number. A number expected by any of the above input functions can be either integer (the first four functions) or real (the last two functions). An integer number begins with a digit or a sign (note that the minus sign is illegal for a *BIG* number) and continues for as long as the subsequent characters are digits. A sign not followed by a digit is not interpreted as the beginning of a number. A real number may additionally contain a decimal point followed by a sequence of digits (the fraction) and/or an exponent. The syntax is essentially as that accepted by the standard UNIX tools for parsing numbers, e.g., *strtod*. A decimal point encountered in an expected integer number stops the interpretation of that number, i.e., the next number will be looked up starting from the first character following the decimal point.

The size of an integer number depends on the size of the object being read. In particular, the range of variables of type *BIG* (*TIME*) may be very big, depending on the declared precision of this type (Sect. 3.1.3).

Three simple features help organize complex data files in a more legible way. If a number read by one of the above functions is immediately (no space) followed by a sign (+ or −) and another number, then the two numbers are combined into one. For example, $120 + 90$ will be read as 210; similarly, $1 - 0.5$ will be read as 0.5, if a real number is expected, or as 1 $(1 - 0)$ otherwise. The rule applies iteratively to the result and thus $1 + 2 + 3 + 4 - 5$ represents a single number (5). Another occasionally useful feature is the symbolic access to the last-read number. The character % appearing as the first (or as the only) item of a sequence of numbers separated by signs stands for the "last-read value." For example, 15, % − 4 will be read as 15, 11. If the type of the expected number is different from the type of the last-read number, the value of % is undefined. One more feature provides an abbreviation for multiple consecutive occurrences of the same number. If a number *m* is immediately followed by a slash (/), followed in turn by an unsigned integer number *n*, the entire sequence is interpreted as *n* occurrences of number *m*.[6] Again, all the expected numbers should be of the same type, otherwise the results are unpredictable.

The fact that only numerical data (together with a few other characters) are expected, and everything else is skipped, makes it easy to include comments in the input data file, but one must be careful not to put a number into a comment by accident. It is possible to insert explicit and unambiguous comments by starting them

---

[6]The original rationale for this feature was entering link distance matrices (Sect. 4.3.3) that, in many cases, consisted of a large number of identical entries.

with # or * (a hash mark or an asterisk). Such a comment extends until the end of the current line.

Below is a function for extracting sets of numbers, e.g., tables. from a character string. While the function is not obviously compatible with the ones listed above, it is useful for reading numerical data from XML files (see Sect. 3.2.3).

int parseNumbers (const char *str, int n, nparse_t *res);

where *nparse_t* is a structure declared as follows:

```
typedef struct {
        int type;
        union {
                double DVal;
                LONG LVal;
                int IVal;
        };
} nparse_t;
```

The first argument points to the character string from where the numbers will be extracted, the second argument tells how many numbers we expect to read, and the third one is an array of records of type *nparse_t* with at least *n* elements. The function will try to fill up to *n* entries in *res* with consecutive numbers located in the input string *str*. It will skip all useless characters in front of a number, and will stop interpreting the number as soon as it cannot be correctly continued. The exact format of an expected number depends on *type* in the current entry of *res*, which should be preset, before the function is called, to one of the following values:

TYPE_double	The expected number is *double*. When found, it will be stored in *DVal*
TYPE_LONG	The expected number is of type *LONG*. When found, it will be stored in *LVal*. If the number begins with *0x* or *0X*, possibly preceded by a sign, it will be decoded as hexadecimal
TYPE_int	The expected number is of type *int*. When found, it will be stored in *IVal*. If the number begins with *0x* or *0X*, possibly preceded by a sign, it will be decoded as hexadecimal
TYPE_hex	The expected number is an integer-sized value coded in hexadecimal, regardless of whether it begins with *0x*, *0X*, a digit, or a sign. When found, it will be stored in *IVal*
ANY	If the actual number looks like *double*, but not *LONG*, i.e., it has a decimal point or/and an exponent, it will be stored in *DVal*, and *type* will be set to *TYPE_double*. Otherwise, the number will be decoded as *LONG* and stored in *LVal*, while *type* will be set to *TYPE_LONG*. In the latter case, the number may optionally begin with the hex prefix (*0x* or *0X*) possibly preceded by a sign

The function value tells how many numbers have been located in the string. This can be more than *n*, which simply means that there are more numbers in the string than we asked for. If any of the numbers encountered by the function overflows the limitations of its format, the function returns *ERROR* (-127).

The following functions write simple data items to the results file:

void print (LONG n, const char *h = NULL, int ns = 15, int hs = 0);
void print (double n, const char *h = NULL, int ns = 15, int hs = 0);
void print (BIG n, const char *h = NULL, int ns = 15, int hs = 0);
void print (const char *n, int ns = 0);
void print (LONG n, int ns);
void print (double n, int ns);
void print (BIG n, int ns);

The first argument provides the data item to be written; it can be a number or a character string (the 4th variant). In the most general case (the first three functions), there are three more arguments:

h	A textual header to precede the data item. If this argument is *NULL*, no header is printed
ns	The number of character positions taken by the encoded data item. If the value of this argument is greater than the actual number of positions required, the encoded item will be right-justified with spaces inserted on the left
hs	The number of character positions taken by the header. The header is left-justified and the appropriate number of spaces are added on the right. The total length of the encoded item (together with the header) is *ns + hs*

The last three functions are abbreviations of the first three ones with the second argument equal *NULL* and *hs* equal 0.

3.2.3 The XML Parser

To facilitate interfacing the simulator to external programs that may act as the suppliers of simulation data and, possibly, absorbers of the output, SMURPH has been equipped with a simple XML parser. The document structure is represented as a tree of XML nodes, with every node pointing to a complete XML element (tag) of the document, which may include subtrees, i.e., subordinate tags. The document is stored entirely in memory.

A tag is described by a record whose type is hidden and only available via a pointer of type *sxml_t*. The idea is to never touch the records and only access the relevant items from the document via functions. This way the parser is compatible with plain C.[7] Here is the complete list of functions available from the parser:

sxml_t sxml_parse_str (char *s, size t len);

---

[7]The parser has been adapted from a publicly available (MIT license) code by Aaron Voisine.

Given a string *s* of length *len* representing an XML document, the function transforms it into a tree of nodes and returns a pointer to the root node. For economy, the input string is modified and recycled; therefore it isn't passed as *const*. The tree is stored in dynamically allocated memory, which can be later freed by *sxml_free* (see below).

int sxml_ok (sxml_t node);

The function should be called after a conversion by *sxml_parse_str* to tell whether the operation has succeeded (the function returns *YES*) or failed (the function returns *NO*). The argument should point to the root node of the XML tree as returned by *sxml_parse_str*.

const char *sxml_error (sxml_t node);

If *sxml_ok* returns *NO* on the root node produced by *sxml_parse_str*, this function (invoked on the same node) will return a *NULL*-terminated error description string. The function returns an empty string (not the *NULL* pointer), if there was no error.

char *sxml_txt (sxml_t node);

This function returns the character string representing the textual content of the node. For example, if *node* describes a tag looking like this:

<tolerance quality = "3" dist = "u"> 0.000001 </tolerance>

the function will return the *NULL*-terminated string "0.000001".

char *sxml_attr (sxml_t node, const char *attr);

This function returns the value of the attribute identified with *attr* and associated with the tag pointed to by node. For example, if *node* points to the "tolerance" tag from the previous illustration, *sxml_attr (node, "quality")* will return the string "3". Note that there is a difference between a non-existent attribute and an empty one. In the former case, the function returns *NULL*, while in the latter it returns an empty string.

sxml_t sxml_child (sxml_t node, const char *name);

This function returns the first child tag of *node* (one level deeper) with the given *name*, or *NULL* if no such tag is found. For example, if *node* points to the following tag:

<goodies>This is a list of goodies:
    <goodie>First</goodie>
    <other>Not a goodie</other>
    <goodie>Second</goodie>
    <goodie>Third</goodie>
</goodies>

sxml_child (node, "goodie") will return a pointer to the tree node representing the first < *goodie* > element, i.e., < *goodie* > First < /goodie >.

sxml_t sxml_next (sxml_t node);

This function returns a pointer to the next tag (at the same level) with the same name as the one pointed to by *node*, or *NULL* if there are no more tags. For example, if *node* points to the first "*goodie*" tag in the above document, *sxml_next (node)* will return a pointer to the second "*goodie*" (not to the "*other*" tag). Note that to get hold of "*other*," one would have to call *sxml_child (node, "other")* on the "*goodies*" node.

sxml_t sxml_idx (sxml_t node, int idx);

> This function returns the *idx*'th tag (counting from zero) with the same name as the tag pointed to by *node* and located at the same level. In particular, for *idx* = 0, the function returns *node*. If *idx* is larger than the number of tags with the same name following the one pointed to by *node*, the function returns *NULL*. For example, *sxml_idx (node, 2)* with *node* pointing to the first "*goodie*" in the above example, returns a pointer to the third "*goodie.*"

char *sxml_name (sxml_t node);

> This function returns the name of the tag pointed to by *node*. For example, with *node* pointing to the "*other*" tag in the above example, *sxml_name (node)* returns the string "*other*".

sxml_t sxml_get (sxml_t node, …);

> This function traverses the tree of nodes rooted at *node* to retrieve a specific subitem. It takes a variable-length list of tag names and indexes, which must be terminated by either an index of −1 or an empty tag name. For example,
>
> *title = sxml_get (library, "shelf", 0, "book", 2, "title", −1);*
>
> retrieves the title of the 3rd book on the 1st shelf of *library*. The function returns *NULL* if the indicated tag is not present in the tree.

const char **sxml_pi (sxml_t node, const char *target);

> This function returns an array of strings (terminated by *NULL*) consisting of the XML processing instructions for the indicated target (entries starting with "<?"). Processing instructions are not part of the document character data, but they are available to the application.

The above functions extract components from an existing XML structure, e.g., built from a parsed input string. It is also possible to create (or modify) an XML structure and then transform it into a string, e.g., to be written to the output file. This can be accomplished by calling functions from the following list. Some of these functions exist in two variants distinguished by the presence or absence of the trailing "_d" in the function's name. The variant with "_d" copies the string argument into new memory, while the other variant uses a pointer to the original string, thus assuming that the string will not be deallocated while the resultant XML tree is being used.

sxml_t sxml_new[_d] (const char *name);

> This function initializes a new XML tree and assigns the specified name to its root tag. The tag has no attributes and its text is empty.

sxml_t sxml_add_child[_d] (sxml_t node, const char *name, size_t off);

> The function adds a child tag to *node* and assigns a *name* to it. The *off* argument specifies the offset into the parent tag's character content. The returned value points to the child

node. The offset argument can be used to order the children, which will appear in the resulting XML string in the increasing order of offsets. This will also happen if the character content of the parent is shorter than the specified offset(s).

*sxml_t sxml_set_txt[d] (sxml_t node, const char *txt);*

This function sets the character content of the tag pointed to by *node* and returns *node*. If the tag already has a character content, the old string is overwritten by the new one.

*sxml_t sxml_set_attr[_d] (sxml_t node, const char *name, const char *value);*

The function resets the tag attribute identified by *name* to *value* (if already set) or adds the attribute with the specified *value*. A *value* of *NULL* will remove the specified attribute. Note the difference between attribute removal and setting it to an empty string. The function returns *node*.

sxml_t sxml_insert (sxml_t node, sxml_t dest, size_t off);

This function inserts an existing tag node as a child in *dest* with *off* used as the offset into the destination tag's character content (see *sxml_add_child*).

sxml_t sxml_cut (sxml_t node);

This function removes the specified tag along with its subtags from the XML tree, but does not free its memory, such that the tag (whose pointer is returned by the function) is available for subsequent operations.

sxml_t sxml_move (sxml_t node, sxml_t dest, size_t off);

This is a combination of *sxml_cut* and *sxml_insert*. First the tag (node) is removed from its present location and then inserted as for *sxml_insert*.

void sxml_remove (sxml_t node);

As *sxml_cut*, but the memory occupied by the tag is freed.

*char *sxml_toxml (sxml_t node);*

The function transforms the XML tree pointed to by node into a character string ready to be written to a file.

These two operations complete the set:
void sxml_free (sxml_t node);

The function deallocates all memory occupied by the XML tree rooted at *node*. It only makes sense when applied to the root of a complete (stand-alone) XML tree.

*sxml_t sxml_parse_input (char del = '\0', char **data = NULL, int *len);*

The function reads the standard data file (Sect. 3.2.2) and treats it as an XML string to be parsed and transformed into a tree. It returns the root node of the tree, which should be checked for a successful conversion with *sxml_ok* and/or *sxml_error* (see above). If the first argument is not '\0', its occurrence as the first character of a line will terminate the interpretation of the input file. If the second argument is not *NULL*, then the (null-terminated) string representing the complete data file (before XML parsing) will be stored at the pointer passed in the argument. The string will be complete in the sense that the contents of any included files (see below) will have been inserted in their respective places, i.e., there will be no "include" tags in the string. If the third argument is not *NULL* (and the second argument is not *NULL* as well), then the length of the string returned via the second argument will be stored at the pointer passed in the third argument.

The last function is the only one that interprets *includes* in the parsed file, i.e., tags of these forms:

<xi:include href= "filename"> ... </xi:include>
<xi:include href= "filename"/>

where *filename* is the name of the file whose contents are to be inserted into the current place of the processed XML data, replacing the entire *<xi:include>* tag. The tag may have a text content, which will be ignored, but is not allowed to have sub-tags. This feature implements a subset of the full capabilities of the official XML *<xi:include>* tag, in particular, *href* is the only expected and parsed attribute. The included file is interpreted as XML data (being a straightforward continuation of the source file), which is allowed to contain other *<xi:include>* tags. In the SMURPH variant, the tag's keyword can be abbreviated as *include*, e.g., these two constructs:

<xi:include href= "somefile.xml"/>
<include href= "somefile.xml"> </include>

are equivalent.

Unless the name of the included file begins with a slash (/), it is first looked up in the current directory, i.e., in the directory in which the simulator has been called. If that fails, and if data include libraries are defined (see Sect. B.2), the library directories are then searched in the order of their declarations.

3.2.4   Operations on Flags

In several situations, it is desirable to set, clear, or examine the contents of a single bit (flag) in a bit pattern. SMURPH defines the type *FLAGS*, which is equal to *Long* or *LONG*,[8] and is to be used for representing 32-element flag patterns (or sets). A collection of flags is associated with packets (Sect. 6.2). The following functions[9] implement elementary operations on binary flags:

FLAGS setFlag (FLAGS flags, int n);
FLAGS clearFlag (FLAGS flags, int n);
int flagSet (FLAGS flags, int n);
int flagCleared (FLAGS flags, int n);

---

[8]The minimum guaranteed size of *FLAGS* is 32 bits.
[9]Implemented as C macros.

The first two functions respectively set and clear the contents of the *n*-th bit in flags. The updated value of *flags* is returned as the function value. Bits are numbered from 0 to 31, 0 being the index of the rightmost (least significant) bit. The last two functions examine the contents of the *n*-th bit in *flags* and return either 0 or 1 depending on whether the bit is 0 or 1, respectively, for *flagSet*; and 1 and 0, for *flagCleared*.

3.2.5 Type Boolean

SMURPH defines type *Boolean* (alternatively named *boolean*) as *unsigned char*. This type is intended to represent simple binary flags that can have one of two values: 0 represented by the symbolic constant *NO* (standing for "false") and 1 represented by the symbolic constant *YES* (which stands for "true"). Clearly, a variable of type *Boolean* can hold values other than 0 and 1. Anything nonzero is interpreted as "true," unless the variable is directly compared against *YES*. Value 2 (constant *YESNO*) is sometimes considered special and intended to mark a *Boolean* variable as "uninitialized." Being nonzero, it formally evaluates as "true."

3.2.6 Pools

In several (mostly internal) places, SMURPH uses the so-called *pools*, which are (possibly ordered) sets of items stored in doubly-linked lists. As pools may be generally useful (for implementing sets of non-standard objects) the requisite tools have been made available to user programs. To be managed as a pool element, an object should define two attributes: *prev* and *next* to be used as the links, for example:

```
struct my_pool_item_s {
        struct my_pool_item_s *prev, *next;
                ... other stuff ...
};
typedef struct my_pool_item_s my_pool_item;
```

A pool of such objects is represented by a single pointer, e.g.,

```
my_pool_item *Head;
```

which should be initialized to *NULL* (for an empty pool). The following operations (macros) are available:

```
pool_in (item, head);
```

to add the new item pointed to by the first argument (an item pointer) to the pool represented by the second argument (the head pointer);

pool_out (item);

to remove the indicated item from the pool (*item* is a pointer);

for_pool (item, head)

to traverse the pool as in this example:

```
my_pool_item *find (int ident, my_pool_item *Head) {
        my_pool_item *el;
        for_pool (el, Head) {
                if (el->Ident == ident)
                        return el;
        }
        return NULL;
}
```

and finally,

trim_pool (item, head, cond);

to remove from the pool all items satisfying the specified condition, e.g.,

trim_pool (el, Head, el->Expiry>Time);

The pool is organized in such a way that new items (added with *pool_in*) are stored at the head. The list is not looped: the *prev* pointer of the first element points to the head (or rather to a dummy item whose *next* pointer coincides with the head pointer. The *next* pointer of the last (oldest) element in the pool is *NULL*.

3.2.7 Error Handling

SMURPH is equipped with a standard error handling mechanism, which can also be invoked from the user program. That simple mechanism offers the following three functions:

*void excptn (const char *string)*
*void assert (int cond, const char *string);*
*void Assert (int cond, const char *string);*

The first function is used to terminate the experiment due to a fatal error condition. The text passed as *string* is written to the standard output, standard error, and the results file. When the run is aborted by *excptn*, SMURPH prints out a description of the context in which the error has occurred.

A call to *assert* or *Assert* is formally equivalent to:

if (!cond) excptn (string);

The difference between *assert* and *Assert* is that when the program is created with the −*a* option (see Sect. B.5), all references to *assert* are disabled (removed from the program), while references to *Assert* are always active.

3.2.8   Identifying the Experiment

The following declarative operation can be used to assign a name to a SMURPH experiment:

identify name;

where *name* can be any piece of text that does not contain blanks. This text will be printed out in the first line of the results file, together with the current date and time (Sect. 3.2.9).

If the experiment identifier contains blanks, it should be encapsulated in parentheses, e.g.,

identify (Expressnet version B);

or in quotes, e.g.,

identify "Conveyor belt driver";

A parenthesis within an identifier encapsulated in parentheses or a quote within a quoted string can be escaped with a backslash.

3.2.9   Telling the Time and Date

The following function returns the number of seconds of the CPU time used by the program from the beginning of the current run:

double cpuTime ();

When the program has been compiled with −*F* (no floating point, see Sec. B.5), the type of *cpuTime* is *Long*, and the function returns an entire number of seconds.

To get the current date, the program can invoke this function:

*char *tDate ();*

which returns a pointer to the character string containing the date in the standard format *www mmm dd hh:mm:ss yyyy*, e.g.,

Tue Apr 2 13:07:31 2019

Date/time in this format is included in the header of the results file, together with the experiment identifier (Sect. 3.2.8).

The real-time execution of a control program can be offset based on the time stamp of a journal file (see Sect. 9.5.2). This is (mostly) intended for playing back control scenarios that took place in the past. In such a case, the value returned by *tDate* reflects the playback time.

3.2.10 Decibels

In wireless communication, when dealing with RF channel models (Sects. 2.4.5 and 4.4.1), we often deal with values that are most naturally expressed on a logarithmic scale, i.e., in decibels (dB). Such values typically represent ratios of signal levels. Sometimes a ratio expressed this way can be interpreted as an absolute power setting, by assuming a specific reference. Thus, for example, 1dBm is a logarithmic measure of power read as 1 dB relative to 1 mW.

When programming RF channel models (see Sect. 2.4.5), as well as protocol programs interacting with such models, one often has to deal with the direct (linear) representations of signal ratios and signals, because this is how SMURPH handles those measures internally. Thus, it is sometimes necessary to convert from one representation to the other. This is because the logarithmic scale is more natural to the human experimenter, when specifying data and interpreting the output, so it would be difficult and unnatural to postulate that the linear scale be used in all circumstances.

SMURPH defines the following two functions for converting between the linear and logarithmic scales:

double linTodB (double vlin);
double dBToLin (double vlog);

The conversion formulas implemented by the functions are:

$$v_{dB} = \log_{10}(v_{lin}) \times 10$$

and

$$v_{lin} = 10^{\frac{v_{dB}}{10}}$$

respectively. The first method checks whether the argument is strictly greater than zero and triggers an error otherwise.

3.3 SMURPH Types

By a SMURPH type we mean a compound, predefined, user-visible type declared (internally) as a C++ class with some standard properties. We conveniently assume that *BIG* (and *TIME*), although sometimes declared as a class, is a simple type. In this section we will be concerned with the SMURPH types only, in the sense of the above definition. The words "class" and "type" will be used interchangeably. We mildly prefer to avoid the word "class" when referencing SMURPH *class* types, because those types form a family with some characteristic syntactic and semantic features.

3.3.1 Type Hierarchy

Figure 3.1 presents the hierarchy of built-in, basic SMURPH types. The dummy root of the tree emphasizes the fact that all those types belong to SMURPH, sharing some common features, which we shall discuss in this section. They are all compound types described by C++ classes.

All objects exhibiting a dynamic behavior belong to the internal type *Object*, which is not visible directly to the user. Its role is to bind together its descendant types by furnishing them with a unifying collection of common attributes and methods that each *Object* is expected to have. Type *EObject*, which is an actual, visible type, can be used to prefix user-defined subtypes of *Object* unrelated to any built-in *Object* subtypes.

The most relevant property of an *Object* (or *EObject*) is that it can be *exposed*. By exposing an object (see Sect. 12.1) we mean presenting some information related to the object in a standard way. This information can be "printed," i.e., included in the output file produced by the SMURPH program, or "displayed," i.e., shown in a DSD window (Sec. C.1).

Timer and *Client* (which we first met in Sects. 2.1.1 and 2.2.3, respectively), and also *Monitor* (Sect. 5.4), stand for specific objects rather than types. These objects represent some important elements of the protocol environment (see Sects. 5.3 and 6.1) and are static, in the sense that they are not created by the user program and they exist throughout its entire execution, each in exactly one copy. Therefore, their types are uninteresting and hidden from the user, and we can informally (in Fig. 3.1) equivalence those types with the objects themselves. Other *Objects* may exist in multiple copies; some of them may be dynamically created and destroyed by the protocol program.

Generally, all objects rooted at *AI*, which is another internal type binding the so-called *activity interpreters*, are models of some entities belonging to the protocol environment. They are responsible for modeling or perceiving the flow of time, which is discussed in (Sect. 5.1.1). In other words, they generate *events* needed for progress in the evolution of the modeled system.

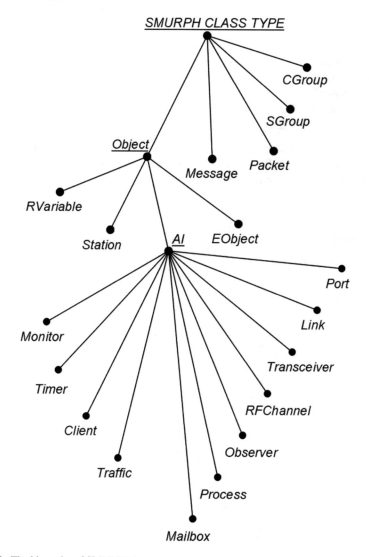

Fig. 3.1 The hierarchy of SMURPH types

3.3.2 Object Naming

Each *Object* has several attributes that identify it from the outside and a number of methods for accessing these attributes. The need for so many identifiers mostly stems from the fact that the dynamic display applet (DSD) responsible for exposing *Objects* "on screen" (Sect. C.1) must be able to identify individual *Objects* and recognize some of their general properties (Sect. C.4). The following seven identification attributes are associated with each Object:

- the class identifier
- the serial number
- the type name
- the standard name
- the nickname
- the base standard name
- the output (print) name

The *class identifier* of an *Object* is a number (represented by a numerical constant) identifying the base SMURPH type to which the *Object* belongs. This attribute can be accessed by the parameter-less method *getClass*, e.g.,

int cl;

...

cl = obj->getClass ( );

which returns the following values:

AIC_timer	if the object is the *Timer* AI
AIC_client	if the object is the *Client* AI
AIC_monitor	if the object is the *Monitor* AI
AIC_link	if the object is a *Link*
AIC_port	if the object is a *Port*
AIC_rfchannel	if the object is an *RFChannel*
AIC_transceiver	if the object is a *Transceiver*
AIC_traffic	if the object is a *Traffic*
AIC_mailbox	if the object is a *Mailbox*
AIC_process	if the object is a *Process*
AIC_observer	if the object is an *Observer*
AIC_rvariable	if the object is an *RVariable*
AIC_station	if the object is a *Station*
AIC_eobject	if the object is an *EObject*

The serial number of an *Object*, also called its *Id*, tells apart different *Objects* belonging to the same class. This attribute is accessed via the method *getId*, e.g.,

int id;

...

id = obj->getId ( );

Except for *Ports*, *Transceivers*, and *Mailboxes*, all dynamically created *Objects* are assigned their *Ids* in the order of creation, such that the first created *Object* in its class is numbered 0, the second object is numbered 1, etc. The numbering of *Ports*, *Transceivers*, and *Mailboxes* is described in Sects. 4.3.1, 4.5.1, and 9.2.1, respectively.

The *Id* attribute of *Timer*, *Client*, and *Monitor* (which never occur in multiple copies) is equal to *NONE* (-1).

The *type name* of an *Object* is a pointer to a character string storing the textual name of the most restricted type (C++ class) to which the object belongs. This pointer is returned by the method *getTName*, e.g.,

```
char *tn;

...

tn = obj->getTName ( );
```

For example, the *getTName* method of the *AckQueue* mailbox from Sect. 2.4.3 will return the string "*AckQueue*".

The purpose of the *standard name* is to identify exactly one *Object*. The standard name is a character string consisting of the object's type name concatenated with the encoded serial number, e.g. *MyProcess 245*. The two parts are separated by exactly one space.

For a *Port*, a *Transceiver*, and a station *Mailbox*, the standard name contains as its part the standard name of the *Station* owning the object (see Sects. 4.3.1, 4.5.1, and 9.2.1). For an object that always occurs in a single instance, the numerical part of the standard name is absent, i.e., the standard names of *Timer*, *Client*, and *Monitor* are just "*Timer*," "*Client*," and "*Monitor*," respectively.

The pointer to the object's standard name is returned by the method *getSName*, e.g.,

```
char *sn;

...

sn = obj->getSName ( );
```

The *nickname* is an optional character string which can be assigned to the object by the protocol program. This assignment is usually made when the object is created (see Sect. 3.3.8). The nickname pointer is returned by the method *getNName*, e.g.,

```
char *nn;

...

nn = obj->getNName ( );
```

If no nickname has been assigned to the object, *getNName* returns *NULL*.

The nickname is the only name that can be changed after it has been assigned (all the other names are assigned by the system and they cannot be changed). The following method of *Object* assigns or changes the object's nickname:

*void setNName (const char *nn);*

where *nn* is the assigned nickname. If the *Object* already had a nickname assigned, the old nickname is discarded and the new one becomes effective. A copy of the string passed as the argument to *setNName* is stored with the object, so the original can be safely deallocated.

The *base standard name* is a character string built in a similar way to the standard name, except that the name of the base SMURPH type, from which the type of the object has been derived, is used instead of the object's (derived) type name. For example, assume that *ms* points to an *Object* of type *MyStation* which has been derived directly or indirectly from the built-in SMURPH type *Station*. The (regular) standard name of this object is "*MyStation n*", while its base standard name is "*Station n*". The pointer to the base standard name is returned by *getBName*, e.g.,

```
char *bn;

...

bn = obj->getBName ( );
```

The base standard names of objects that are direct instances of base types are the same as their (regular) standard names.

The *output name* of an *Object* is a character string which is the same as the nickname, if one is defined; otherwise, it is the same as the standard name. When an *Object* is exposed "on paper" (see Sect. 12.1), its output name is used as the default header of the exposure. The method *getOName* returns a pointer to the object's output name, e.g.,

```
char *on;

...

on = obj->getOName ( );
```

The character strings returned by *getSName*, *getBName*, and *getOName* are not constants. Thus, if a string returned by one of these functions is to be stored, it must be copied to a safe area. On the other hand, the strings pointed to by *getTName* and *getNName* are constants and they do not change for as long as the object in question is alive (or, in the case of nickname, the name is not changed).

3.3.3   Type Derivation

Some of the types (classes) presented in Sect. 3.3.1 are not usable directly. For example, type *Process* is merely a frame for defining process types needed by the protocol program. One element of a process class that must be provided by the user (to make instances of that class runnable as processes), is a method outfitting the process with code.

SMURPH defines its own way of deriving new types from the built-in ones. Here is the simplest syntax of the operation serving this end:

```
declarator typename {

        ...

        attributes and methods

        ...

};
```

where "declarator" is a keyword corresponding to a base SMURPH type (Sect. 3.3.1) and "typename" is the name of the newly defined type. For example, the following declaration:

```
packet Token {
        int status;
};
```

defines a new packet type called *Token*. This type is established as an extension of the standard type *Packet* and contains one user-defined attribute: the integer variable *status*.

The declarator keyword is obtained from the corresponding base SMURPH type by changing the first letter to the lower case. The base types that can be extended this way are: *Message*, *Packet*, *Traffic*, *Station*, *Link*, *RFChannel*, *Mailbox*, *Process*, *Observer*, and *EObject*.

It is possible to define the new type as a descendant of an already defined type derived from the corresponding base type. In such a case the name of the parent type should appear after the new type name preceded by a colon, i.e.,

```
declarator typename : itypename {

        ...

        attributes and methods

        ...
};
```

where "itypename" identifies the inherited type, which must have been declared previously with the same declarator. A declaration without itypename can be viewed as an abbreviation for an equivalent declaration in which itypename identifies the base type. For example, the previous declaration of type *Token* is equivalent to

```
packet Token : Packet {
        int status;
};
```

For base types *Traffic*, *Process*, and *Mailbox*, a subtype declaration may optionally specify one or two argument types (Sects. 6.6.1, 5.1.2 and 9.2.1). If present, these argument types are provided in parentheses preceding the opening brace. Thus, the format of a subtype definition for these three base types is:

```
declarator typename : itypename (argument types) {

        ...

        attributes and methods

        ...
};
```

where the parts ": itypename" and "(argument types)" are independently optional. For example, the following process type declaration:

```
process Guardian (Node) {
    ...
};
```

defines a process type *Guardian* descending directly from the base type *Process*. The
argument type should refer to a defined (or announced, see Sect. 3.3.6) station type; it
identifies the type of stations that will be running processes of type *Guardian*. Given
the above declaration, a new process type can be defined like this:

```
process SpecializedGuardian : Guardian (MyNode) {
    ...
};
```

as a descendant of *Guardian*. Processes of this type will be owned by stations of type
MyNode. The latter need not be a descendant of *Node*, but it must be a known station
type.

The exact semantics of the SMURPH type extension operations will be described
individually for each extensible base type. Such an operation is expanded into the
definition of a class derived either from the corresponding base type or from the
specified (already defined) descendant type of the base type. Some standard attributes
(mostly invisible to the user) are automatically associated with that class. These
attributes are used by SMURPH to keep track of what happens to the objects created
from the defined type.

The inherited supertype class is made formally *public* in the subclass. Following
the opening brace of the type definition, the user can define attributes and methods of
the new type. By default, all these attributes are *public*. The C++ keywords *private*,
protected, and *public* can be used to associate specific access rights with the user-
defined attributes.

3.3.4   Multiple Inheritance

Sometimes it is convenient to define a SMURPH subtype as a direct descendant of
more than one supertype. This is especially useful when creating libraries of types.
The standard concept of multiple inheritance (present in C++) is pretty much naturally
applicable in such situations; however, SMURPH adds to the issue a specific flavor
resulting from the following two problems:

- It is meaningless to define types derived simultaneously from different base types.
  Thus, there should be a way of controlling whether a SMURPH type extension
  makes sense.
- An object of a type derived from multiple supertypes must have exactly one frame
  of its base type.

The first fact is rather obvious: the base types have been designed to be orthogonal,
and, e.g., an object that is both a *Station* and a *Message* at the same time doesn't make

a lot of sense. The second requirement is a consequence of the way objects derived from the base SMURPH types are handled by the SMURPH kernel. Internally, such an object is represented by various pointers and tags kept in its base type segment. Having two or more copies of this segment, besides wasting a substantial amount of space, would have the confusing effect of perceiving a single object under multiple internal views.

In C++ it is possible to indicate that the frame corresponding to a given class occurs exactly once in the object frame, even if the class occurs several times in the object's type derivation. This is achieved by declaring the class as *virtual* in the derivation sequence.

Thus, one possible solution to be adopted in SMURPH would be to force each base type to occur as *virtual* in the derivation list of a user-defined type. This solution, however, would be costly and clumsy in typical situations, amounting to 99.9% of all cases of SMURPH type inheritance. To make it work, one should always have to make the base types *virtual*, even if no multiple inheritance were used.[10] That would be expensive, because the frame of a virtual subclass is referenced via a pointer, which substantially increases the cost of such a reference. Therefore, a lighter solution has been adopted.

If the name of a newly defined SMURPH supertype is followed by *virtual*, e.g.,

station VirPart virtual {

    ...

};

it means that formally the type belongs to the corresponding base type, but the frame of the base type is not to be attached to the defined type's frame. The "itypename" part of a SMURPH subtype definition (Sect. 3.3.3) may contain a sequence of type names separated by commas, e.g.,

packet DToken : AToken, BToken, CToken {

    ...

};

Such a sequence represents the inheritance list of the defined subtype. The following rules apply:

- The inheritance list may contain at most one non-virtual type name (in the sense described above). If a non-virtual type name is present, it must appear first on the list.
- If the name of the defined type is followed by *virtual* (this keyword must precede the colon), the inheritance list cannot include a non-virtual type.
- If the inheritance list contains virtual types only, but the keyword *virtual* does not follow the name of the defined type, the defined type is not virtual, i.e., the base type is implicitly added to the inheritance sequence (as the first element). Thus, only the types explicitly declared as *virtual* end up being virtual.

---

[10]A SMURPH program may be contained in several files and the fact that no multiple inheritance is used in one file does not preclude it from being used in another one.

One disadvantage of this solution is that a *virtual* (in the SMURPH sense) type cannot reference attributes of the base type. This is not a big problem, as very few of these attributes may be of interest to the user. Virtual functions can be used to overcome this little deficiency, should it become a problem.

3.3.5 Abstract Types

A SMURPH type can be explicitly declared as incomplete (and thus unusable for creating objects without being extended) by putting the keyword *abstract* into its declaration, e.g.,

```
traffic MyTrafficInterface abstract {
    ...
};
```

An abstract type can be derived from other (possibly non-abstract) types, but it cannot be virtual (Sect. 3.3.4). Thus, the keywords *abstract* and *virtual* cannot both occur in the same type declaration.

If an abstract type is derived from other types, then those types must be listed after a colon following the *abstract* keyword, e.g.,

```
traffic MyTrafficInterface abstract : XTraffic (Msg, Pkt) {
    ...
};
```

Essentially, an abstract type has all the features of a regular type, except that *create* for such a type (Sect. 3.3.8) is illegal. Moreover, a type *must* be declared *abstract*, if it declares *abstract* virtual methods. For example, this declaration:

```
traffic MyTrafficInterface abstract (Msg, Pkt) {
    virtual Boolean check it (Pkt *p) = 0;
    ...
};
```

would trigger errors if the *abstract* keyword following the traffic type name were absent.

3.3.6 Announcing Types

A SMURPH subtype can be announced before it is completely defined. This may be needed, if the name of the type is required (e.g., as an argument type, Sect. 3.3.3) before the full definition can take place. The following type declaration format is used for this purpose:

declarator typename;

where "declarator" has the same meaning as in Sect. 3.3.3. The type name can be optionally followed by *virtual* in which case the type is announced as virtual in the sense of Sect. 3.3.4. Note that a full type definition must precede the usage of such a type in the inheritance sequence of another type.

3.3.7 Subtypes with Empty Local Attribute Lists

There are many cases when the local attribute list of a SMURPH subtype is empty, e.g.,

traffic MyTPattern (MyMType, MyPType) {};

This situation occurs most often for traffic patterns (subtypes of *Traffic*) and mailboxes (subtypes of *Mailbox*). In such cases, it is legal to skip the empty pair of braces, provided that the new type declaration cannot be confused with a type announcement (Sect. 3.3.6). For example, the above definition of *MyTPattern* can be shortened to:

traffic MyTPattern (MyMType, MyPType);

as, due to the presence of the argument types, it does not look like a type announcement. On the other hand, in the declaration:

packet MyPType {};

the empty pair of braces cannot be omitted, as otherwise the declaration would just announce the new packet type without actually defining it. Generally, an abbreviated type declaration (without braces) is treated as an actual type definition rather than an announcement, if at least one of the following conditions is true:

- the declaration contains argument types
- the declaration specifies explicitly the supertype

Thus, the above declaration of *MyPType* can be written equivalently as:

packet MyPType : Packet;

which, however, is hardly an abbreviation. The most common cases when the attribute list of a newly defined type is empty deal with subtypes of *Traffic* and *Mailbox*, and then argument types are usually present.

3.3.8 Object Creation and Destruction

The following operation explicitly creates an object belonging to a SMURPH type (base or derived):

obj =create typename *(*setup args*);*

where "typename" is the name of the object's type. The arguments in parentheses are passed to the object's *setup* method. If the object has no *setup* method, or the list of arguments of this method is empty, the part in parentheses does not occur (the empty parentheses can be skipped as well). The operation returns a pointer to the newly-created object.

The purpose of the *setup* method is to perform object initialization. If needed, it should be declared as:

void setup (...);

within the object type definition. Argument-less C++ constructors can also be used in user-defined derived SMURPH types; however, there is no way to define and invoke upon object creation a constructor with arguments. The role of such a constructor is taken over by *setup*. Non-extensible types (Sect. 3.3.3), and types that need not be extended, define appropriate standard/default *setup* methods. Their arguments and semantics will be discussed separately for each base type.

It is possible to assign a nickname (see Sect. 3.3.2) to the created object. The following version of *create* can be used for this purpose:

obj = create typename, nickname, *(*setup args*);*

where "nickname" is an expression that should evaluate to a character pointer. The string pointed to by this expression will be duplicated and used as the object's nickname. If the "setup args" part does not occur, the second comma together with the parentheses can be omitted.[11]

The typename keyword (in both versions of *create*) can be optionally preceded by an expression enclosed in parentheses, i.e.,

*obj = create (*expr*)* typename *(*setup args*);*

or

*obj = create (*expr*)* typename, nickname, *(*setup args*);*

where "expr" must be either of type *int* (in which case it will be interpreted as a station *Id*) or a pointer to an object belonging to a *Station* subtype. It identifies the station in whose context the new object is created, and is relevant for some object types (e.g., see Sect. 2.2.5). The context of the indicated station is assumed for the *create* operation, as well as any operations triggered by it, most notably the execution of the created object's *setup* method.

---

[11] The comma following "nickname" is needed (it is not a mistake).

Objects belonging to types *SGroup*, *CGroup*, *Station*, *Traffic*, *Link*, *RFChannel*, *Mailbox*,[12] *Port*, and *Transceiver* can only be *created* and never explicitly destroyed. Other dynamically created objects can be destroyed with the standard C++ operator *delete*.

User-defined subtypes of *Object* must belong to type *EObject*, i.e., be declared as derivatives of *EObject* or its descendants. A declaration of such a subtype should start with the keyword *eobject* and obey the rules listed in Sects. 3.3.3–3.3.5. An *EObject* subtype may declare a *setup* method (or a collection of such methods). Object instances of *EObject* subtypes should be generated by *create* and deallocated by *delete*. The *Id* attributes of *EObjects* reflect their order of creation starting from 0. The same, common indexing is shared by all subtypes of *EObject*.

It is illegal to *create* an object of a *virtual* type (Sect. 3.3.4).

References

1. P. Gburzyński, P. Rudnicki, A Better-than-token protocol with bounded packet delay time for Ethernet-type LAN's, in *Proceedings of Symposium on the Simulation of Computer Networks* (Colorado Springs, Co., 1987)
2. P. Gburzyński, P. Rudnicki, Using time to synchronize a token Ethernet, in *Proceedings of CIPS Edmonton '87* (1987)
3. P. Gburzyński, P. Rudnicki, A note on the performance of ENET II. IEEE J. Sel. Areas Commun. **7**(4), 424–427 (1989)
4. L.G. Birta, G. Arbez, *Modelling and Simulation* (Springer, 2013)
5. L.A. Adamic, Zipf, power laws, and Pareto: a ranking tutorial. Xerox Palo Alto Research Center, Palo Alto, CA, http://ginger.hpl.hp.com/shl/papers/ranking/ranking.html (2000)

---

[12]This only concerns those mailboxes that are owned by stations. Mailboxes owned by processes can be destroyed (see Sect. 9.2.1).

Chapter 4
The Virtual Hardware

The logical structure of the modeled/controlled system (for simplicity, we shall call it a network), although static from the viewpoint of the protocol program, is defined dynamically by explicit object creation and calls to some functions and methods. As perceived by SMURPH, a network consists of stations which are conceptually the processing units of the (distributed) system. A station can be viewed as a parallel computer for running collections of processes.

Sometimes (especially in network simulators) stations are interconnected via links or/and radio channels. Links, represented by type *Link* (and its several built-in variants), are models of simple communication channels that can be implemented in real life based on some broadcast-type communication media, e.g., a piece of wire, a coaxial cable, an optical fiber, or even a wireless link (although *RFChannels* are better equipped for this purpose).

Stations are interfaced to links via ports, which can be viewed as specific points (taps) on the links that the stations may use to send and receive information structured into packets. Links introduce propagation delays; therefore, one application of a link in a control program may be to implement a delayed communication path between a pair of stations.

Radio channels, represented by objects of type *RFChannel*, are similar to links, excepts that they allow for more dynamic (possibly mobile) configurations of stations interconnected by them. Also, radio channels are more open-ended: they come with hooks whereby users can describe (program) their favorite characteristics related to transmission power, receiver gain, transmission range, signal attenuation, interference, i.e., all the attributes of real-life radio channels relevant from the viewpoint of modeling signal propagation and reception. Owing to the large and unforeseeable set of possible channel models that one may want to plug into SMURPH simulators, the package does not presume any built-in characteristics, except for providing a generic framework for modeling signal/packet propagation according to the (Euclidean) distance between the concerned parties.

The radio channel equivalent of a port is called a *transceiver*. Like a port, it represents an interface of a station to a (radio) channel. Multiple (independent) radio

© Springer Nature Switzerland AG 2019
P. Gburzyński, *Modeling Communication Networks and Protocols*,
Lecture Notes in Networks and Systems 61,
https://doi.org/10.1007/978-3-030-15391-5_4

channels can coexists within a single network model. Of course, links and radio channels can coexist, too, so hybrid (wired/wireless) networks can be modeled.

4.1 Stations

A station is an object belonging to a type derived from *Station* (see Sect. 3.3.1). It is possible to create a station belonging directly to type *Station*, but such a station is not very useful.

The standard type *Station* defines only two attributes visible to the user, which are practically never accessed directly. These attributes describe the queues of messages awaiting transmission at the station and are discussed in Sect. 6.6.5.

4.1.1 Declaring Station Types

The definition of a new station type starts with the keyword *station* and obeys the rules elaborated on in Sects. 3.3.3–3.3.5. The *setup* method for a station type needs only be defined, if the type declares attributes that must be initialized when a station object of this type is created. For example, the following declaration:

```
station Hub {
        Port **Connections;
        int Status;
        void setup ( ) { Status = IDLE; };
};
```

defines a new station type called *Hub* derived directly from *Station*, with two attributes: an array of pointers to ports and an integer variable. When a *Hub* object is created, the *Status* attribute will be set to *IDLE*. The array of port pointers is not initialized in the *setup* method, so, presumably, it will be created outside the object.

The *Hub* type defined above can be used to derive another station type, e.g.,

```
station SuperHub : Hub {
        TIME *TransferTimes;
        void setup (int hsize) {
                Hub::setup ( );
                TransferTimes = new TIME [hsize];
        };
};
```

The *SuperHub* type inherits all attributes from *Hub* and declares one attribute of its own: an array of *TIME* values. The *setup* method has now one argument. Note the explicit call to the *setup* method of the supertype, to initialize the *Status* attribute.

Stations are the most natural candidates to be defined from virtual supertypes, according to the rules described in Sect. 3.3.4. For example, consider the following declaration:

```
station TwoBusInterface virtual {
        Port *Bus0Port, *Bus1Port;
        void setup (RATE tr) {
                Bus0Port = create Port (tr);
                Bus1Port = create Port (tr);
        };
};
```

which defines a station component. This component consists of two ports created by the setup method (the details are explained in Sect. 4.3.1). The two ports may interface the station to a two-bus network without specifying any other details of the station's architecture. The next declaration:

```
station ThreeBuffers virtual {
        Packet Buf1, Buf2, Buf3;
};
```

can be viewed as a description of the *Client* interface of a station (Sect. 6.3) consisting of three packet buffers. The two interfaces can be used together, e.g.,

```
station MyStationType : TwoBusInterface, ThreeBuffers {
        RVariable *PacketDelay;
        void setup ( ) {
                RATE MyXRate;
                PacketDelay = create RVariable;
                readIn (MyXRate);
                TwoBusInterface::setup (MyXRate);
        };
};
```

as independent building blocks for a complete (non-virtual) station type.

4.1.2 Creating Station Objects

Given a station type, a station object is built by executing *create* (see Sect. 3.3.8). The serial number (the *Id* attribute, see Sect. 3.3.2) of a station object reflects its creation order. The first-created station is assigned number 0, the second one is numbered 1, etc. The global read-only variable *NStations* of type *int* contains the number of all stations that have been created so far. At the end of the network initialization stage

(Sect. 5.1.8), and during the proper execution of the protocol program, *NStations* tells the total number of stations in the network.[1]

As the configuration of the modeled system must be completely determined before the protocol execution is started, it is illegal to create new stations after starting the protocol. Also, once created, a station can never be destroyed. At first sight, this may look like a limitation reducing the expressing power for the model's dynamics, e.g., in ad hoc wireless networks, where one may want to study mobility and varying populations of stations that come and go. Note, however, that a station's behavior can be arbitrarily dynamic and independent of the behavior of other stations. Stations can remain dormant, then wake up, then become dormant again.[2] In a wireless network, they can also change their locations, move out of range, and so on. The operation of pre-building the model's virtual hardware should be viewed as creating a virtual universe. Once created and set in motion, the "substance" of this universe is not allowed to change, but it can behave in every which way allowed by the universe's laws.

There exists one special (and usually inconspicuous) station, which is created automatically before anything else. Its purpose is to run a few internal processes of SMURPH and also the user's *Root* process (see Sect. 5.1.8). This station is pointed to by the global variable *System* whose type (a subtype of *Station*) is hidden from the user. The *Id* attribute of the *System* station is *NONE* and its type name is *SYSTEM*. The System station does not count to the pool of "normal" stations, and, in particular, it does not contribute to *NStations*.

Sometimes it is convenient and natural to reference a station by its serial number, i.e., the *Id* attribute, rather than the object pointer. For example, station numbers are used to identify senders and receivers of messages and packets (see Sect. 6.2). Quite often, especially in the initialization stage, one would like to perform some operation(s) for all stations, or perhaps, for their significant subsets. In such a case, it may be natural to use a loop of the following form:

for (Long i = 0; i < NStations; i ++)...

To make such operations possible, the following function converts a station *Id* into a pointer to the station object:

Station *idToStation (Long id);

The function checks whether the argument is a legal station number and, if it is not the case, the execution is aborted with an error message.

For symmetry, the following macro looks like a function that returns the *Id* attribute of a station (or any *Object*) whose pointer is specified as the argument:

Long ident (Object *a);

This macro expands into a call to *getId* (see Sect. 3.3.2).

---

[1]Minus the *System* station (see below).

[2]When we say that a station becomes dormant (or active), we mean the processes running at the station.

4.1.3 The Current Station

At any moment during the execution of the protocol program, whenever a process is awakened in one of its states, it is always clear which station is the one being currently active. That (current) station is pointed to by the global variable *TheStation* of type *Station* which belongs to the so-called process environment (Sect. 5.1.6).

One can say that anything that ever happens in a SMURPH program always happens within the context of a specific station. When SMURPH starts, and the user program is given control for the first time, in the first state of the *Root* process (Sect. 5.1.8), *TheStation* points to the *System* station. Then, whenever a new "regular" station is created, *TheStation* is set to point to that newly-created station. Many other objects that get created along the way (e.g., ports, transceivers, processes) must belong (be owned) by specific stations. They are assigned by default to the station pointed to by *TheStation* at the time of their creation. Sometimes it may make sense to assign *TheStation* explicitly, to assume the context of some specific station and make sure that the object (or objects) subsequently created will be assigned to the right owner. There exists a version of *create* with an extra argument specifying the station that should be made current (assigned to *TheStation*) before the object is created (Sect. 3.3.8). That assignment is reverted when *create* completes.

4.2 Links

Links are objects belonging to type *Link*, which exists in three standard subtypes: *BLink*, *ULink*, and *PLink*. The primary purpose of a link is to model a simple wired communication medium. Two types of wired channel models, broadcast and unidirectional links, are built into SMURPH, with the latter type occurring in two slightly different flavors.

4.2.1 Propagation of Signals in Links

A broadcast link is a uniform signal passing medium with the property that a signal entered into any port connected to the link reaches all other ports of the link in due time. A typical real-life representative of the broadcast link model is a cable (with multiple taps). In a unidirectional link, signals travel in one direction only: for any two ports A and B on such a link, there is either a path from A to B or from B to A, but never both. A typical physical implementation of a unidirectional link is a single, segmented fiber-optic channel.[3]

---

[3]Technically, a segmented, unidirectional fiber optic cable can be implemented in SMURPH using repeater stations (i.e., explicit models of the interconnections of essentially bidirectional segments).

Irrespective of the link type, the geometry of the link is described by the link's distance matrix[4] which specifies the propagation distance between every pair of ports connected to the link. The propagation distance from port A to port B is equal to the (integer) number of *ITUs* required to propagate a signal from A to B. SMURPH models of signals in a link are called (link) *activities*. Activities are started at originating ports and they propagate to other ports according to the distance matrix. If an activity, typically a packet transmission (see Sect. 7.1.2), is started on port A at time t, it will arrive at port B at time $t + D[A, B]$, where D is the link's distance matrix. For a broadcast, bidirectional link (type *BLink*), the propagation distance is independent of the direction, i.e., the propagation distance from A to B is the same as the propagation distance from B to A. In other words, the distance matrix of a broadcast link is symmetric.

For a unidirectional link (types *ULink* and *PLink*), the order in which ports have been connected to the link (see Sect. 4.3.2) determines the propagation direction, i.e., if port A was connected before B, then signals can propagate from A to B, but not from B to A. It means that an activity inserted at time t into A will reach B at time $t + D[A, B]$, as for a broadcast link, but no activity inserted into B will ever reach A. Thus, the distance matrix of a unidirectional link is triangular with the "absent" entries counted as infinities.

If a unidirectional link has a strictly linear topology, i.e., for every three ports A, B, C created in the listed order $D[A, C] = D[A, B] + D[B, C]$, the link can be declared as a *PLink*, which type can only be used to represent strictly linear, unidirectional links. Semantically, there is no difference between types *PLink* and *ULink* used to represent such a link; however, the *PLink* representation is usually much more efficient, especially when the link is very long compared to the duration of a typical activity (packet).

Type *BLink* is equivalent to the simplest generic link type *Link*. More information about links, including a detailed description of their functionality, will be given in Sect. 7.1.

4.2.2 Creating Links

Link objects are seldom, if ever, referenced directly by the protocol program, except for the network initialization stage, where the links must be built and configured. It is technically possible (but seldom useful) to define a new link type with the following operation:

The built-in unidirectional link model provides a convenient shortcut whose implementation is incomparably more efficient.

[4]Distance matrices do not occur explicitly (as matrices): they are distributed into distance vectors associated with ports rather than links.

link newtypename : oldtypename {

    ...

};

One use for this possibility would be to augment the standard methods for collecting link performance data by user-defined functions (see Sect. 10.3), or to declare a non-standard link method for generating errors in packets (Sect. 7.3). If the "oldtypename" part is absent, the new link type is derived directly from *Link*.

A link object is created by *create*, like any other *Object* (Sect. 3.3.8). The standard link *setup* method is declared with the following header (the same for all link types):

void setup (Long np, RATE r, TIME linger, int spf);

The first argument, *np*, specifies the number of ports that will be connected to the link. The second argument, *r*, provides the default transmission rate for a port connected to the link. That rate will be assigned to every port connected to the link that doesn't directly specify a transmission rate at its creation (Sect. 4.3.1). The third argument, *linger*, gives the link linger (or archival) interval in *ITUs*. The attribute determines for how long descriptions of activities that have formally disappeared from the link (having reached and crossed the most distant ports) are going to be kept in the link data base, because they may still be needed by the protocol program (see Sects. 7.1.1 and 7.2.2). Value 0 (or *TIME_0*) means that any activity leaving the link is immediately deleted and forgotten (as it happens in real life). The last argument indicates whether standard performance measures are to be calculated for the link. *ON* (1) means that the standard performance measures will be computed; *OFF* (0) turns them off.[5]

There is an alternative (shortcut) *setup* method for *Link* declared as:

void setup (Long np, TIME linger = TIME_0, int spf = ON);

which translates into this call:

setup (np, RATE_0, linger, spf);

and, basically, ignores the default port rate argument, setting it to zero (note that such a rate is not usable for transmission, Sect. 7.1.2). Thus, ports are then expected to take care of their transmission rates individually, all by themselves. With the two *setup* variants for *Link*, the *create* operation expects at least one and at most four arguments.

The *linger* attribute can also be set dynamically, after the link has been created, by invoking this *Link* method:

void setPurgeDelay (TIME linger);

---

[5]If non-standard performance measuring methods have been declared for a link (Sect. 10.3), then they will be invoked, even if the link has been created with *spf* = *OFF*. The flag only applies to the standard measures. The non-standard methods must explicitly check the flag, if they want to properly respond to its setting.

Note that the default value of the attribute is zero *ITUs*. The interpretation of the zero *linger* value is that any activity inserted into the link will be removed at the time when its last bit passes the most distant port on the link, as seen from the port of its insertion (Sect. 7.1.1).

A newly-created link is *raw*, in the sense that it is not configured into the network. The value of *np* specifies the number of slots for ports that will be (eventually) connected to the link, but the whereabouts of those ports are not known yet. The link's geometry is not known either; as we said in Sect. 4.2.1, it is determined by distances between pairs of ports. Those distances must be assigned explicitly after the port slots in the link have become filled (Sect. 4.3.3).

The *Id* attributes of links (and their standard names, see Sect. 3.3.2) reflect the order in which the links have been created. The first created link receives number 0, the second link number 1, etc. In many cases when a link has to be referenced, it is possible to use either a pointer to the link object or the link number (its *Id* attribute). The following function:

Link *idToLink (Long id);

converts a numerical link identifier to the *Link* pointer. The global variable *NLinks* of type *int* stores the number of all links created so far. As links cannot be created once the protocol program has started (Sect. 5.1.8), following the initialization stage, *NLinks* shows the immutable, total number of links in the network.

4.3   Ports

Ports are objects of the standard type *Port*, which is not extensible by the user. Therefore, there is no *port* operation that would declare a *Port* subtype.

4.3.1   Creating Ports

There are two ways to create a port. One is to use the standard *create* operation (Sect. 3.3.8). The port *setup* method for this occasion is declared as follows:

void setup (RATE rate = RATE_inf);

where the optional argument defines the port transmission rate (*Port* attribute *XRate*)[6] as the number of *ITUs* required to insert a single bit into the port. This attribute is only relevant, if the port will ever be used for transmission (Sect. 7.1.2) and determines the amount of time required to transmit a sequence of bits (a packet) through the port. The default value, *RATE_inf*, is interpreted as "none" and means that the default

---

[6]Strictly speaking, this is the reciprocal of the rate.

transmission rate associated with the link (Sect. 4.2.2) should be applied to the port. That rate will be copied by the port from the link at the time when the port becomes connected to the link (Sect. 4.3.2).

Ports are naturally associated with stations and links. From the viewpoint of the protocol program, the association of a port with a station is more important than its association with a link, because links are not perceived directly by the protocol processes. Therefore, a port can also be declared as a class variable within a station class, i.e.,

Port portname;

Then, it will be created automatically along with the station. Such a port cannot be explicitly assigned a transmission rate, so it will inherit the rate from the link (Sect. 4.3.2). The rate can also be specified later (and redefined at any time) by the *setXRate* method of *Port*. For illustration, in the following code: *setNName*

```
station BusStation {

    ...

    Port P;

    ...

    void setup ( ) {

        ...

        P.setXRate (myXRate);
        P.setNName ("Port P");

        ...

    };

    ...

};
```

port *P* (declared as a *Port* class variable within *BusStation*) is assigned a transmission rate as well as a nickname in the *setup* method of its station. Note that it doesn't matter whether the port is connected to a link at this time or not, as long as the specified rate (*myXRate*) is different from *RATE_inf* (which would mean "undefined"). Any definite (different from *RATE_inf*) rate explicitly assigned to a port takes precedence over the default rate brought in later by the link. Note that the zero value of the transmission rate is formally legal (and interpreted as a definite rate). It indicates that the port is not to be used for transmission, but it can be used for reception. An attempt by such a port to transmit will trigger an error.

In addition to setting the transmission rate of its port, *setXRate* returns the previous setting of the rate. The method is declared with the following header:

RATE setXRate (RATE r = RATE_inf);

When called with *RATE_inf* (or with no argument at all), the method will cause the port to be assigned the default transmission rate of its link, but that will only work if the port is already connected, Sect. 4.3.2). Any other value is interpreted as the port's new, private, and specific transmission rate.

This method:

RATE getXRate ();

returns the current setting of the port's transmission rate.[7]

No matter which way a port has been created, from the beginning of its existence, it is associated with a specific station. If the port has been declared as a (non-pointer) class variable of a station, the situation is clear: as soon as the station has been created, the port comes into existence and is automatically assigned to the station. When a port is created by *create*, it is assigned to the current station pointed to by the environment variable *TheStation* (Sect. 4.1.3).

All ports belonging to the same station are assigned numerical identifiers starting from 0 up to $n-1$, where n is the total number of ports owned by the station. This numbering of ports reflects the order of their creation and is used to construct the *Id* attributes of ports (Sect. 3.3.2). The complete *Id* attribute of a port consists of two numbers: the *Id* attribute of the station owning the port and the station-relative port number determined by its creation order. These numbers are shown separately in the port's standard name (Sect. 3.3.2) which has the following format:

Port pid *at* stname

where "pid" is the relative number of the port at its station, and "stname" is the standard name of the station owning the port (note that it includes the station's numerical *Id*).

The following *Station* method:

Port *idToPort (int id);

converts the numerical station-relative port identifier into a pointer to the port object. An alternative way to do the same is to call a global function with the identical header as the above method, which assumes that the station in question is the current station.

The following two *Port* methods:

int getSID ();
int getYID ();

return the numerical *Id* of the port's owning station and the station-relative port *Id*, respectively.[8]

4.3.2   Connecting Ports to Links

A port that has just been created, although it automatically and necessarily belongs to some station (Sect. 4.3.1), is not yet configured into the network. Two more steps are

---

[7] As a rule, most "set" methods have their "get" counterparts.

[8] *SID* stands for station *Id*, while *YID* means "your" (station-relative) *Id*.

required to specify the port's place in the overall structure. The first of these steps is connecting the port to one of the links. The link to which the port is to be connected must have been created previously.

One of the following two *Port* methods can be used to connect a port to a link:

*void connect (Link *l, int lrid =NONE);*
void connect (int lk, int lrid = NONE);

In the first variant of *connect*, the first argument points to a link object. The second variant allows the user to specify the link number (its *Id* attribute, Sect. 3.3.2) instead of the pointer.

A port connected to a link is assigned a link-relative numerical identifier which, in most cases, is of no interest to the user. That identifier is a number between 0 and $n-1$, where n is the total number of ports connected to the link, and normally reflects the order in which the ports were connected. For a unidirectional link (types *ULink* and *PLink*), the link-relative numbering of ports determines the link direction. If A and B are two ports connected to the same unidirectional link, and the link-relative number of A is smaller than that of B, then there is a path from A to B, but not in the opposite direction.

It is possible to assign to a port being connected to a link an explicit link-relative number by specifying this number as the second argument of *connect*. By default, this argument is *NONE* (-1), which means that the next unoccupied link-relative number is to be assigned to the port. The user can force this number to be any number unused so far, but the resulting numbering of all ports that eventually become connected to the link must be continuous and start from 0.

When a port is being connected to a link, and its transmission rate is undefined at the time of connection (equal to *RATE_inf*), the port's transmission rate is copied from the link's default (Sect. 4.3.1).

4.3.3 Setting Distance Between Ports

One more operation required to configure the network is assigning distances to all pairs of ports that have been connected to the same link. This can be done with the following global functions:

*void setD (Port *p1, Port *p2, double d);*
void setD (Port p1, Port p2, double d);

which set the distance between the ports represented by *p1* and *p2*. The distance *d* is expressed in distance units (*DUs*, Sect. 3.1.2) and is converted internally into *ITUs*. One should remember that although the distance is specified as a floating-point value, the internal representation of the distance is always an integer number. For example, with the default setting of 1 *DU* = 1 *ITU*, any fractional components of *d* are rounded (not truncated) to the nearest integer value (see Sect. 3.1.2).

The order in which the two ports occur in the argument list of *setD* is immaterial. For a bidirectional link (types *Link*, *BLink*), the distance from *p1* to *p2* must be the same as the distance from *p2* to *p1*. For a unidirectional link (types *ULink* and *PLink*), only one of these distances is defined (and this is the one that will be set), as determined by the port connection order (see Sect. 4.3.2).

SMURPH offers more functions for setting distance between ports. The following ones specify only one port each; the other port is implicit:

void setDFrom (Port *p, double d);
void setDFrom (Port p, double d);
void setDTo (Port *p, double d);
void setDTo (Port p, double d);

Each of the two variants of *setDFrom*, sets the distance from port *p* to the port that was last used as the second argument of *setD*. For a bidirectional link, the direction is immaterial: the two-way distance between the two ports is set to *d*. For a unidirectional link, SMURPH checks whether the direction coincides with the link-relative ordering of the two ports, and an error is signaled if it doesn't. Similarly, the two variants of *setDTo* set the distance to port *p* from the port that last occurred as the first argument of *setD*. The rules are the same as for *setDFrom*.

The four functions listed above also exist as methods of type *Port*. For such a method, the implicit port is the one determined by *this*, e.g.,

myPort->setDTo (otherPort, 12.45);

sets the distance from *myPort* to *otherPort* according to the same rules as for the global function *setDTo*.

It is OK to define the same distance twice or more times, provided that all definitions specify the same value. For a strictly linear unidirectional link (type *Plink*, Sect. 4.2.1), there is no need to define distances between all pairs of ports. As soon as enough distances have been provided to describe the entire link, SMURPH is able to figure out the "missing" distances on its own.

It is possible to look up the distance between two ports (connected to the same link) from the protocol program. The following port method:

double distTo (Port *p);

returns the distance (in *DUs*) from the current (*this*) port to *p*. This distance is undefined (the method returns *Distance_inf*), if the two ports belong to different links, or if the link is unidirectional and *p* is upstream from the current port.

Note that the distance returned by *distTo*, although a floating point number, is derived from the integer number of *ITUs* representing the distance internally. Thus, the value may differ slightly from the one to which the distance was originally set. For illustration, assume the setting:

setDu (10.0);

from the example in Sect. 3.1.2. The distance of 2.34 (meters, according to the user intention) will be stored as 23 *ITUs*, because the *double* value will be multiplied by 10.0 and rounded to an integer. Thus, when subsequently queried with *distTo*, the distance will be returned as 2.3 (meters).

4.4 Radio Channels

In principle, links and ports (Sects. 4.2 and 4.3), in their purely broadcast flavor (type *BLink*) could be used to model radio channels. However, their built-in features miss certain basic aspects of functionality of such channels, which would require the user to construct too much of their models by hand. Two problems with links are particularly difficult to overcome in this context:

- There is no built-in notion of neighborhood (implied by the propagation range), which means that all stations (ports) connected to the link always see all traffic. While the protocol program can ignore its parts based on a programmed notion of range, sensitivity, and so on, the model of a large network with narrow neighborhoods would incur a lot of overhead.
- It is difficult to implement any kind of node mobility. The problem is that once the topology/geometry of a link is determined (at the initialization stage), it is meant to remain fixed for the entire simulation experiment.

These problems get in the way of any attempts to tweak the link models to offer the kind of flexibility one would like to exercise with useful models of RF channels. Even ignoring the fact that links lack any explicit tools for modeling signal attenuation and interference (and the programmer would have to be responsible for creating them all), the rigid and global structure of links makes them practically useless as models of wireless links.

Consequently, SMURPH brings in a separate collection of tools for expressing the behavior of wireless channels. Those tools are comprised of radio channels[9] (type *RFChannel*) and transceivers (type *Transceiver*), which can be viewed as the wireless equivalents of links and ports. Many general concepts applicable to links and ports directly extend over their wireless counterparts.

---

[9]As we noticed in Sect. 2.4.5, the word "channel" is a bit unfortunate in this context, because we may want to model channels within a channel (Sect. A.1.2). Perhaps "medium" or "band" would better fit the role of *RFChannel*, except that the term "channel model" is rather well established in the literature in reference to what we have in mind.

The role of a radio channel in SMURPH is to interconnect transceivers, which (like ports with respect to links) provide the interface between stations and radio channels. In contrast to a link, a radio channel need not guarantee that a signal originating at one transceiver will reach all other transceivers.[10] Another important difference is that the built-in channel type *RFChannel* is incomparably more open-ended than *Link* (or any of its built-in subtypes). Although it does provide a complete functionality of sorts, that functionality is practically useless. This is because the issue or modeling wireless channels (with a meaningful fidelity) is considerably more complex than modeling a piece of wire. Owing to the proliferation of wireless channel models, it would be unwise to restrict our model to one (or some) of them. Consequently, *RFChannel* does not provide a complete channel model, but rather the generic, common content of all such models. It is thus a parent type for building actual channel types whose exact behavior is specified by a collection of virtual methods programmed by the user (Sect. 4.4.3).

One attribute of a transceiver is its (geographical) position in the form of two or three Cartesian coordinates specified in *DUs* and stored internally in *ITUs*. The dimensionality of the network's deployment area is selected at compilation time (Sect. B.5). As many practically interesting cases are expressible in two dimensions, the default is 2d, which simplifies network parameterization and reduces the complexity of the model. With the 3d option, certain methods receive different headers, to accommodate one more coordinate needed to express locations.

The following global function calculates the distance between a pair of points:

double dist (double x0, double y0, double x1, double y1);

where $(x0, y0)$ and $(x1, y1)$ are the coordinates of the two points. With the 3d option, the function's header looks like this:

double dist (double x0, double y0, double z0, double x1, double y1, double z1);

i.e., each point is described by three coordinates.

Note that positions are attributed to transceivers rather than stations (nodes). Normally the position of a node is that of its transceiver. If a station has more than one transceiver, then the multiple transceivers are positioned independently and their locations need not coincide. Stations are not active objects (they are not activity interpreters, Sect. 5.1.1), so their role is to formally glue things together with no physical implication. Unlike ports (Sect. 4.3.3), where positions are irrelevant and only distances matter (and those distances must be specified directly), the distances among transceivers are derived from their positions (using the straightforward, Euclidean notion of distance). The explicit positions of transceivers are useful for

---

[10]We remember from Sect. 2.2.5 that (wired) links can be made faulty, so some bona-fide packets may not make it to destinations because of spontaneous errors. The issue of spontaneous faults in links has been handled in a simple way (see Sect. 7.3), because such faults are usually not the most interesting part of the model. In contrast, the primary challenge in authoritative modeling of wireless channels is precisely the high-fidelity expression of errors caused by signal attenuation and interference.

mobility models and also for advanced RF propagation models, e.g., based on samples collected from real-life environments. In contrast to the distance matrix of a link (Sect. 4.2.1), the distance matrix of a radio channel is flexible (and is not really a matrix). Transceivers can change their positions in a practically unrestricted way. A natural way to model node mobility is to have a special process that modifies the locations of nodes/transceivers according to some prescription. Even though the coordinates are de facto discrete (stored internally in *ITUs*, Sect. 3.1.2), the *ITU* can be chosen to represent an arbitrarily fine grain of distance. This is important because the coordinate change is always instantaneous: a node is simply teleported from one location to another. There is no problem with this approach, if the mobility pattern is realistic, i.e., the nodes move in reasonably small steps at a reasonably low (non-relativistic) speed. It is up to the user to maintain the sanity of the mobility model.

4.4.1 Signals, Attenuation, Interference

The present section presents a brief introduction to the mechanics of a channel model, so we can understand the configuration parameters of *RFChannels* and *Transceivers*. A more detailed explanation of those mechanics, supplementing the information from the present section, will be given in Sect. 8.1, where we shall also discuss the interface offered by *RFChannels* and *Transceivers* to the protocol program.

The user-provided methods of a subtype of *RFChannel*, replacing the virtual stubs in *RFChannel* and implementing a specific channel model according to the user's prescription, are referred to as *assessment methods*. They provide formulas to calculate the impact of signals on other signals and decide whether those signals trigger events on transceivers that would result in packet reception. We saw examples of most of them in Sect. 2.4.5, in a less formal context.

An *RFChannel* carries RF signals which represent packets or their preambles. Like the activities in links, those signals propagate through the channel with the timing induced by the coordinates of the transceivers involved in their generation and perception. In fact, we sometimes use the term "activity" for a signal, but, in this section, we shall stick to "signals," because we are primarily interested in their RF properties. From this point of view, an RF signal, as represented in the model, is characterized by two attributes encapsulated into this standard structure:

```
typedef struct {
        double Level;
        IPointer Tag;
} SLEntry;
```

The first attribute, *Level*, is a nonnegative floating-point number typically interpreted as the signal's power or strength; the other attribute, *Tag*, describes the signal's *features*. The meaning of *Level* is obvious. The role of *Tag* is to account for all other properties of the signal that may affect its propagation and impact on other signals. For example, signals transmitted at slightly different frequencies (corresponding to different, selectable "channels" within the same RF medium), or using different codes (in CDMA) will interfere with each other according to some rules (crosstalk) whose precise formulation involves factors other than just the signal strength. For another example, in a frequency-hopping schemes, the *Tag* may describe a pattern of frequencies that must be compared with the patterns of interfering activities to assess the impact of their interference.

The core (built-in) part of the model takes care of timing the signals as they propagate through the wireless medium, but it does not care what *Level* and *Tag* precisely mean. When a signal arrives at a transceiver, the core model invokes assessment methods to evaluate the signal's impact on other signals perceived by the transceiver at the same time, and/or decide whether the signal should trigger any of the events that the protocol program may be awaiting on the transceiver. The interpretation of *Level* and *Tag* is thus up to the assessment methods.

The headers of the assessment methods are listed in Sect. 4.4.3. They are intentionally orthogonal, in the sense that they don't get in each other's way (the questions they answer are well encapsulated and isolated), and complete, in the sense that they can (in principle) describe any conceivable RF channel. By wrapping the user-level description of a channel model into a relatively small collection of orthogonal functions, SMURPH reduces the complexity of the model specification to filling in a few blanks with prescriptions how to answer a few straightforward questions. Those prescriptions have the appearance of static formulas completely parameterizing the model, while also completely hiding its rather complex dynamics. The configuration of assessment methods (making them simple and few) was the primary challenge in developing the core wireless model [1].

Some assessment methods take signals as arguments, i.e., pointers to structures of type *SLEntry* (see Sect. 4.4.3). Recall from Sect. 3.1.1 that *IPointer* (which is the type of *Tag* in *SLEntry*) is the smallest integer type capable of accommodating a pointer; thus, *Tag* can point to a larger structure, if the set of signal features cannot be described directly in the *Tag's* (integer) value. In addition to representing signals in transit, *SLEntries* may also refer to the settings of transmitters and receivers. For example, the transmit power setting of a transceiver involves a numerical value interpreted as the signal strength of a departing packet, but the full setting may also involve a *Tag* to be assigned to the packet, i.e., the set of features selected for the transmission. For a receiver, the *Level* attribute is typically interpreted as the (antenna) gain (to be applied as a multiplier to a received signal level). Of course, the signal features (referring to the current tuning of the receiver to the specific channel or code) may also apply. We should emphasize it once again that the interpretation of all those attributes is up to the assessment methods. For example, in a simple model with featureless signals, the *Tag* attribute may not be used at all (Sect. 2.4.5).

The assessment methods (their headers and roles in the model) are geared for analyzing sets of signals perceived by transceivers at interesting moments to evaluate the extent to which they interfere, and transform those interferences into chances for

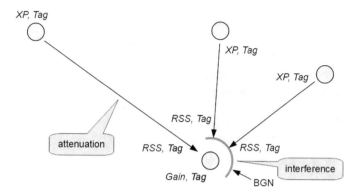

Fig. 4.1   The signal transformation model

successful packet reception. Figure 4.1 sketches the way signals are processed by the core model. The circles represent transceivers. A signal (representing a packet or its preamble) is transmitted at some transmit power *XP* with some configuration of features represented by *Tag*. At another transceiver, *XP* gets transformed into the received strength (*RSS*) of the signal according to the attenuation function implemented by one of the assessment methods. The set of features of the signal (*Tag*) is not meant to change along the way. If some transceiver, say the one at the bottom of Fig. 4.1, receives a packet (represented by some signal), then the remaining signals audible at the transceiver, plus the background noise, will combine into an interfering signal competing with the received packet. Some other assessments methods will be invoked to add the interfering signals, calculate the interference, and ultimately come up with a decision regarding the success (or failure) of the packet reception. To make the model as realistic and as open as possible (while keeping the collection of assessment methods easily comprehensible and small) there is no single reception event, but several assessment points in the packet's life, which facilitates high-fidelity models of the various real-life techniques of packet reception.

Since different fragments of packets (or preambles) can suffer different levels of interference (something we already mentioned in Sect. 2.4.5), the core model implements a special data structure (class) for representing the interference experienced by a signal "in time," i.e., the history of the interference to which the signal has been exposed. That data structure is called the interference histogram and described by type *IHist* (see Sect. 8.1.5). Interference histograms are maintained internally by the core model and made available to those assessment methods that may care about them. The raw contents of interference histograms are not interesting to the protocol program: the data structure is queried via its methods described in Sect. 8.1.5.

4.4.2   Creating Radio Channels

A single network model can feature several radio channels. Separate radio channels
are independent: if the user wants to model their cross-interference, the virtual methods describing the conditions for packet reception must account for that. A more
natural approach is to have a single radio channel and use *Tags* to identify different
"channels" (bands) within it (see Sect. A.1.2).

A radio channel type is defined in the following way:

```
rfchannel newtypename : oldtypename {

    ...

    attributes and methods

    ...

};
```

If the "oldtypename" part is absent, the new radio channel type is derived directly
from *RFChannel*. It is possible to use a proper subtype of *RFChannel* as the ascendant
of another subtype; in such a case, the ascendant type should appear as "oldtypename"
in the declaration.

The *setup* method of *RFChannel* has the following header:

```
void setup (Long nt, RATE r, int pre, double xp, double rs, TIME linger = TIME_1,
    int spf = ON);
```

where *nt* is the total number of transceivers that will be interfaced to the channel.
Arguments *r* through *rs* provide the default values for the respective attributes of
those transceivers (Sect. 4.5.1). Those attributes can also be specified on a per-
transceiver basis (when the transceiver is created), in which case they will take
precedence over the channel's defaults (Sects. 4.5.1 and 4.5.2). There exist more
attributes with this property, i.e., the de facto *Transceiver* attributes whose values
can be set from defaults associated with the *RFChannel* (to which the transceiver
eventually becomes interfaced). Even though those attributes do not occur on the
setup argument list, they can be set (and queried) with some methods of *RFChannel*
and *Transceiver*. We discuss them all in Sect. 4.5.2, where we talk about configuring
the transceivers.

The *linger* argument defines the so-called linger or purge delay (expressed in
ITUs) and referring to the delay in removing the activity from the channel's database
following its formal disappearance from the channel's "ether." By that we mean the
moment when the activity has passed the most distant (potential) destination of its
neighborhood (Sect. 8.2.1). This is similar to the *linger* attribute of a link (Sect. 4.2.2).

The last argument of the *setup* method, *spf*, plays the same role as the corresponding *Link setup* argument (Sect. 4.2.2), and is used to switch on or off the automatic
collection of standard performance measures for the rfchannel.

There exists an abbreviated version of the *setup* method for *RFChannel* which skips the default attributes of transceivers, setting them to generic (mostly unusable) values. That way it assumes that the transceivers will take care of those attributes all by themselves, e.g., setting them at the time of their creation (Sect. 4.5.1). The header of the abbreviated variant looks like this:

void setup (Long nt, TIME linger= TIME_1, int spf =ON);

and its effect is equivalent to:

setup (nt, RATE_0, 0, 0.0, 0.0, linger, spf);

Like for a link, the linger delay can also be set (or reset) at any time, after the rfchannel has been created, by invoking this method of *RFChannel*:

void setPurgeDelay (TIME linger);

Similar to links, the *Id* attributes of radio channels (and their standard names, Sect. 3.3.2) reflect the order in which the channels have been created (the first channel is assigned number 0, and so on). In those few cases when a radio channel must be specified as an argument of a function/method, it is possible to use either a pointer to the *RFChannel* object, or the channel's numerical *Id* attribute. The following function:

*RFChannel *idToRFChannel (Long id);*

converts a numerical channel *Id* to the *RFChannel* pointer. The global variable *NRFChannels* of type *int* stores the number of all radio channels created so far.

Like for links, the geometry of a radio channel is determined by the configuration of its transceivers. The transceivers are assigned positions represented by pairs (or triplets) of coordinates (Sect. 4.5.1). The distance between a pair of transceivers is interpreted as the Euclidean distance in two or three dimensions and expressed in *DUs*.

4.4.3   The List of Assessment Methods

The primary reason for extending the base type *RFChannel* is to define the *assessment methods*, i.e., the virtual functions completing the generic core model into a working instance. Below we list the headers of the assessment methods, together with their default contents, and briefly describe their roles. The detailed explanation of the way those methods determine the fate of transmitted packets is given in Sect. 8.1. The methods execute in a certain environment (Sect. 8.1.7), which means that the information available to them consists of more than just the arguments.

Attenuation

```
double RFC_att (const SLEntry *sl, double d, Transceiver *src) {
      return sl->Level;
};
```

This method calculates signal attenuation depending on the distance and possibly other attributes that can be extracted from the signal *Tag* as well as the source and destination transceivers. The first argument refers to the original (departing) signal (corresponding to *XP* and *Level* in Fig. 4.1), the second argument is the distance in *DUs* separating the source and destination transceivers. The source transceiver is pointed to by *src*; the destination transceiver can be referenced via the global (environment) variable *TheTransceiver* (see Sect. 8.1.7). The method is expected to return the attenuated level of the original signal, i.e., its level as perceived by the destination transceiver (*RSS* in Fig. 4.1).

The default implementation of the method does not attenuate the signal at all, i.e., it returns the original signal level.

Addition of signals

```
double RFC_add (int n, int followed, const SLEntry **sl, const SLEntry *xm) {
      double tsl;
      tsl = xm->Level;
      while (n--)
            if (n != followed)
                  tsl += sl [n] -> Level;
      return tsl;
};
```

This method prescribes the formula for calculating the aggregation of multiple signals arriving at the receiver at the same time. The first argument, *n*, specifies the number of entries in *sl*, which is an array of signals representing all the activities currently perceived by the transceiver. The last argument, *xm*, pertains to the signal being issued (transmitted) by the present transceiver. If the transceiver is currently transmitting, then the *Level* attribute of *xm* is nonzero and represents the transmitted signal level (*XP* in Fig. 4.1). In all circumstances, the *Tag* attribute of *xm* stores the current value of the transceiver's transmit tag (Sect. 4.5.2). If *followed* is not *NONE* (−1), it gives the index of one entry in *sl* that should be ignored. The same method can be invoked to calculate the level of interference affecting one selected signal from the pool of signals perceived by the transceiver, e.g., when the transceiver attempts to receive a packet, while some other signals interfere with the reception. In such a case, the signal index of the packet being received is presented in *followed*; that signal should be excluded from the calculation.

Like in the case of *RFC_att*, the environment variable *TheTransceiver* points to the transceiver for which the assessment is being carried out. The default version of the method simply adds the signal levels ignoring their tags (and skipping the *followed* signal, if any).

Activity sense

```
Boolean RFC_act (double sl, const SLEntry *s) {
        return sl > 0.0;
};
```

The role of this method is to tell whether the transceiver senses any signal at all (carrier sense) based on the total combined signal level *sl* arriving at the transceiver at the time (the transceiver is pointed to by the environment variable *TheTransceiver*). The second argument provides the receiver's tag as well as its gain (sensitivity) stored in the *Level* component of *s*. The method is invoked to tell when the channel is perceived busy or idle (and when to trigger the events *ACTIVITY* and *SILENCE*, see Sect. 8.2.1).

The total signal level sensed by the transceiver is calculated by calling *RFC_add*. The default version of *RFC_act* returns *YES* (true) for any nonzero received signal level.

Beginning of packet

```
Boolean RFC_bot (RATE r, const SLEntry *sl, const SLEntry *rs, const IHist *ih) {
        Long nbits;
        if ((nbits = h->bits (tr)) < 8)
                return NO;
        if (h->max (tr, nbits - 8, 8) > 0.0)
                return NO;
        return YES;
};
```

The responsibility of this method is to decide whether the beginning of a packet arriving at a transceiver is recognizable as such, i.e., the packet has been formally detected and stands a chance of being received. The first argument is the transmission rate at which the packet was transmitted (in *ITUs*), *sl* refers to the received signal of the packet, i.e., the *(RSS, Tag)* pair in Fig. 4.1, *rs* describes the receiver's gain and tag setting, and *ih* points to the interference histogram storing the history of the interference experienced by the packet's preamble.

The method is invoked at the time when the preamble ends and the proper packet begins, i.e., the *BOT/BMP* event becomes eligible (Sect. 8.2.1). It has to say *YES* or *NO* deciding whether the event will be triggered.

The default method illustrates a simple usage of the interference histogram pertaining to a packet preamble. Note that the entire interference history of the preamble is available when the method is called, because the preamble has just finished. The *bits* method of *IHist* returns the number of bits (i.e., the length) of the preamble, which is derived from the total duration of the preamble's history divided by the rate (the amount of time corresponding to one bit). If the preamble is less than 8 bits long, the method says *NO* immediately. Otherwise, it invokes another method of the histogram, *max*, which (with the given configuration of arguments) returns the maximum level of interference experienced by the last 8 bits of the preamble (Sect. 8.1.5). If that level is nonzero (any interference at all), the method says *NO*, otherwise, the event is approved.

In a nutshell, the *BOT/BMP* event is approved, if the preamble preceding the packet is at least 8 bits long and there has been absolutely no interference within the last 8 bits. This does not look like a very realistic assessment, but the default method is not meant to be good for anything. The example we saw in Sect. 2.4.5 was much more sensible, but we must wait until Sect. 8.1 before it can be fully explained.

As in all cases when the assessment relates to a specific transceiver, the environment variable *TheTransceiver* points to the transceiver in question. Another environment variable, *ThePacket*, points to the packet whose *BOT/BMP* event is being assessed.

End of packet

```
Boolean RFC_eot (RATE r, const SLEntry *sl, const SLEntry *rs, const IHist *ih) {
        return (h->max ( ) == 0.0);
};
```

This method decides whether the end of a packet arriving at a transceiver should trigger *EOT/EMP* (Sect. 8.2.1). The arguments (and environment variables) are exactly as for *RFC_bot*, except that the interference histogram refers to the packet rather than the preamble.

The default version of the method approves the event, only if the total interference experienced by the packet is zero.

Cutoff

```
double RFC_cut (double sn, double rc) {
        return Distance_inf;
};
```

This method returns the cut-off distance (in *DUs*), i.e., the distance after which a signal ceases to be noticeable (causes no perceptible interference). The two arguments specify the transmitted signal level and the receiver's gain. The method is used to keep track of node neighborhoods, i.e., groups of nodes being affected by other nodes. Note that it does not take tags into consideration and bases its estimate solely on numerical signal levels. Conservative values returned by *RFC_cut* are formally safe, but they may increase the simulation time owing to unnecessarily large neighborhoods.

The default version of the method returns infinity, which means that all signal levels are always relevant, and all transceivers interfaced to the channel belong to one global village.

This method has no access to environment variables.

Transmission time

```
TIME RFC_xmt (RATE r, Packet *p) {
        return (TIME) tr * p->TLength;
};
```

This method calculates the amount of time needed to transmit the packet pointed to by the second argument at rate r by the current transceiver (*TheTransceiver*). The reason why an assessment method is provided for this operation (instead of, say, assuming that the result should be a simple product of the number of bits and the transmission rate) is that the encoding scheme of the model may render the issue more complex. For example, the logical (payload) bits may have to be transformed into physical bits (e.g., encoding four-bit nibbles into six-bit symbols). In even more complex cases, different types and/or portions of packets may be encoded using different schemes. Also, if there are any extra framing bits (like start/end symbols) added to physical packets, the method may want to account for them. Note that the preamble is a separate issue (Sect. 8.1.1), and its transmission is not subjected to assessment by *RFC_xmt*.

Error bits

```
Long RFC_erb (RATE r, const SLEntry *sl, const SLEntry *rs, double il, Long nb) {
        no default implementation
};
```

The method returns the number of error bits (which is typically randomized) within the sequence of nb bits received at rate r, for the received signal described by sl, with receiver parameters rs, and at the steady interference level il. Its role is to distribute bit errors for a given signal level and interference, e.g., by applying the concept of a SIR-to-BER function (Sect. 2.4.5). The method needs to be defined, if the protocol program (or other assessment methods) calls *errors* or *error*, which occur in several variants as methods of *RFChannel* and *Transceiver* (see Sect. 8.2.4).

The method answers a relatively straightforward question which is this: "given a signal level and an interference level, plus a sequence of bits received steadily under the same conditions, how many of those bits have been received in error?" Being able to answer a question like this, the core model can answer more complex questions, like: "given an interference histogram of a sequence of bits received at some signal level, what is the number of error bits in that sequence?"

The method has no access to environment variables and no default implementation. An attempt to invoke it triggers an execution error.

Number of bits until error condition

Long RFC_erd (RATE r, const SLEntry *sl, const SLEntry *rs, double il, Long nb) {
     no default implementation
};

The first four arguments are as for *RFC_erb*. The method returns the randomized number of received bits (under the conditions described by the first four arguments, exactly as for *RFC_erb*) preceding the occurrence of a user-definable "expected" configuration of bits described by the last argument. Typically, that configuration (the amount of time until its occurrence expressed in bits) is strongly related to the distribution of bit errors (thus, *RFC_erd* is closely correlated with *RFC_erb*). For example, the method may return the (randomized) number of bits preceding the nearest occurrence of a run consisting of *nb* consecutive bit errors. The sole purpose of the method is to trigger *BERROR* events, whose interpretation is up to the protocol program (Sect. 8.2.4). If the program is not interested in perceiving those events, the method need not be defined.

A typical application of *RFC_erd* is to generate randomized intervals until the first error bit, e.g., to cater to packet reception models where the first erroneous symbol aborts the reception (Sects. 8.2.4 and A.2.2). The method has no access to environment variables and has no default implementation. An attempt to invoke it triggers an execution error.

4.5 Transceivers

Transceivers are represented by type *Transceiver*, which, like *Port*, is not extensible by the user. There are more similarities between transceivers and ports: the two types of objects play conceptually the same role of interfacing communication channels to stations and processes.

4.5.1 Creating Transceivers

One natural way to create a transceiver is to resort to the standard *create* operation (Sect. 3.3.8). The *setup* method of *Transceiver*, which determines the configuration of arguments for *create*, has the following header:

void setup (RATE r = RATE_inf, int pre = −1, double xp = −1.0, double gain = −1.0,
     double x = Distance_0, double y = Distance_0);

All arguments are optional and have the following meaning:

r	This is the transmission rate in *ITUs* per bit, analogous to the corresponding *Port* attribute (Sect. 4.3.1). The default setting of *RATE_inf* formally stands for "none" and means that the rate should be copied from the rfchannel to which the transceiver will be interfaced (Sect. 4.4.2)
pre	The preamble length in "physical" bits to be inserted by the transmitter before an actual packet. The default (negative) value formally stands for "none" and means that the attribute should be copied from the rfchannel
xp	Transmission power, i.e., the initial setting of the signal level for a transmitted packet (see Sect. 4.4.1). The default (negative) value is treated as "none" meaning that the attribute should be copied from the rfchannel
gain	Receiver gain (sensitivity). The default (negative) value is interpreted as "none" meaning that the attribute should be copied from the rfchannel
x and *y*	These are the planar coordinates of the transceiver expressed in *DUs*. By default, all transceivers initially pile up at the same location with coordinates (0, 0). There is no default transceiver location attribute associated with the *RFChannel* object that could be inherited by transceivers (it wouldn't make a lot of sense)

The list of *setup* arguments for *RFChannel* does not cover all the *Transceiver* attributes whose values can be modified and queried by the protocol program. For example, it does not include the transmitter and receiver tags, which accompany the transmit power and receiver gain to form complete descriptions of the signal features as *SLEntry* objects (see Sect. 4.4.1). The tags can be set by *Transceiver* methods (see Sect. 4.5.2); their initial values can also be inherited from the *RFChannel* to which the transceiver is connected. Initially, when a *Transceiver* is created (but not yet connected), both tags are set to the special value *ANY* (all ones or -1) interpreted as "unassigned." When the transceiver is subsequently connected to an rfchannel, the tags (both of them) will be set to the default values associated with the rfchannel. Those values are initially zeros, but the rfchannel defaults can be changed before transceivers are interfaced to the rfchannel, or even half way through (see Sect. 4.5.2). If this sounds complicated, then please bear in mind that, if used at all by the channel model, the tags are likely to be a dynamic feature, i.e., one that will have to be reset very often, possibly on every transmission. Thus, their default (or semi-default) presets are probably not that important.

If a transceiver is to be used for transmission, its rate attribute must be definite, because its role is to determine the transmission time as a function of the packet's length. Note that the duration of a packet transmission is calculated by an assessment method (Sect. 4.4.3), which, technically, is free to ignore the rate. However, the rate is still needed in calculating the amount of time needed to transmit the preamble (which is handled by the core model). Thus, when we say that the rate applies to "physical" bits, we mean the bits as they are counted within the preamble (note that preamble length is specified in bits). *RFC_xmt* receives that "physical" rate as its argument and is expected to produce the packet's transmission time, which also accounts for any physical/logical transmutation of bits that may be needed along the way. As every packet is preceded by a preamble, a transceiver used for transmission must define a positive preamble length, as well as a positive rate.

Transceiver locations are settable dynamically and can be changed at any time, so node mobility can be modeled. The *Transceiver setup* method, in its version presented above, applies to the 2d variant of the model. With the 3d variant, selectable with a compilation option (Sect. B.5), the list of coordinates receives one more element, so, the method's header becomes:

void setup (RATE r = RATE_inf, int pre = –1, double xp = –1.0, double gain = –1.0,

        double x = Distance_0, double y = Distance_0, double y = Distance_0);

Note that a transceiver can only be used for anything related to transmission or reception after it has been interfaced (*connected*) to an rfchannel (Sect. 4.5.2). Thus, the default "unassigned" values of transceiver attributes will always be overridden by the respective channel defaults before the transceiver is used for a transmission or reception. To be able to transmit anything, a transceiver must specify a positive rate different from 0 and *RATE_inf*, as well as a preamble strictly greater than zero. Moreover, transmitting at power zero or/and receiving at sensitivity zero is probably not going to succeed, but that is up to the assessment methods to decide.

Similar to a port (Sect. 4.3.1), a transceiver can be declared as a class variable within a station type, e.g.,

```
station SensorNode {

    ...

    Transceiver RFI;

    void setup ( ) {

        ...

        RFI.setXRate (myXRate);

        RFI.setPreamble (32);

        RFI.setNName ("SensorInterface");

        RFI.setLocation (24.1, 99.9);

        RFI.setMinDistance (0.1);

        ...

    };

    ...

};
```

Such a transceiver may obtain some attributes from the channel (at the time of connection); however, its location must be assigned explicitly by *setLocation*, unless we really want the transceiver to end up at the default position (0, 0), or (0, 0, 0), in the 3d case.

Like a port, a transceiver must always be associated with a specific station. If the transceiver has been declared as a class variable within a station class, it comes into existence together with the station object and is automatically assigned to it. When a transceiver is created by *create*, it is assigned to the current station pointed to by the environment variable *TheStation* (Sect. 4.1.3).

All transceivers belonging to one station are assigned numerical identifiers starting from 0 up to $n-1$, where n is the total number of transceivers owned by the station. This numbering reflects the order of transceiver creation and is used to construct the *Id* attributes of transceivers (Sect. 3.3.2), which resemble the *Id* attributes of ports (Sect. 4.3.1). Thus, the standard name of a transceiver has the following format:

Tcv tid *at* stname

where "tid" is the relative number of the transceiver at its station, and "stname" is the standard name of the station owning the transceiver.

The following *Station* method:

*Transceiver *idToTransceiver (int id);*

converts the numerical station-relative transceiver identifier into a pointer to the transceiver object. An alternative way to achieve the same effect is to call a global function declared exactly as the above method, which assumes that the station in question is the current station, i.e., the one pointed to by *TheStation*.

The following two *Transceiver* methods:

int getSID ();
int getYID ();

return the numerical *Id* of the transceiver's owning station and the station-relative transceiver *Id*, respectively.

4.5.2 Interfacing and Configuring Transceivers

One important step before a transceiver becomes functional is to interface it to a radio channel. This can be accomplished with the following method of *RFChannel*:

*void connect (Transceiver *tcv);*

or one of these methods of *Transceiver*:

*void connect (RFChannel *rfc);*
void connect (int rfcId);

In the last case, the argument is the numerical *Id* of the radio channel (Sect. 4.4.2).

In contrast to ports and links, there is no need to set up a distance matrix for the radio channel. In the wireless case, the propagation distances between transceivers are their Euclidean distances on the virtual plane or in the virtual 3-space.

When a transceiver is connected to a link, and some of its setup attributes have "unassigned" values (Sect. 4.5.1), those attributes will be copied from the default attributes of the channel (Sect. 4.4.2). The list of those attributes extends beyond the list of arguments of the *setup* methods for *RFChannel* and *Transceiver*. For example, the transmission and reception tags do not appear among the *setup* arguments, but they are handled in the same way as the other attributes.

The following methods of *Transceiver* change transceiver attributes. We list them in the present section, because they can be sensibly called before the protocol program is started, i.e., they can be used for the initial configuration of transceivers replacing or augmenting the *setup* method (Sect. 4.5.1). A more detailed discussion of these and other methods of *Transceiver* will be given in Sect. 8.1.

RATE setXRate (RATE r = RATE_inf);

This method sets (or re-sets) the transmission rate of the transceiver. The transmission rate can be changed at any time, and it immediately becomes effective for all packets transmitted after the change. The method returns the previous setting of the transmission rate.

If the argument is *RATE_inf* (or absent), the method will reset the transceiver's rate to the default rate associated with the rfchannel (Sect. 4.4.2). This is only possible if the transceiver has been already interfaced (connected) to an rfchannel.

Long setPreamble (Long p =–1);

This method sets (or re-sets) the preamble length and returns the previous setting. If the argument is negative (or absent), the method will reset the preamble length to the default length associated with the rfchannel (Sect. 4.4.2). This is only possible if the transceiver has been already interfaced (connected) to an rfchannel.

double setXPower (double xp =–1.0);
double setRPower (double rp =–1.0);

These methods set (or re-set), respectively, the transmission power and the receiver gain of the transceiver, returning the previous setting. If the argument is negative (or absent), the method will reset the respective attribute to the default value associated with the rfchannel (Sect. 4.4.2). This is only possible if the transceiver has been already interfaced (connected) to an rfchannel.

IPointer setXTag (IPointer tag = ANY);
IPointer setRTag (IPointer tag = ANY);
IPointer setTag (IPointer tag = ANY);

These methods set (or re-set) the tags of the transceiver and return the previous setting. The first method sets the transmission tag, the second one sets the reception tag, and the third one sets both tags to the same value. In the last case, the returned previous setting refers to the reception tag.

If the argument is equal to *ANY* (all ones, or −1), the corresponding attribute will be copied from the rfchannel. The default channel setting of both tags is zero. In the case of *setTag*, both tags will be set to the channel defaults.

void setLocation (double x, double y);

This method moves the transceiver to a new location with the specified Cartesian coordinates expressed in *DUs*. If the simulator has been created with the *–3* option (selecting the 3d geometry of the transceiver deployment space, Sect. B.5), the method is declared as:

void setLocation (double x, double y, double z);

The move incurred by invoking *setLocation* is instantaneous (a teleportation). In a realistic mobility model, the transceivers should probably be moved in small steps combining into reasonably smooth trajectories. One exception is the initial configuration of nodes when it makes perfect sense to teleport the transceivers to their starting locations before starting the protocol program.

Recall that positions are attributed to transceivers, not stations. If a single station is equipped with multiple transceivers, all those transceivers should be positioned in unison, unless the station is elastic and, for some (sensible) reason, the transceivers are spatially independent. To facilitate, simplify, and foolproof the most reasonable scenario, there exists a *Station* variant of *setLocation* with the same configuration of parameters. The *Station* method moves all transceivers owned by the station to the indicated location at the same time. Of course, the difference between the *Station* and the *Transceiver* variants of *setLocation* only matters when the station is equipped with multiple transceivers.

One important point regarding transceiver coordinates, which occasionally amounts to a minor nuisance, is that they cannot be negative. This is because those coordinates end up being represented by time values (expressed in *ITUs*), and time in SMURPH cannot be negative. An attempt to use a negative coordinate as an argument of *setLocation*, in any of its standard variants, will be signaled as an error. The nuisance part of this feature surfaces in some mobility models, e.g., when we want the transceivers (nodes) to wander randomly, say, within some area surrounding their initial positions. If the initial position is close to the origin, and the random "delta" turns out negative, the node may end up in a forbidden quadrant of the plane. The nuisance is minor, because it is trivially easy to offset all node locations by some (possibly huge) positive margins to eliminate this problem. For example, in this code:

```
#define  X_OFFSET  10000.0
#define  Y_OFFSET  10000.0
station MovingNode {

    . . .

    void setLocation (double x, double y) {

        . . .

        Station::setLocation (x + X_OFFSET, y + Y_OFFSET);

        . . .

    };

};
```

MovingNode defines its own *setLocation* method which shifts the origin of the network deployment plane away from the boundaries of the allowed quadrant.

void setMinDistance (double d = −1.0);

This method sets the minimum distance between a pair of transceivers. Whenever the model asks for the distance between a pair of transceivers, and the distance resulting from their current coordinates turns out to be less than the declared minimum, then the declared minimum is used instead of the actual distance. This feature is useful, occasionally needed, and natural, considering that the modeled time is discrete and the transceiver locations, despite formally being floating point numbers, are in fact integers aligned to the *ITU* grid (Sect. 4.4). In the real word, the propagation time between two distinct points is never exactly zero, and no two transceivers can possibly occupy the same point in space. If distance zero happens in the model, because of the discretization of locations, then it may be "unphysical," and the model may not be prepared to handle it (e.g., because the action of propagating a signal between two distinct locations would cause no time advance).

Similar to tags, the minimum distance attribute has its channel default, which is applied when the above method is called with no argument, or with a negative argument (interpreted as "unassigned"). This also happens when a transceiver whose minimum distance has never been assigned is being connected to a channel. The default setting of minimum distance, which is good for most occasions, is 1 *ITU* expressed in *DUs*, i.e., the minimum distance that when converted to *ITUs* ends up as nonzero.

Each of the above "set" methods has a corresponding "get" method that just returns the value of the respective attribute without changing it. Except for *getLocation*, such a method takes no arguments and returns a value of the pertinent type. For *getLocation*, the header is:

void getLocation (double &x, double &y);

or

void getLocation (double &x, double &y, double &z);

depending on the selected dimensionality of the network deployment space. The method returns the two or three coordinates of the transceiver (in *DUs*) via the arguments. There exists a *Station* variant of the method (with the same header) which returns the location of the first transceiver of the station. Note that normally all transceivers of the same station are (or should be) at the same location, although this is not formally required.

Owing to the fact that location coordinates are quantized to the *ITU*, in the following sequence:

st->setLocation (x, y, z);
st->getLocation (x, y, z);

the values returned by the second call need not be the same as those passed in the first one. This is because *getLocation* reverses the transformation done by *setLocation*, which has converted the argument coordinates to *ITUs*.

Similar to ports (Sect. 4.3.3), this transceiver method:

double distTo (Transceiver *p);

returns the distance (in *DUs*) from the current (*this*) transceiver to *p*. This distance
is undefined (the method returns *Distance_inf*), if the two transceivers belong to
different rfchannels. This restriction may appear questionable, because, in contrast to
ports, transceiver locations are interpreted independent of their associated rfchannels,
The special value returned by the method when the rfchannels differ can be used as
a predicate to tell whether two given transceivers belong to the same rfchannel.

   In addition to *getPreamble*, this *Transceiver* method:

TIME getPreambleTime ();

provides a shortcut producing the amount of time (in *ITUs*) needed to transmit the
preamble, based on its current length, as defined for the transceiver. This is equivalent to multiplying the value returned by *getPreamble* by the transceiver's (current)
transmission rate.

   To determine the total amount of time needed to transmit a given packet (without
actually transmitting it), use this *Transceiver* method:

TIME getXTime (Packet *p);

   The returned value tells the number of *ITUs* needed to transmit the packet pointed
to by *p*. Note that the calculation is (at least potentially) more complex than in the
case of *Port* (where it boils down to a simple multiplication of the packet's length
(*TLength*) by *XRate*, Sect. 7.1.2) and involves *RFC_xmt* (Sect. 4.4.3).

   Many attributes of transceivers can be set globally, i.e., to the same value for all
transceivers interfaced to the same rfchannel, by calling the corresponding method of
RFChannel rather than the one of *Transceiver*. Except for *setLocation*, all the "set"
methods listed above have their *RFChannel* variants. Such a variant can be used to
one of these ends:

1. To set the attribute in all transceivers connected to the channel to the same value,
   and make that value a new channel default. This happens when the channel
   method is invoked with a "definite" argument.
2. To set the attribute in all transceivers to the current channel default (without
   changing that default). For this, the channel method should be called with no
   argument, or with the "unassigned" value of the argument.

   For example, if *CH* points to an *RFChannel*, this invocation:

CH->setPreamble (18);

will set the preamble length for all transceivers interfaced to the channel to 18 bits
and redefine the channel's default for preamble length to 18. When, subsequently,
the same method is called with no argument:

CH-> setPreamble ();

it will set the preamble length in all transceivers connected to the channel to 18 (leaving the channel default intact). The same idea applies to these methods: *setXRate*, *setXPower*, *setRPower*, *setXTag*, *setRTag*, *setTag*,[11] *setMinDistance*, which were discussed above, as well as *setErrorRun* and *setAevMode*, to be introduced in Sects. 8.2.1 and 8.2.4. The *RFChannel* variants of the "set" methods return no values (they are declared as *void*).

One method of *RFChannel*, potentially useful in the network configuration stage, which has no *Transceiver* counterpart, is:

double getRange (double &lx, double &ly, double &hx, double &hy);

and its 3d version:

double getRange (double &lx, double &ly, double &lz, double &hx, double &hy, double &hz);

The 2d variant returns via the four arguments the lower-left and upper-right coordinates of the rectangle encompassing all the transceivers interfaced to the channel, according to their current locations. The coordinates are in *DUs* and represent the extreme coordinates over all transceivers. Thus, *lx* is the smallest *x* coordinate, *ly* is the smallest *y* coordinate, *hx* is the highest *x* coordinate, and *hy* is the highest *y* coordinate. The 3d variant adds the third coordinate to those values, in the obvious way. The method returns as its value the length of the rectangle's diagonal (it is the space diagonal of the rectangular cuboid in the 3d case).

Following the initialization stage, SMURPH (the core model) builds the neighborhoods of transceivers based on the indications of *RFC_cut*. The neighborhood concept is defined in the transmitter-to-receiver direction. By the neighborhood of a transceiver *T*, we mean the population of all transceivers that lie within the cutoff distance of *T*. Intentionally, the cut-off distance is a (conservative) estimate of the maximum distance on which a transmitting transceiver can affect (not effect) a reception at a receiving transceiver. The primary purpose of this concept is to reduce the computational complexity of the model. If not sure, one can start with large neighborhoods (e.g., a trivial *RFC_cut* returning the constant value *Distance_inf*) and then narrow them down experimentally checking how this reduction affects the observed behavior of the network. In any case, the only penalty for large neighborhoods is a longer execution time and, possibly, increased memory requirements of the simulator.

Reference

1. P. Gburzyński, I. Nikolaidis, Wireless network simulation extensions in SMURPH/SIDE, in *Proceedings of the 2006 Winter Simulation Conference (WSC '06)*, Monterey, 2006

---

[11]The method is a shortcut for *setXTag* + *setRTag* executed in sequence with the same argument.

Chapter 5
The Execution Framework

In this chapter, we introduce the essential tools of SMURPH, agnostic to the intended applications of the package, and constituting the base (or the common denominator) of all SMURPH programs. These tools define an execution framework where the flow of time is marked by responding to events.

5.1 Processes

A program in SMURPH takes the form of a collection of cooperating threads called *processes*. Each process can be viewed as an event handler: its processing cycle consists of being awakened by some event, responding to the event by performing some *activities*, and going back to sleep to await the occurrence of another event.

5.1.1 Activity Interpreters: The Concept

An activity interpreter, or AI for short, is an object belonging to a (possibly derived) SMURPH type rooted under *AI* in Fig. 3.1. Objects of this type are responsible for triggering events that can be perceived by processes, and for advancing the virtual time. To await an event, a process issues a *wait* request addressed to some *AI*. Such a request specifies a class of events that the process would like to respond to. The occurrence of an event from this class will wake the process up.

A typical example of an *AI* is a port. A process, say P_1, may issue a wait request to a port, say p_1, e.g., to be awakened as soon as a packet arrives at p_1. Another process, P_2, may start a packet transmission on port p_2 connected to the same link. This *activity* exhibited by P_2 will be transformed by the link into an event that will wake up process P_1 when, according to the propagation distance between p_1 and p_2, the packet has materialized at p_1.

© Springer Nature Switzerland AG 2019   207
P. Gburzyński, *Modeling Communication Networks and Protocols*,
Lecture Notes in Networks and Systems 61,
https://doi.org/10.1007/978-3-030-15391-5_5

Different *AIs* use different means (methods) to absorb the relevant activities exhibited by processes, but each of them makes essentially the same single method available to the protocol program as a tool for awaiting events. Its header looks like this:

void wait (etype *event, int state, LONG order = 0);*

where *event* identifies an event class (the type of this argument is *AI*-specific), and *state* indicates the process state to be assumed when the awaited event occurs. The last argument (*order*) is optional and seldom used. Its role is to assign priorities to awaited events. When multiple events occur at the same time (within the same *ITU*), those priorities determine the order in which the events are presented. The third parameter of *wait* can only be specified if the program has been compiled with the *–p* option (Sect. B.5).[1]

5.1.2   Declaring Process Types

A process is an object belonging to type *Process*. A process consists of its private data (attributes) and a possibly shared code[2] describing the process's behavior. To create a non-trivial process, i.e., one that has some code to run, one must define a subtype of *Process*. Such a definition must obey the standard set of rules (Sects. 3.3.3 and 3.3.4) and begin with the keyword *process*. Typically, it has the following layout:

```
process ptype : itypename ( ostype, fptype ) {

    ...

    local attributes and methods

    ...

    states { list of states };

    ...

    perform {

        the process's code

    };

};
```

where "ostype" is the type name of the station owning the process and "fptype" is the type name of the process's *father*, i.e., the creating process (see Sect. 5.1.3). Every process is created by some other (previously existing) process, deemed the process's father, and belongs to some station. Two standard process attributes, F and S, point to the process's father and to the station owning the process. They are implicitly declared within the process type as:

---

[1]Because the order argument is seldom used, "normally" compiled code does not implement it for the reasons of efficiency. According to what we said in Sect. 1.2.3, the most natural treatment of multiple events scheduled at the same time slot is to process them in a nondeterministic order.

[2]By multiple processes/instances of the same process type.

ostype *S;
fptype *F;

i.e., the arguments "ostype" and "fptype" determine the type of the attributes F and S. If the parenthesized part of the process type declaration header is absent, the two attributes are not defined. This is legal if the process makes no reference to them.[3] If only one argument is specified within parentheses, it is interpreted as the station type. In such a case, the S pointer is defined, while F is not.

A *setup* method can be declared for a process type (Sect. 3.3.8) to initialize the local attributes upon process creation. The default *setup* method provided in type *Process* is empty and takes no arguments.

The definition of the process code starting with the keyword *perform* resembles the declaration of an argument-less method. The code within the method is partitioned into states whose names must be announced with the *states* declaration in and can be any C++ identifiers. These identifiers are formally interpreted as values (enumeration constants) of type *int*.

As for a regular method, the body of a code method can be defined outside the process type, in the following way:

```
ptype::perform {

   ...

};
```

In such a case, the process type declaration must contain an announcement of the code method in the form:

perform;

The reason why an announcement is needed is that not every process is obliged to have a code method. Sometimes a process type is introduced solely as a basis to define other process types, ones that will bring in their own code methods. In particular, a process type declared as *virtual* (Sect. 3.3.4) is not even allowed to offer a code method.

A process type T_2 derived from another process type T_1 naturally inherits all the attributes and methods of T_1, including the *setup* method(s) and the code method. Of course, these methods can be redefined in T_2, in which case they override the corresponding methods of T_1. Although an overridden *setup* method of the parent type T_1 can be referenced from T_2, there is no natural way to reference the overridden code method of the parent. A process without a code method can be run, provided that a code method is declared somewhere in its supertype. That method may use *virtual* functions adapting its behavior to the code-less subtype (see Sect. 2.3.2 for examples).

---

[3] While many processes reference their stations, few of them reference their fathers.

5.1.3   Creating and Terminating Processes

A process is created like any other *Object* (Sect. 3.3.1) with the *create* operation
(Sect. 3.3.8). All activities of the protocol program are always carried out from within
a process; therefore, a (user-defined) process is always created by another (known)
process. This creating process becomes the father of the created process and can
be referenced via its *F* attribute (Sect. 5.1.2). Similarly, every process belongs to a
specific station, which is pointed to by *TheStation* (Sect. 4.1.3) whenever the process
is run. The contents of *TheStation* at the time of a process creation determine the
station that will own the new process. That station can also be referenced by the
process's *S* attribute (Sect. 5.1.2). The latter way of referencing the station owning
the process is usually more convenient, because *S* can be declared to be of the proper
station subtype, while *TheStation* is of the generic type *Station* and (usually) must
be explicitly cast to be useful. Note, however, that the types of *S* and *F* are declared
statically along with the process type declaration, and their declared types are not
enforced at process creation. For example, processes of the same type can be created
by fathers of multiple types and belong to stations of multiple types. Arguably, such
scenarios are seldom useful, but they are not completely unheard of. *S* and *F*, if
defined at all, always point to the station owning the process and to the process's
creator, but whether the types of these pointers agree with those of the underlying
objects is up to the program to ensure.

One more way to get hold of the station owning a process is to invoke this *Process*
method:

*Station *getOwner ();*

which returns a pointer to the station that owns the process. The method is not very
useful from within the process itself (*S* and *TheStation* provide better access to the
same object), but can be applied when a process pointer is available, and the reference
is from another process.

A process can terminate itself by calling this method:

void terminate ();

or by executing

terminate;

There is no formal difference when the operations are carried out from the process
that wants to terminate itself, but the first option (being a method) can be also invoked
to terminate another (any) process, given its pointer, e.g.,

p->terminate ();

One more option is to call a global function with this header:

*terminate (Process *p);*

which can also be used to terminate the invoking process in this somewhat contrived
way:

::terminate (this);

A terminated process ceases to exist. If the termination is by the process itself, then any statements in the process code following the termination statement are ignored (the call to *terminate* does not return). If the process is terminated by another process, then, by the very nature of process execution in SMURPH, the process being terminated must be inactive waiting for some event. Such a process simply disappears as an object and its lists of pending wait requests is erased.

SMURPH maintains a hierarchy of objects, based on their ownership/creation context, whose most important role is to make the objects locatable by DSD for exposures (Sect. C.5.1). It may happen that a terminated process has a non-empty set of descendants, which may include children processes and/or other exposable objects that have been created by the process during its lifetime. Such objects are linked to the process through the "father-child" relationship (attribute *F*) and show up as its descendants in the hierarchy of exposable objects (Sect. C.5.1). When a process whose descendant list is nonempty is being terminated, the descendants are moved to the process's father, i.e., the process originally responsible for its creation. Then, if the descendant is a process, its father (the *F* attribute) is set to the new ancestor. This may cause problems, because the type of the new ancestor need not be the same as the type of the previous (original) one, and it may not agree with the declared type of *F*. Therefore, if such situations can legitimately arise at all, the program must take precautions to prevent confusion.

A non-process descendant of a terminated process (e.g., an *RVariable* or *EObject*) is simply moved to the new ancestor. Normally, a process that terminates itself would take precaution to clean up its dependents (the objects it has created), although sometimes it may be difficult, especially when those dependents are processes. While it is illegal to use standard C++ constructors for a process, standard destructors are OK. A process that expects to be terminated unexpectedly (by some other process) can use its destructor to make sure that the unnecessary dependents and other dynamic objects are always cleaned up properly.

There exists a way to terminate all processes owned by a specific station. This is accomplished by executing *terminate* as a *Station* method, e.g., as in the following sequence:

...

```
state Reset:
    S->terminate ( );
    sleep;
```
...

The role of this operation is to offer a way to reinitialize (reset) a station by erasing all its processes and starting from a clean slate. When a process is terminated this way, all its non-process descendants are deallocated, and its children processes are moved to *Kernel* (Sect. 5.1.8) as its new father. This is usually very temporary: such children are typically owned by the same station, so they will be terminated as well; however, it is technically possible for a process running at station *A* to create a process at station *B* (e.g., by explicitly indicating the new owner, Sect. 3.3.8). Such tricks may lead to a confusion, so they should be applied with diligent care.

As the deallocation of the descendants of a process killed by the *Station* variant of *terminate* follows the execution of the process's destructor, there will be no confusion when the process itself deallocates all or some of them upon termination (from its destructor). Note that the process executing *S->terminate ()* will be terminated as well by this operation. As in many cases the operation must be followed by other statements, e.g., ones that start up a new set of processes, it does not automatically return from the process (in contrast to the *Process* variant of *terminate*). One should remember, however, that following *terminate*, the current process has been deallocated as a C++ object and formally is no more (even, if it still can execute statements). Any references to its methods and/or attributes may result in errors. The sequence of operations that the process can safely execute after *S->terminate ()* is thus restricted. Preferably the process should invoke some method of its station, or a global function, to carry out any non-trivial actions required by the reset. But note that the process should not try to reference *S* (even though a reference to an attribute of the just-deallocated process may statistically succeed, it is formally dangerous); thus, the only recommended, safe way to access the process's owning station is via *TheStation*.

There is a way to locate processes by their type names and owning stations, which feature is not used very often, but it does provide a powerful last-resort means to find processes and account for them all, e.g., to terminate them as part of a cleanup operation, even if no explicit pointers to them are available. This operation:

Long getproclist (Station *s, ptype, Process **PL, Long Len);

stores in the array of pointers represented by the third argument the list of pointers to processes described by the first two arguments. The last argument gives the length of the array of pointers. The operation looks like a global function, but, strictly speaking, it isn't a C++ function, because the second argument is the name of a process type. The operation is intercepted by *smpp* (Sects. 1.3.2 and B.1) and transformed into a call to some internal and rather obfuscated function.

If the first argument is not *NULL*, then it should point to a station. The processes located by the operation will belong to that station. The second argument can be absent, in which case the operation's header will look like this:

Long getproclist (Station *s, Process **PL, Long Len);

and the operation will return all processes owned by the station pointed to by the first argument. Finally, if the first argument is also *NULL*, it can be skipped like this:

Long getproclist (Process **PL, Long Len);

This variant is truly desperate, because it will return pointers to all processes run by the simulator.

The array of pointers is filled up to the specified length (*Len*). The number of process pointers stored in it is returned as the function value. If that number is equal to *Len*, then there may be more processes to be returned. For illustration, here is the (not very efficient but effective) code to kill all copies of process *Consumer* run by the current station:

Process *PT [1];

...

while (getproclist (TheStation, Consumer, PT, 1))

      PT [0] -> terminate ( );

    A call to *getproclist* is rather costly, as the operation may end up traversing a significant part of the entire object ownership hierarchy (Sect. C.5.1) before it locates all the indicated processes.

5.1.4 Process Operation

Each process operates in a cycle consisting of the following steps:

1. the process is awakened by one of the awaited events
2. the process responds to the event, i.e., it performs some operations prescribed by its code
3. the process puts itself to sleep.

    Before a process puts itself to sleep, it usually issues at least one wait request (Sect. 5.1.1), to specify the event(s) that will wake it up in the future. A process that goes to sleep without specifying a waking event is terminated. The effect is the same as if the process issued *terminate* as its last statement (Sect. 5.1.3).

    To put itself to sleep, a process can execute the following statement:

sleep;

or simply exit by falling through the end of the list of statements at the current state (or by the closing brace of the code method).

    By issuing a wait request (using the *wait* method of an *AI*, Sect. 5.1.1), a process identifies the following four elements:

- the activity interpreter responsible for triggering the event
- the event type
- the process state to be assumed upon the occurrence of the event
- the order (priority) of the awaited event.

    The first three items are mandatory. The last one can be left unspecified (as it mostly is) in which case the default order of 0 is assumed. Note that the specified order can be negative, i.e., lower than the default.

    If before suspending itself (i.e., going to sleep) a process issues more than one wait request, the multiple wait requests are interpreted as an alternative of waking conditions. It means that when the process becomes suspended, it will be waiting for a collection of event types, possibly coming from different *AIs*, and the occurrence of the earliest of those events will resuscitate the process. When a process is restarted, its collection of awaited events is cleared, which means that in each operation cycle these events are specified from scratch.

A process may undo its already issued collection of wait requests by calling this function:

void unwait ();

This operation is only useful in special circumstances and must be applied with care, as some wait requests may be implicit (and they will be also undone by *unwait*). For example, undoing the implicit wait request issued by *transmit* (Sect. 7.1.2) seldom makes sense. We saw a useful example of *unwait* in Sect. 2.4.3.

It is possible that two or more events, from the collection of events awaited by the process, occur at the same time (within the same *ITU*). In such a case, the event with the lowest order is selected and this event restarts the process. If multiple events occurring at the same time have the same lowest order (e.g., the default order of 0), the waking event is chosen nondeterministically from amongst them.[4] In any case, whenever a process is awakened, it is always due to exactly one event.

Most protocols can be programmed without assuming any presumed ordering of events that may be scheduled for the same *ITU*. Generally, the nondeterminism inherent in such scenarios is useful and tends to enhance the model's fidelity by statistically exploring all the potential paths of execution (see Sect. 1.2.4). Therefore, for the sake of efficiency, the possibility to specify the third argument of *wait* is switched on with the *–p* compilation option of *mks* (Sect. B.5). Without this option, *wait* accepts only two arguments, and all events (except for the cases mentioned in Sects. 5.2.2 and 9.2.4) have the same order (equivalent to 0).

On the other hand, it is also possible to force a completely deterministic processing of multiple events scheduled at the same *ITU*. This is accomplished by setting the *–D* option of *mks* (Sect. B.5). When this option is in effect, all events of the same order occurring at the same *ITU* are additionally ordered by the absolute timing (ordering) of the corresponding wait requests (first awaited events receive lower order). This feature is usually selected together with the real mode of operation (option *–R*) to make sure that a program controlling a real physical system behaves deterministically.

The state identifier specified as the second argument of a wait request is an enumeration object of type *int*. When the process is awakened (by one of the awaited events), its code method is simply called (as a regular function) and the state identifier is presented in the global variable:

int TheState;

which can be used by the process to determine what has happened.

A newly created process is automatically and immediately restarted (within the *ITU* of its creation) with the value of *TheState* equal 0. Value 0 corresponds to the first symbolic state declared with the *states* command (Sect. 5.1.2).[5] This first waking event is assumed to have arrived from the process itself (Sect. 5.1.7). Its default

---

[4]Exceptions are discussed in Sects. 5.2.1, 7.1.5, 8.2.2 and 9.2.4.

[5]Note that this is the first item on the *states* list, not the first *state* listed in the code method.

order is 0; therefore, multiple processes created within the same *ITU* may be run in an unpredictable order. If truly needed, the following *Process* method can be used to make this order predictable:

void setSP (LONG order);

The method can be called by the creator immediately after executing *create* (which returns a *Process* pointer) or, preferably, from the created process's *setup* method. The argument of *setSP* specifies the order of the first waking event for the new process. The program must have been compiled with −*p* (the order attribute must be enabled) for *setSP* to be available.

5.1.5   The Code Method

A process code method is usually programmed in a way resembling a finite state machine. An awakened process gets into a specific state and performs a sequence of operations associated with that state. By issuing wait requests, the process dynamically specifies the transition function, which tells it where to go from the current state upon the occurrence of interesting events.

The standard layout of the process code method is this:

```
perform {
       state S₀:

             ...

       state S₁:

             ...

       state Sₙ₋₁:

             ...

};
```

Each *state* statement is followed by a symbol identifying one state. This symbol must be declared in the process's *states* list (Sect. 5.1.2). The dots following a *state* statement represent instructions to be executed when the process wakes up in each state. The interpretation of the contents of *TheState* (Sect. 5.1.4) is automatic. When the list of instructions associated with the given state is exhausted, i.e., the code method attempts to fall through the state boundary, the process is automatically put to sleep.

The code method does not absolutely have to organize itself the standard way. For example, a single-state code method, like the one of the Callback process in the PID controller (Sect. 2.3.2) can be simplified with no reference to states at all, e.g.,

```
process Callback abstract (Plant) {
        states { Loop };
        virtual void action (int) = 0;
        perform {
                action (Loop);
                sameas Loop;
        };
};
```

at some (arguable) loss of legibility, like the apparent irrelevance of the state name
(*Loop*). Any sequence of actual *state* statements in a code method can be replaced
by a sequence of *if* statemens, or by a *case* statement, referencing *TheState* and
comparing it to the state names (interpreted as integer constants).

Keyword *transient* can be used instead of *state* to allow the state to be entered
directly from the preceding state. In other words, if the list of statements of the state
preceding a *transient* state is exhausted (and it does not end with sleep), the *transient*
state will be entered, and its statements will be executed.

Here is a sample process declaration:

```
process AlarmClock {
        TIME delay;
        Mailbox *sgn;
        setup (TIME d, Mailbox *m) {
                delay = d;
                sgn = m;
        };
        states { Start, GoOff };
        perform {
                state Start:
                        Timer->wait (delay, GoOff);
                state GoOff:
                        sgn->put ( );
                        proceed Start;
        };
};
```

This process can be viewed as an alarm clock that sends out a signal every *delay*
time units (*ITUs*). The semantics of operations *Timer->wait*, *sgn->put*, and *proceed*
will be discussed in Sects. 5.3.1 and 9.2.3.

The code method of *AlarmClock* can be rearranged into:

```
perform {
        state GoOff:
                sgn->put ( );
        transient Start:
                Timer->wait (delay, GoOff);
};
```

to (essentially) retain the original semantics, while making the code a bit shorter (no *proceed* statement is needed). Even though *GoOff* now precedes *Start* in the above code, *Start* is still the initial state, because it occurs first on the *states* list in the declaration of *AlarmClock*.

Strictly speaking, the above code is not *exactly* equivalent to the original, because the removal of *proceed* eliminates some non-determinism in responding to the events (see Sect. 5.3.1). Guessing at the purpose of *AlarmClock*, and its possible context within the entire program, the simplification is probably OK.

Here is yet another way to program the code method:

```
perform {
        if (TheState == GoOff)
                sgn->put ( );
        Timer->wait (delay, GoOff);
};
```

Note that the statement *sgn->put ()* must be executed on every turn of the code method (at its every run), except for the first one in state *Start*. More formally, any statements appearing before the first *state* statement are executed always, i.e., on every invocation of the code method, regardless of the value of *TheState*. Following the first *state* command, the rest of the code method is forced into a sequence of states automatically selected by *TheState* (effectively amounting to a *case* statement).

5.1.6 The Process Environment

By the process environment we understand a collection of variables readable by the process code method and describing some elements of the process context, or carrying some information related to the waking event. Two such variables: *TheStation* (Sect. 4.1.3) and *TheState* (Sect. 5.1.4) have already been discussed. Two more variables in this category are:

Time

of type *TIME*. This variable tells the current virtual time measured in *ITUs*.

TheProcess

of type *Process*. This variable points to the object representing the process currently active. From the viewpoint of the process code method, the same object is pointed to by *this*.

In many cases, when a process is awakened by an event, the event carries some information (besides just being triggered). For example, for an event caused by a packet, it would be reasonable to expect that the packet in question will be available to the awakened process. To this end, there exist additional environment variables which are set by some events, before those events are presented to processes, to convey to the process event-specific information.

As the *actual* execution of code methods by the SMURPH kernel is sequential, the single (global) set of event-related environment variables is OK, as long as those variables are set to the right values before a process is awakened by a specific event. Technically, there are exactly two actual environment variables of this kind declared as:

*void *Info01, *Info02;*

Note that they are generic pointers, and they are mostly used to pass pointers to objects related to the event, e.g., a packet pointer for an event triggered by a packet. The two variables always suffice, which means that the amount of (extra) information presented to the process by any event is limited to what can be stored in two pointers.

The above variables are practically never used directly; there exist numerous aliases (macros) that cast them to the proper types, corresponding to the types of the data items returned by specific event types. For example, the following standard macro declares an alias for the occasion of receiving a packet pointer along with an event:

#define ThePacket ((Packet)Info01)*

This, in effect, brings in an environment variable, *ThePacket*, available to a process code method whenever it receives an event triggered by a packet. This and other environment variables will be discussed in the context of the respective activity interpreters and their events. We can generally forget that they are macros aliasing just two actual variables, because multiple aliases pointing to the same variable never collide within the realm of a single wake-up scenario. One should remember, however, that the values returned by event-related environment variables are only guaranteed to be valid immediately after the respective event is received (i.e., the process wakes up in the corresponding state). This also applies to the dynamic objects pointed to by the variables, e.g., the packet object for a packet-related event (Sects. 7.1.6 and 8.2.3).

5.1.7   Process as an Activity Interpreter

Type *Process* is a subtype of *AI* (Sect. 3.3.1) which suggests that a process can appear as an activity interpreter to another process. For one facet of this feature, it is possible for a process, say P_1, to issue a wait request to another process, say P_2, to be awakened when P_2 enters a specific state. The first argument of *wait* in this case is an integer enumeration value identifying a process state. The process P_1 restarted by P_2 entering the state specified in P_1's wait request is awakened *after* P_2 is given control, i.e., when P_2 completes its state and puts itself to sleep, but still within the same *ITU*. The environment variables *Info01* and *Info02* (Sect. 5.1.6) presented to P_1 look exactly as they were when P_2 was completing its state. Note that P_2 can modify them to pass a message to P_1.

A process P_1 can await the termination of another process P_2 by waiting for the *DEATH* event[6] on P_2. With the possibility of awaiting the termination of another process, one can easily implement a subroutine-like scenario in which a process creates another process and then waits for the termination of the child. For example, with the following sequence of statements:

pr = create MyChild;
pr->wait (DEATH, Done);
sleep;

the current process spawns a child process and goes to sleep until the child terminates. When it happens, the parent will resume its execution in state *Done*.

In addition to *DEATH* of a specific process, a process can await the termination of any of its immediate children by issuing a wait request to itself with *CHILD* as the first argument. Here is a sample sequence to make sure that all the current process's children have terminated before proceeding:

```
...
state WaitChildren:
        if (children ( ) != 0) {
                wait (CHILD, WaitChildren);
                sleep;
        }
...
```

The *children* method of *Process* returns the number of processes appearing in the descendant list of the process. This can be viewed as the formal definition of the children set. There are circumstances (see Sect. 5.1.3) when a process that has not been explicitly created by a given process may nonetheless show up as its descendant. Note that any process can await the termination of any process's child. For *DEATH*, as well as for *CHILD*, the environment variables *Info01* and *Info02* presented to the awakened process look as they were last seen by the process that has triggered the

---

[6]Non-state events are represented by negative numbers, e.g., *DEATH* is defined as -1, so they are never confused with state events.

event. A process terminated by the *terminate* method of its owning station (aka the station reset, Sect. 5.1.3), triggers neither a *DEATH* nor a *CHILD* event.

The first waking event for a process, generated after the process's creation (Sect. 5.1.3), to run the process for the first time in its initial state, is formally issued by the process itself. This event is called *START*, and it cannot be awaited explicitly.

Following up on the scenario discussed at the beginning of the present section, it is possible for a process to wait for itself entering a specific state, although this possibility may appear somewhat exotic. For example, consider the following fragment of a code method:

...

```
state First:
        TheProcess->wait (Second, Third);
        proceed Second;
state Second:
        sleep;
state Third:
```

...

When the process wakes up in state *Second*, it apparently dies, as it does not issue any wait requests in that state (Sect. 5.1.4). However, in the previous state, the process declared that it wanted to get to state *Third* as soon as it entered *Second*. Thus, after leaving state *Second* the process will wake up in state *Third*, within the current *ITU*. The effect is as if an implicit *proceed* operation to state *Third* were issued by the process at *Second*, just at the moment that state is entered.

Note that *proceed* is just a special case of a wait request (Sect. 5.3.1) and, if the process issues other wait requests that are fulfilled within the current *ITU*, the (implicit or explicit) *proceed* may be ineffective, i.e., the process may end up in another state. The order argument of the state wait request can be used to assign a higher priority to this transition (should the problem become relevant).

The range of the wait request issued in state *First* is just one state ahead. According to what we said in Sect. 5.1.4, all pending wait requests are cleared whenever a process is awakened, so the state wait request to itself is an exception in some sense.

In the code fragment presented above, the call to *wait* at state *First* does not have to be preceded by *TheProcess->*, as the wait request is addressed to the current process. It is suggested, however, to reference all *AI* methods with explicit remote access operators. In the above example, *this* used instead of *TheProcess* would have the same effect.

5.1.8   The Root Process

From the user's perspective, all activities in SMURPH are processes. One system process is created by SMURPH immediately after the protocol program is run, and exists throughout its entire lifetime. This process is pointed to by the global variable

Kernel of type *Process* and belongs to the *System* station (Sect. 4.1.2). Before the execution is started, *Kernel* formally creates one process of type *Root* which must be defined by the user. This process is the root of the user process hierarchy. The *Root* process is always created with an empty *setup* argument list.

The *Root* process is responsible for building the network and starting the processes of the protocol program. Then it should put itself to sleep awaiting the end of execution which is signaled as the termination of *Kernel*. For a control program (as opposed to a simulator) the end of execution event may never be triggered, so the *Root* process is free to assume other duties (or even terminate itself). For example, the *Root* process of the PID controller discussed in Sect. 2.3.2 issues no wait request in its only (startup) state; thus, it effectively terminates itself (it will never run again).

As a side note, there is a subtle difference between an actual termination (with *terminate*, Sect. 5.1.3) and an "effective" termination (by failing to issue a wait request). In the latter case, the process is formally retained as an object and keeps its place in the ownership tree (Sect. C.5.1).

The following structure of the *Root* process is recommended for a typical simulator:

```
process Root {
        states { Start, Done };
        perform {
                state Start:
                        ... read input data ...
                        ... create the network ...
                        ... define traffic patterns ...
                        ... create protocol processes ...
                        Kernel->wait (DEATH, Done);
                state Done:
                        ... print output results ...
        };
};
```

Of course, the individual initialization steps listed at state *Start* can be carried out by separate functions or methods, e.g., see Sect. 2.2.5.

The termination of *Kernel* that indicates the end of the simulation run is fake; in fact, *Kernel* doesn't die and, for example, can be referenced (e.g., exposed, see Sect. 12.5.10) in state *Done* of the above *Root* process, after its apparent death. On the other hand, *Kernel* can only be killed once, and its *DEATH* event can only happen at most once.

The user can define multiple termination conditions for the run (Sect. 10.5). Meeting any of those condition will cause *Kernel* to be formally killed effecting the soft termination of the simulation experiment. It is also possible to do that directly from the program by executing:

Kernel->terminate ();

i.e., by formally killing the *Kernel* process.

One more potentially useful event that can be awaited on the *Kernel* process (and only on that process) is *STALL*, which it is triggered when the simulator runs out of events. All processes waiting for *STALL* on *Kernel* will then be restarted within the current *ITU*. The *STALL* event is not available in the real mode (because SMURPH cannot possibly know whether and when an event will arrive from an external device), and in the visualization mode (because the lack of progress in the model is then visualized).

The two-state layout of *Root* is often natural, but not the only possible one. Sometimes there is no pronounced stopping action, e.g., in a control program (see Sect. 2.3.2), which operates like a never-ending daemon. This can also be the case in a simulator whose purpose is to visualize the system in action, rather than collect some performance measures from a run of a definite duration. Sometimes, as mentioned in Sect. 2.1.3, a third state becomes useful, reflecting the need to mark a steady state of the model for starting the collection of performance measurements. One important constrain is that the initialization (at least its first step taking place in *Root's* first state) must be done by the *Root* process itself (which, e.g., cannot start another process and delegate that step to it), because by completing its first state (going to sleep for the first time) *Root* declares that it is done with the construction of the virtual hardware. This is exactly when the simulator decides that the virtual hardware has been built: when the *Root* process goes to sleep for the first time. If *Root* created another process to take care of the job, then it would necessarily leave its first state and go to sleep before that other process got a chance to enter its initial state.

The function invoked by SMURPH when the *Root* process completes its first state is available to the protocol program, which may decide to call it before its implicit automatic invocation. The (global) function is:

void rootInitDone ();

and its call can appear anywhere within the sequence of statements issued in the first state of *Root*. Following its execution, any additions to the system configuration, like creating new stations, links, ports, rfchannels, transceivers, and traffic patterns, are illegal. Note that it only makes sense to make the call from the first state of *Root*; once the function has been invoked, all its subsequent invocations are void (and it is always invoked automatically at the end of *Root's* first state).

The reason why the function is available to the user program is that it marks the moment when the system configuration operations issued in the first state of *Root* become truly effective. Some elements of the network configuration and the traffic generator (the *Client*) are not completely built until *rootInitDone* gets called. The final touch of their construction is postponed, even though all the requisite operations have been issued (thus creating a somewhat premature sense of completion), because SMURPH wants to have a full view of the network configuration before finalizing the details of the virtual environment. For example, the operations defining links (Sect. 4.2.2), rfchannels (Sect. 4.4.2), and (at least some) traffic patterns (Sect. 6.6.2) are incremental. Ports are added to the link one-by-one, the elements of the distance matrix are set, and so on, the link's status remaining in a dangling state until it is clear that no further changes (additions) will be made. New senders and/or receivers can be added to the sets of stations associated with traffic patterns until someone says that they are ready to go. On occasion, one may want to put into the initialization code, executed in the first state of *Root*, operations assuming that things have been fully set up, e.g., to start a packet transmission.[7] In a situation like this, the completion of the network setup can be forced, via *rootInitDone*, before the *Root* process exits its first state.

5.2 Signal Passing

In one of its guises as an activity interpreter, a process can also be viewed as a repository for signals providing a simple means of process synchronization. Other, more powerful, inter-process communication tools are mailboxes described in Sect. 9.1.

5.2.1 *Regular Signals*

In the simplest case, a signal is sent to (we say deposited at) a process by calling the following *Process* method:

*int signal (void *s = NULL);*

where the argument can point to additional information to be passed along with the signal. If the argument is absent, no signal-specific information is passed. It is then assumed that the occurrence of the signal is the only event of interest. Formally, the signal is deposited at the process whose signal method is invoked. Thus, by executing:

prcs->signal ();

the current process deposits a signal at the process pointed to by *prcs*, whereas each of the following two calls:

signal (ThePacket);
TheProcess->signal ();

deposits a signal at the current process.

A deposited signal can be perceived by its recipient via wait requests addressed to the process. A process can declare that it wants to await the occurrence of a signal by executing the following method:

prcs->wait (SIGNAL, where);

---

[7]Inserting a token packet into a token ring is an example that comes to mind.

where *prcs* points to the process at which the awaited signal is expected to arrive, and *where* identifies the state where the waiting process wants to be awakened upon the signal arrival. Note that the signal does not have to be deposited at the waiting process. For example, it can be deposited at the sender or even at a third party which is neither the sender nor the recipient of the signal.

If the signal is pending at the time when the wait request is issued (i.e., it has been deposited and not yet absorbed), the waking event occurs within the current *ITU*. Otherwise, the event will be triggered as soon as a signal is deposited at the corresponding process.

When a process is awakened due to a signal event, the signal is removed from the repository. This only happens when the waking event is in fact a signal event. Thus, when multiple events occur (and are scheduled) at the same *ITU*, and one of them happens to be a signal event, the signal repository will be cleared only if the signal event ends up being the one actually presented to the waking process. Two environment variables (Sect. 5.1.6) are then set: *TheSender* (of type *Process**) points to the process that has sent (deposited) the signal, and *TheSignal* (of type *void**) returns the value of the argument passed to *signal* by the sender.

It is possible that the process responsible for sending a signal is no longer present (it has terminated itself or has been terminated by some other process) when the signal is received. In such a case, *TheSender* will point to the *Kernel* process, which can be used as an indication that the signal sender has terminated. Generally, any event triggered by a process that terminates before the event is presented is assumed to have been triggered by *Kernel*.

Only one signal can remain pending at a process at a time, i.e., multiple signals are not queued. The *signal* method returns an integer value that tells what has happened to the signal. The following return values (represented by symbolic constants) are possible:

ACCEPTED	The signal has been accepted and there is a process already waiting for the signal. The signal has been put into the repository and a waking event for the awaiting process has been scheduled. Note, however, that this event is not necessarily the one that will wake up the process (the process may have other waking events scheduled at the same *ITU*)
QUEUED	The signal has been accepted, but nobody is waiting for it at the moment. The signal has been stored in the repository, but no event has been triggered
REJECTED	The repository is occupied by a pending (not yet accepted) signal, and the new signal has been rejected. The *signal* operation has no effect in this case

It is possible to check whether a signal has been deposited and remains pending at a process. The following *Process* method can be used to this end:

int isSignal ();

The method returns *YES* (1) if a signal is pending at the process, and *NO* (0) if the process's signal repository is empty. In the former case, the environment variables *TheSender* and *TheSignal* are set as described above. The signal is left in the repository and remains pending. The following process method:

int erase ();

behaves like *isSignal*, except that the signal, if present, is removed from the repository.

A process can also wait for a signal repository to become empty. This is accomplished with the following wait request:

prcs->wait (CLEAR, where);

where, as before, *prcs* identifies the owner of the repository. The *CLEAR* event is triggered by *erase*, but only if the repository is nonempty when *erase* is invoked.

The flexibility in selecting the signal repository makes it possible to create signal-based communication scenarios that are best suited for the application. For example, in a situation when a single process P_0 creates several child processes $P_1, \ldots, P_k$ and wants to pass a signal to each of them, it is natural for P_0 to deposit each signal at the corresponding child. On the other hand, if each of the child processes $P_1, \ldots, P_k$ is expected to produce a message for P_0, the most natural place for a child process to deposit a signal carrying such a message is in its own repository. In the first case, the signal operation issued by P_0 will be addressed to a child and the child will issue a wait request to itself; in the second case, the child will execute its own *signal* method and P_0 will issue a signal wait request to the child.

5.2.2 Priority Signals

Sometimes one would like to give a signal event a higher priority, to make sure that the signal is received immediately, even if some other waking events are scheduled at the same *ITU*.[8] While nondeterminism in handling multiple events scheduled for the same time slot is generally beneficial from the viewpoint of modeling (Sect. 1.2.4), signal passing is often used for programming convenience (to coordinate control switching among processes) rather than for expressing events in the modeled system. Thus, one may want to treat (some) signals as special events triggered and accepted beyond the scope of the model. One way to accomplish this is to assign a very low order (Sect. 5.1.1) to the wait request for the signal event. This solution requires a cooperation from the signal recipient.[9] It is also possible for the signal sender to indicate that the signal should be received immediately, irrespective of the order of the corresponding wait request at the recipient. Such a communication mechanism can be interpreted as a "dangling branch" from one process to another and is called a *priority event*. A signal priority event is triggered by calling the following method:

---

[8] Similar problems are discussed in Sects. 7.1.5, 8.2.2 and 9.2.4.

[9] And the program must have been compiled with the *-p* option (Sect. B.5).

int signalP (void *s = NULL);

where the meaning of *s* is the same as for the regular *signal* operation (Sect. 5.2.1). Like *signal*, *signalP* deposits a signal at the process (in the same single-signal repository). The following additional rules apply to *signalP*:

- At most one *signalP* operation can be issued by a process from the moment it wakes up until it goes to sleep. In fact, a somewhat stronger statement is true. Namely, at most one operation generating a priority event (including the *putP* operation introduced in Sect. 9.2.4) can be issued by a process in any single state. In other words: at most one priority event can remain pending at any time.
- At the time when *signalP* is executed, the process's signal repository should be empty, and there must be exactly one process waiting for the signal event. In other words: the exact deterministic continuation of the process generating the signal must be known at the time when *signalP* is issued.

The situation when nobody is waiting for a priority signal (the second condition does not hold) is not treated as a hard error. The value returned by *signalP* indicates whether the operation has succeeded (the method returns *ACCEPTED*), or failed, i.e., nobody has been waiting for the signal (in which case the method returns *REJECTED*). On the other hand, if the target signal repository is nonempty when the operation is issued, the error is hard, and the program is aborted.

As soon as the process that executes *signalP* is suspended, the waiting recipient of the signal is restarted. No other process is run in the meantime, even if there are some other events scheduled at the current *ITU*.

Note that the recipient of a priority signal does not declare in any special way that it is awaiting a priority event: it just executes a regular signal wait request (Sect. 5.2.1) with any (e.g., default) order. The fact that the transaction is to be processed as a priority event is indicated solely by *signalP*, and the order of the matching wait request is irrelevant. In Sect. 9.2.4, we give an example of a situation where priority events can be useful.

5.3   The Timer AI

Although formally the *Timer AI* occurs in a single copy, it in fact provides an unlimited number of independent alarm clocks that can be set and responded to by protocol processes. This way the processes can explicitly advance the internal notion of time (wait until the clock moves by the prescribed amount). For simulation (running in the virtual mode) this advancement is conceptual; in the real (or visualization) mode, this operation involves actual, real-time delays.

5.3.1   Wait Requests

By invoking the following method of *Timer*:

void wait (TIME delay, int where);

or

void wait (TIME delay, int where, LONG order = 0);

the process declares that it wants to be awakened in state *where*, *delay ITUs* after the current moment. The delay can also be specified in *ETUs* with this method used instead of *wait*:

void delay (double etus, int state);
void delay (double etus, int state, LONG order = 0);

Timer ->delay (e, st) is equivalent to *Timer->wait (etuToItu (e), st)*.

   Implicit wait requests to the *Timer AI* are issued by some operations, those resulting in events scheduled after calculable intervals (e.g., see Sect. 7.1.2). One important example is this operation:

proceed newstate;

used to transit directly (without awaiting any explicit event) to the state indicated by the argument. This operation is basically equivalent (with a minor exception explained below) to:

Timer->wait (TIME_0, newstate);
sleep;

Thus, the transition is performed in response to an event that wakes up the requesting process after the delay of zero *ITUs*, i.e., within the current *ITU*. This is not the same as a straightforward "go to" the indicated state, because other events scheduled at the same current *ITU* can preempt the transition. If the order option (*–p*) is compiled in (Sect. B.5), then the order of the *Timer* wait request implied by *proceed* is 0. Any wait request with a negative order scheduled within the current *ITU* will take precedence over *proceed*, and any request with the same order (scheduled for the current *ITU*) stands a chance to preempt it at random. The function of *proceed* can be described as an event-less, zero-delay, state transition accounting for the non-determinism in processing actions occurring within the indivisible grain of virtual time (Sect. 2.1.4). Even if the process executing the *proceed* statement does not await any (other) events that may occur within the current *ITU*, other processes whose events are scheduled within the current *ITU* may be run (nondeterministically, or based on the *order* attribute of their requests) before the transition takes place.

   If the program has been compiled with *–p*, the transition implied by *proceed* can be assigned an explicit order different from the default of zero, e.g.:

proceed newstate, order;

where *order* can be any expression evaluating to type *LONG*. The value of that expression will be used as the third argument of the implicit *Timer* wait request issued by the operation. The command syntax admits parentheses around the arguments, i.e.:

proceed (newstate);
proceed (newstate, order);

Another operation resembling *proceed* is:

skipto newstate;

which is equivalent to:

Timer->wait (TIME_1, newstate);
sleep;

and performs a transition to *newstate* with the minimum possible, nonzero delay of 1 *ITU*. This operation is useful for aging out events that persist until time is advanced (see Sect. 2.2.3). The syntax of *skipto* is like that of *proceed*, e.g., the command accepts the order argument, if the program has been compiled with –*p*, as well as the optional parentheses around the argument(s).

If the nondeterminism/uncertainty of *proceed* is not acceptable, unnatural, or not needed, then this statement:

sameas newstate;

carries out a truly atomic and perfectly straightforward, immediate transition to the indicated state, effectively amounting to a "go to." Almost the same effect can be obtained (albeit in a slightly less elegant way) by inserting a label immediately after the *state* statement of the destination state and replacing *sameas* with *goto* specifying that label as the target. Thus, a sequence like this:

```
...
state TargetState:
        ... do something ...
state AnotherState:
        ... do something else ...
        sameas TargetState;
...
```

is (almost) equivalent to:

```
...
state TargetState:
label_1:
        ... do something ...
state AnotherState:
        ... do something else ...
        TheState = TargetState;
        goto label_1;
...
```

We say "almost," because there are two additional actions performed by *sameas* (as opposed to *goto*):

1. setting *TheState* to the target state
2. undoing any wait request issued by the process before executing *sameas* (see the *unwait* operation in Sect. 5.1.1).

The extra actions make the operation appear like a formal state transition (in response to some hypothetical event). In particular, the process starts in the new state with a clean slate of pending wait requests.

 For compatibility with *proceed* and *skipto*, *sameas* accepts optional parentheses around its (single) argument, but accepts no order attribute (which would make no sense). The compatibility has its limitation: while *proceed* and *skipto* can be invoked from a function or method called by the process (with the state passed as an argument), *sameas* can only be executed directly from the process's code method.

 When a process is restarted by a *Timer* event, the two environment variables (Sect. 5.1.6) are both set to *NULL*, i.e., *Timer* events carry no environment information. When a process wakes up in response to *proceed*, the environment variables have the same contents as at the end of the previous state (at the time when *proceed* was executed). This is the only difference between *proceed* and a *Timer wait* for *TIME_0*. However, for a transition by *skipto*, the environment variables are nullified, as for a straightforward *Timer* event. This is because even the tiniest time advance may cause object pointers that were valid one *ITU* ago (and were accessible via the environment variables) not to be valid any more. The environment variables are not affected at all by *sameas*.

5.3.2 Operations

Issuing a *Timer* request (Sect. 5.3.1) can be viewed as setting up an alarm clock or a timer. Unless the process is awakened by an earlier (or simultaneous) event, the alarm clock will make sure to restart the process after the specified delay.

 A process can inquire about an alarm clock set up by another process and also reset that alarm clock to a different delay. For this purpose, alarm clocks are identified by their target states.[10] The following method of *Timer*:

*TIME getDelay (Process *prcs, int st);*

returns the delay of the timer set up by the indicated process that, when (and if) it goes off, will move the process to the state indicated by the second argument. If no such alarm clock is currently pending for the process (the process does not await a *Timer* event with a transition to the specified state), the method returns *TIME_inf* (i.e., "undefined", Sect. 3.1.3). The value returned upon success gives the current

---

[10]A single process may await several different *Timer* events at the same time specifying different target states (regardless how little sense that makes), but only one event per target state.

(i.e., remaining) delay of the timer, rather than the original value for which the alarm clock was set.

A timer can be reset with the following *Timer* method:

*int setDelay (Process *prcs, int st, TIME value);*

where the third argument specifies the new delay value. The method will succeed (and return *OK*) only if an alarm clock restarting the process in the indicated state is set for the process. Otherwise, the method returns *ERROR* and has no effect. This means that one cannot introduce new timers this way, i.e., an alarm clock can be reset from the outside, but it cannot be set up. There is no variant of *setDelay* that would accept the delay argument in *ETUs*.

The state argument of *getDelay* and *setDelay* must specify a symbolic state name of the process indicated by the first argument (not the numerical value of the state enumeration constant). The first argument must be a process pointer of the proper type, i.e., it cannot be a generic *Process* pointer, because the state name can only be interpreted within the context of a specific process type. These rules imply that *getDelay* and *setDelay* are not straightforward C++ methods. Their invocations are intercepted by *smpp* (Sects. 1.3.2 and B.2) and transformed into calls to some obfuscated internal functions.

If the original alarm clock was set with the variant of *wait* specifying the order of the timer event (Sect. 5.3.1), this order will be retained when the alarm clock is reset by *setDelay* (the original order setting cannot be changed).

5.3.3   Clock Tolerance

There exists only one global *Timer AI*, which would suggest that all stations in the network use the same notion of time and their clocks run in a perfectly synchronized fashion. This may make sense in a control program driving a real physical system (where the different stations represent some conceptual blocks of a de facto centralized system), but is not very realistic in a simulation model (where the stations act as independent components of some hardware). In a realistic simulation model of a physical system, we often need to simulate the limited accuracy of local clocks, e.g., to study race conditions and other phenomena resulting from imperfect timing. Thus, SMURPH makes it possible to specify the *tolerance* (variability) for clocks used by different stations. This idea boils down to randomizing and optionally offsetting all delays specified as arguments of *Timer* wait requests according to the declared *tolerance* parameters. These adjustments also affect implicit wait requests to the *Timer*, e.g., issued by *transmit* (Sect. 7.1.2).

The extent to which the indications of timers (delays measured by them) used by different stations can differ from the absolute delays (as measured by the central, undisturbed, event clock of the simulator) is controlled by three parameters:

Deviation	This is the maximum fraction by which the effective delay calculated by the *Timer* may differ from the specified (absolute) delay. The difference is randomized and can be applied in both directions
Quality	This parameter affects the distribution of *deviation*, as explained below
Drift	This is a constant bias factor included in the effective delay, accounting for the clock's drift

Suppose that a *wait* request is issued to the *Timer* for *delay ITUs*. The effective delay (expressed in absolute time) is calculated as:

$$effective\_delay = delay * (1.0 + dRndTolerance\ (-deviation, deviation, quality) + drift);$$

The above expression is a bit simplified (compared to one actually applied by the *Timer*), because of the need to account for the difference in types (*TIME, double*) used in the expression in the most efficient and safe way, but it captures the essence. Both *deviation* and *drift* are fractional (extra) contributions to the effective delay, with the actual deviation generated as a β-distributed random number between $-deviation$ and $+deviation$ (see Sect. 3.2.1).

For illustration, suppose that *deviation* is 0.00001, corresponding to 10 ppm (parts per million) and *drift* is 0. When a process running at the station executes:

Timer->wait (2000000, 0);

the effective delay will be anything between 1,999,980 and 2,000,020, because 10 ppm of 2,000,000 is 20. If *drift* is set to 0.000005 (i.e., 5 ppm), the effective delay will be between 1,999,990 and 2,000,030, because 5 ppm of 2,000,000 is 10, and this value is added to the delay after accounting for the deviation. The way *dRndTolerance* (and β distribution) works (Sect. 3.2.1), the value of *quality* does not affect the range of the effective delay, but it influences the shape of the distribution. With higher values of *quality*, the randomized deviation tends to be more narrowly focused around zero.

Owing to the discrete nature of time, small delays may be affected in a crudely approximate way, or not affected at all, by the tolerance adjustments. For example, with the same *deviation* of 10 ppm, a delay of less than 50,000 *ITUs* will not be disturbed at all. This is because 10 ppm of 50,000 is 0.5, i.e., less than one *ITU*. The adjustment involves rounding, so anything above 0.5 will end up as 1, but any delta below 0.5 *ITU* will be ignored. This is yet another argument for a fine *ITU*. If the experiment calls for a high fidelity study of phenomena depending on the variability of clocks, the *ITU* must be defined in such a way that the shortest relevant delays issued by the program will be realistically (up to the model's expectations) affected by the finite accuracy of independent clocks (see Sect. 1.2.4). Again, the imprecise terms "relevant" and "expectations" can only be made more precise in the context.

The tolerance parameters can be specified globally, during the network initialization stage (Sect. 5.1.8), before creating stations. Each call to this global function:

void setTolerance (double deviation = 0.0, int quality = 2);

(re)defines *deviation* and *quality* to be applied to all stations created henceforth, until overridden by a new definition, i.e., another call to (global) *setTolerance*. Zero *deviation* results in non-randomized clocks. This is the default assumed before *setTolerance* is called for the first time. Note that nonzero *drift* applied together with zero *deviation* will cause a fixed fractional offset applied to all delays.

The *drift* parameter can only be defined on a per-station basis, i.e., its default value is always zero. This makes sense, because clocks that exhibit the same drift do not drift with respect to each other, and the only interesting aspect of a (constant) drift is that it is different for different clocks. The clock tolerance parameters can be defined individually for a given station by calling this *Station*method:

void setTolerance (double deviation = −1.0, int quality = −1, double drift = 0.0);

The negative (default) value of any of the first two arguments indicates that the present setting of the corresponding parameter should not be changed. To only set the drift, this shortcut is recommended:

void setDrift (double drift);

which is formally equivalent to *setTolerance (−1.0, −1, drift)*, but slightly more efficient. In some high-fidelity models, it may be useful to model dynamic drifts, e.g., changing as a function of temperature. This can be accomplished by having a dedicated process invoking periodically *setDrift* according to some dynamic rules.

The current setting of the global tolerance parameters can be obtained with this global function:

double getTolerance (int *quality = NULL);

which returns the current global *deviation* setting. If the argument is not *NULL*, the *quality* parameter is stored under the indicated location. The corresponding *Station* method is declared as:

double getTolerance (int *quality = NULL, double *drift = NULL);

and offers one more pointer to return the station's drift setting.

5.3.4   The Visualization Mode

When a simulation experiment is run in the visualization mode, the virtual timing of events is synchronized to real time, such that the user can witness the model's real-time behavior, possibly scaled down (slow motion) or up (accelerated motion). The visualization mode is available to any simulation program that has been compiled with −W (Sect. B.5). It must be explicitly entered by the program and can be exited at any time. When the visualization mode is not active, the simulator executes in a purely virtual manner, with time being entirely abstract and bearing no relationship to the running time of the program.

The visualization mode is entered by calling this (global) function:

void setResync (Long msecs, double scale);

where *msecs* is the so-called resync interval (expressed in milliseconds of real time) and *scale* maps that interval to virtual time. For example, if *msecs* is 500 and scale is 0.25, the resync interval is 500 ms (0.5 s) and that interval will map to 0.25 *ETUs* of simulated time. If, say, 1 *ETU* corresponds to 1 virtual second, the visualization will be carried out in slow motion with the factor of 2 (it will take 0.5 s of real time to present 0.25 s of virtual time).

The way it works is that every resync interval the simulator compares its virtual clock (variable *Time*, Sect. 5.1.5) with the real time using the scale factor for conversion. If the virtual time tries to advance faster than real time, the simulator will hold (sleep internally) until the real time catches up. The program cannot do much if the real (execution) time advances faster than the simulated time. In such a case, the visualization may be poor or even completely misleading.

Note that the simulator uses its absolute virtual clock for resyncing to real time. Individual (station) clocks running with nonzero tolerance (Sect. 5.3.3) operate with relation to the absolute time, with the notion of global absolute time thus being always well defined, e.g., as the base for resyncing.

Visualization can be stopped at any time by calling *setResync* with the first argument equal zero (the value of the second argument is then irrelevant). It can be stopped and resumed many times with no effect on the model, unless the model accepts input from the outside. Like in the real mode, it is possible to feed a visualized model with external data, or make it generate data feeding external programs, e.g., through bound mailboxes (Sect. 9.4). Since such a program pretends to run in real time, it can drive graphic displays and/or react with the environment as a make-believe replica of the real system. For all this to appear realistic, the real-time execution of the simulator must be faster (or at least not slower) than the scaled progress of the model in virtual time.

5.4 The Monitor AI

Like *Timer*, the *Monitor AI* occurs in a single instance pointed to by the global constant *Monitor*. It provides a simple, crude, and generic inter-process synchronization tool, primarily intended for synchronizing processes operating "behind the scenes" of the protocol program, e.g., library modules, complex traffic generators, or traffic absorbers. Its true object-oriented counterpart, recommended for higher-level solutions, is provided by barrier mailboxes (Sect. 9.3).

The *Monitor* makes it possible to await and signal simple events represented by pointer-sized numbers (bit patterns). The range of those events is global. This means that any process can await any number, and any process can also signal any number. When a number is signaled, all the processes waiting for it are restarted. To be restarted by a signaled number, a process must be waiting for it already when the number is signaled (*Monitor* signals are never queued).

5.4.1   Wait Requests

There is a single type of wait request that identifies the awaited event. The *Monitor* *wait* method has this header:

void wait (void *event, int where);

or

void wait (void *event, int where, LONG order = 0);

depending on whether the program has been compiled without or with –*p* (Sect. B.5). The type of the event argument is *void**, because one important application of monitor signals is to implement semaphores, and it is natural to use addresses of the "critical" variables or data structures as the event identifies. Note that those identifiers are global, so the uniqueness and consistency of events intentionally restricted to the context of, say, a single station can only be maintained by using different numbers. To this end, SMURPH defines this macro:

#define MONITOR_LOCAL_EVENT(o) ((void*)(((char*)TheStation) + (o)))

to be used for creating event identifiers intended to be local with respect to the current station. For example, by declaring the following values:

#define EVENT_XMIT MONITOR_LOCAL_EVENT(0)
#define EVENT_RCV MONITOR_LOCAL_EVENT(1)
#define EVENT_DONE MONITOR_LOCAL_EVENT(2)

one can use them as three different event identifiers guaranteed to be unique within the context of the current station, at least in all those decent situations where the number of the different events (the argument of the macro) is not larger than the size of the station object.

5.4.2   Signaling Monitor Events

The operation of triggering a monitor event looks like sending a signal to a process, except that the signal is addressed to the (global) monitor. This is accomplished by invoking this *Monitor* method:

int signal (void *event);

where the argument identifies the event being triggered. All the processes waiting for the specified event number will be restarted within the current *ITU*. The method returns the number of processes that have been waiting for the signal. Zero means that the operation has been void.

When a process is resumed due to the occurrence of a monitor event, its environment variable *TheSignal* (of type *void**) returns the event number, while *TheSender* (of type *Process*) points to the process that has sent the signal (executed the *signal* operation). This is the same as for process signals (Sect. 5.2.1).

Similar to the priority signals for processes (Sect. 5.2.2), this *Monitor* method:

int signalP (void *event);

issues the signal as a priority event. The semantics is exactly as described in Sect. 5.2.2. The method returns *ACCEPTED*, if there was a (single) process awaiting the signal, and *REJECTED* otherwise. The situation when more than one process awaits the signal is treated as an error.

One use of monitor signals is to implement locks or semaphores, i.e., a mechanism for forcing one process to wait for something to be accomplished by another process. A typical scenario involves shared resource usage. For illustration, suppose that there is a variable shared by two processes (e.g., declared at their owning station) whose value indicates the availability of a shared resource. A process trying to get hold of the resource may execute this code:

```
...
state GetResource:
            if (S->lock)
      {
            Monitor->wait ((void*)&(S->lock), GetResource);
            sleep;
      }
      S->lock = 1;
      ...
```

and the operation of releasing the resource may look like this:

```
S->lock = 0;
Monitor->signal ((void*)&(S->lock));
...
```

SMURPH defines two macros wrapping the above sequences into handy one-liners:

varwait (var, old, new, st);
varsignal (var, old);

The first macro checks if the value of variable *var* equals *old*. If that is the case, it issues a monitor wait request for the variable's address to state *st* and sleeps the process. Otherwise, it sets the variable to *new* and continues. The second macro sets the variable to *old* and issues a monitor signal with the variable's address used as the event.

Chapter 6
The Client

The *Client AI*, which can be viewed as a union of all *Traffic AIs*, is the agent modeling the network's application and responsible for providing it with traffic (at the sender's end) and absorbing the traffic (at the recipient's end). It is intended practically exclusively for simulators, possibly including some simulated components in hybrid systems controlled by SMURPH programs.

6.1 General Concepts

The traffic in a modeled network consists of *messages* and *packets*. A message represents a logical unit of information that arrives to a station from the application and must be sent to some destination. From the viewpoint of the protocol program, a message is characterized by the sender (i.e., the station at which the message arrives for transmission), the receiver (i.e., the station to which the message is to be sent), and the length (i.e., the number of bits comprising the message).

The process of message arrival is defined by the user as a collection of traffic patterns. Each traffic pattern is described by the distribution of senders and receivers and the distribution of message interarrival time and length. A single simulation experiment may involve multiple and diverse traffic patterns.

Each traffic pattern is managed by a separate activity interpreter called a *Traffic*. That *AI* is responsible for generating messages according to the distribution parameters specified by the user, queuing them at the sending stations, responding to inquiries about those messages, triggering waking events related to the message arrival process, and absorbing messages at their final destinations. The *Client AI* is a union of all

© Springer Nature Switzerland AG 2019
P. Gburzyński, *Modeling Communication Networks and Protocols*,
Lecture Notes in Networks and Systems 61,
https://doi.org/10.1007/978-3-030-15391-5_6

Traffic AIs. For illustration, assume that a protocol process P at some station S wants to check whether there is a message queued at S that belongs (has been generated according) to traffic pattern T. This kind of inquiry will be addressed to one specific traffic pattern (a *Traffic AI*). By issuing the same kind of inquiry to the *Client AI*, P will check whether there is any message (belonging to any traffic pattern) awaiting transmission at S. Similarly, P may decide to go to sleep awaiting the arrival of a message of some specific pattern, or it may await any message arrival at all. In the former case, the wait request will be addressed to the respective *Traffic* object, which will be also responsible for delivering the awaited event. In the latter case, the wait request will be addressed to the *Client*, and the waking event will also arrive from there.

Each *Traffic AI* has its private message queue at every user-defined station. A message arriving to a station is queued at the end of the station's queue handled by the *Traffic AI* that has generated the message. Thus, the order of messages in the queues reflects their arrival time.

If the standard tools provided by SMURPH for defining the behavior of a *Traffic AI* are insufficient to describe a refined traffic pattern, the user can program this behavior as a collection of dedicated processes.

Before a message can be transmitted over a communication channel (a link or an rfchannel), it must be turned into one or more packets. A packet represents a low-level unit of information (a frame) that, besides the payload bits inherited from a message, usually carries some additional information required by the protocol (headers, trailers, etc.).[1]

A typical execution cycle of a protocol process responsible for transmitting packets at a station consists of the following steps:

1. The process checks if there is a message to be transmitted, and if so, acquires a packet from that message and proceeds to 3.
2. If there is no message awaiting transmission, the process issues a wait request to the *Client* or some specific *Traffic AI*, to await a message arrival, and puts itself to sleep. When the waking event is triggered, the process wakes up at 1.
3. The process transmits the packet (having obeyed the rules prescribed by the protocol).
4. When the transmission is complete, the process continues at 1.

The operation of acquiring a packet for transmission consists in turning a message (or just a portion of it) into a packet. It may happen that the protocol imposes some rules as to the maximum (and/or minimum) packet length. In such a case, a single long message may have to be split into several packets before it can be transmitted.[2]

---

[1]This interpretation assumes that messages and packets are used to model traffic in a communication network. It is possible to use messages and packets to model other things, e.g., boxes traveling through conveyor belts. In such scenarios, the headers and trailers (as well as some other message/packet attributes) may be irrelevant and need not be used.

[2]This also applies to many nonstandard interpretations of messages and packets, say, a message representing a truckload of boxes (packets) to be transported via a conveyor belt.

SMURPH collects some standard performance measures, separately for each *Traffic* and globally for the *Client* (i.e., for all traffic patterns combined). The user can declare private performance-measuring functions that augment or replace the standard ones.

6.2   Message and Packet Types

SMURPH defines two standard types *Message* and *Packet* that provide bases for defining protocol-specific message and packet types. These types are *not* subtypes of *Object* (Sect. 3.3.1).

Message types are defined according to the rules described in Sects. 3.3.3 and 3.3.4. The declaration of a message type starts with the keyword *message*. For example, the following declaration:

```
message RMessage {
        int Route [N], RouteLength;
};
```

defines a message type called *RMessage*, which is derived directly from the base type *Message*. Messages are very seldom generated "by hand" in the protocol program, although it is possible. In such a case, the message type may declare a *setup* method and the regular *create* operation can be used to generate an instance of the message type. Note that the version of *create* that assigns a nickname to the created object is not applicable to a message (or a packet), because nicknames (and, generally, standard identifiers) only apply to *Objects* (see Sects. 3.3.1 and 3.3.2). The default *setup* method defined for class *Message* takes no arguments and its body is empty.

The declaration of a new packet type obeys the standard set of rules (Sects. 3.3.3 and 3.3.4) and starts with the keyword *packet*. For example, the following declaration defines a packet type that may correspond to the message type *RMessage* declared above:

```
packet RPacket {
        int Route [N], RouteLength;
        void setup (RMessage *m) {
                for (RouteLength = 0; RouteLength < m->RouteLength; RouteLength++)
                        Route [RouteLength] = m->Route [RouteLength];
        };
};
```

The *setup* method copies the *Route* attribute of the message from which the packet is built into the corresponding packet attribute. A packet *setup* method specifying a *Message* subtype pointer as its argument type is treated in a special way (see below).

Typically, packets are created automatically from messages by the *Client* when a protocol process acquires a new packet. Sometimes it is desirable to create a nonstandard packet directly. In such a case, the regular *create* operation (without a nickname) can be used to explicitly create a packet object. For such occasions, a packet type may declare a *setup* method (or a collection of *setup* methods) to initialize its attributes upon creation. For standard packets, i.e., those built automatically by the *Client*, one *setup* method has a special meaning. A user-defined packet type being a non-trivial extension of *Packet* (i.e., specifying some additional attributes) may want to declare the following *setup* method:

void setup (mtype **m);*

where "mtype" is a message type, i.e., a subtype of *Message*. This method will be called whenever a packet is acquired from a message, and its role will be to set the non-standard attributes of the packet based on the relevant (non-standard) attributes of the message. The argument *m* will then point to the message from which the packet is being extracted.

The standard type *Message* declares a few public attributes that may be of interest to the user, because they are sometimes useful for programming non-standard traffic generators. Here is the set of *public* attributes of the *Message* class:

Message *next, *prev;
Long Length;
IPointer Receiver;
int TP;
TIME QTime;

Attributes *next* and *prev* are used to store the message in a queue at the sending station. They have been made publicly visible to facilitate programming non-standard ways of putting messages into queues, extracting them, and turning them into packets. Attribute *next* points to the next message in the queue, and *prev* points to the previous message in the queue. For the last message in the queue, *next* contains *NULL*. The *prev* attribute of the first message in the queue points to a nonexistent dummy message whose next attribute coincides with the queue pointer (see Sect. 6.6.5).

The *Length* attribute stores the message length. For the standard interpretation of messages and packets, this length is in bits. The *Length* of a message generated by the standard *Client* is guaranteed to be a strictly positive multiple of 8.

Attribute *Receiver* contains the *Id* (Sect. 4.1.2) of the station to which the message is addressed. For a broadcast message (Sects. 6.4 and 6.5), *Receiver* points to a data structure representing a group of stations. One special value of the *Receiver* attribute is *NONE*: its interpretation is that the message is not explicitly addressed to any station in particular.

TP is the *Id* attribute of the traffic pattern (the *Traffic* object) that has created the message (see Sect. 6.6.2). The attribute is set automatically for messages generated in the standard way. It is recommended to use negative *TP* values for messages that are generated explicitly (by *create*), to prevent confusing them with messages issued by the *Client*.

QTime tells the time when the message was generated and queued at the sender. This attribute is used in calculating message delay (see Sect. 10.2.1).

The public part of the *Packet* type declaration contains the following attributes:

Long ILength, TLength, Sender;
IPointer Receiver;
int TP;
TIME QTime, TTime;
FLAGS Flags;

Attributes *Receiver*, *QTime*, and *TP* are directly inherited from the message from which the packet has been acquired. *ILength* stores the length of the *information* part of the packet, i.e., the part that comes from the message and carries "useful" information. This part is also called the packet's payload. *TLength* is always greater than or equal to *ILength* and stores the total length of the packet, including any headers and trailers required by the protocol on top of the payload.[3] Both these lengths are in bits, or, more generally, in the same discrete units as the message length.

The *Sender* attribute contains the *Id* of the station sending the packet. Note that for a message, such an attribute would be redundant, as a message is always queued at a specific station, and only that station can be the sender of the message. On the other hand, a packet propagates among stations (often making several hops on its path to the destination), which makes the issue of identifying its sender less obvious.

TTime stores the packet's "top time," i.e., the time when the packet became ready for transmission. This time is used for measuring packet delay (Sect. 10.2.1).

Attribute *Flags* amounts to a collection of bits describing various elements of the packet status. Five bits, numbered 27–31, have a predefined meaning and are discussed elsewhere (Sects. 6.3, 7.1.6 and 7.3). The remaining flags are left for the user. Symbolic constants *PF_usr0*, …, *PF_usr26*, representing integer values from 0 to 26 are recommended for accessing those flags (Sect. 3.2.4).

Non-standard packets, i.e., packets created outside the traffic patterns and the *Client*, should have their *TP* attributes negative or greater than the maximum valid traffic pattern *Id* (Sect. 6.6.2). These two *Packet* methods:

int isStandard ();
int isNonstandard ();

tell a standard packet from a non-standard one. A packet is assumed to be standard (*isStandard* returns 1 and *isNonstandard* returns 0), if it has been obtained from a message generated by one of the traffic patterns (i.e., if its *TP* attribute contains a valid traffic pattern *Id*). Otherwise, the packet is considered non-standard and i*sNonstandard* returns 1, while *isStandard* returns 0.

If the protocol program has been compiled with the –g (or –G) option of *mks* (debugging, see Sect. B.5), each packet carries one additional, public attribute of type *Long* called *Signature*. For packets acquired from the *Client* (via *getPacket*, see Sect. 6.7.1) this attribute is set (globally) to consecutive integer values, starting from 0, which are different for different packets. Intentionally, packet *Signatures* are to be used for tracking packets, e.g., with *observers* (Sect. 11.3). The *Signature* attribute of

---

[3]For example, the packet length determines the amount of time needed to transmit the packet, i.e., insert it into a port or a transceiver at a given transmission rate (see Sect. 7.1.2).

a non-standard packet is *NONE* (if needed, it should be set explicitly by the protocol program).

6.3   Packet Buffers

Typically, when a protocol process wants to acquire a packet for transmission, it polls the *Client* or one of its *Traffic* objects. If a suitable message is queued at the station, the *AI* will build a packet and return it to the process. The packet is stored in a buffer provided by the polling process.

Packet buffers are associated with stations. A packet buffer is just an object of a packet type (i.e., a subtype of *Packet*) declared as a class variable within a station class. For illustration, in the following sequence:

```
packet MyPType {

        ...

};
    ...
station MySType {

        ...

        MyPType Buffer;

        ...

};
```

Buffer is declared as a packet buffer capable of holding packets of type *MyPType*. Packet buffers declared as packet class variables within a station class are viewed as the station's "official" buffers and, e.g., are exposed by the station's buffer exposure (Sect. 12.5.11). In principle, it is also legal, although not very popular, to use as a packet buffer a packet created dynamically (with the *create* operation). When a packet buffer is created this way, the *setup* argument list should be empty.

As a side note, it is illegal to declare a packet class variable (not a packet pointer) outside a station class. A packet class variable can only be declared within a station class, because the packet buffer that it amounts to must be automatically attributed, at the time of its declaration, to some station.

A packet buffer can be either empty or full, which is determined by the contents of the *PF_full* flag (bit number 29) in the buffer's *Flags* attribute (Sect. 6.2). This flag is irrelevant from the viewpoint of the formal packet contents: its meaning is restricted to the interpretation of the packet object as a buffer.[4] These two *Packet* methods:

---

[4]This minor difference between a packet buffer and the packet itself did not seem to warrant introducing a separate data type for representing packet buffers.

```
int isFull ( ) {
        return (flagSet (Flags, PF_full));
};
int isEmpty ( ) {
        return (flagCleared (Flags, PF_full));
};
```

act as predicates telling whether a packet buffer is full or empty. There exist other methods for examining other flags of packets internally used by SMURPH which, for all practical purposes, completely disguise those flags under a friendlier interface.

In most cases, a packet buffer is filled by calling *getPacket* (Sect. 6.7.1); then, the buffer's *PF_full* flag is set automatically. The standard way to empty a packet buffer, is to call the *release* method of *Packet*, as in:

buf->release ();

where *buf* points to a packet. The purpose of *release* is to mark the buffer as empty and to update some standard performance measures related to the completion of the packet's handling by its sender (Sect. 10.2). The method should be called as soon as the contents of the buffer are no longer needed, i.e., the packet is done from the viewpoint of its sender, e.g., it has been completely transmitted, dispatched, acknowledged, discarded, etc., as prescribed by the protocol.

There exist two *Client* versions of *release* which perform the same action as the corresponding *Packet* method. The above call is equivalent to:

Client->release (buf);

or

*Client->release (*buf);*

i.e., the second version takes an object (reference) as the argument rather than a pointer.[5] Finally, similar two methods are available from traffic patterns. If *TPat* points to a traffic pattern, then each of the following two calls:

TPat->release (buf);
*TPat->release (*buf);*

is equivalent to *buf->release ()*. As an added value, they implement assertions to verify that the buffer being released holds a packet belonging to the traffic pattern pointed to by *TPat*.

It is possible to fill the contents of a packet buffer explicitly, without acquiring into it a packet from the *Client*. This is useful for special packets that may be needed by the protocol, which do not originate in the network's application (like acknowledgments). The following *Packet* method serves this end:

*void fill (Station *s, Station *r, Long tl, Long il = NONE, Long tp = NONE);*

---

[5]It is often more natural, especially when issued from a method of a *Station* subtype, because packet buffers are most often declared "statically" (as class variables) at stations.

The first two arguments point to the stations to be attributed at the packet's sender and receiver, respectively. The third argument will be stored in the *TLength* attribute: it contains the total length of the packet. The next, optional, argument gives the length of the packet's "information" part, i.e., the payload. If unspecified (or equal *NONE*), it is assumed that the payload length (attribute *ILength* of the packet) is the same as the total length. It is checked whether the payload length, if specified, is not bigger than the total length. Finally, the last argument will be stored in the *TP* attribute of the packet. It cannot be equal to the *Id* of any defined traffic pattern (it is illegal for a non-standard packet to pose for a *Client* packet).

There exists an alternative version of *fill* which accepts station *Ids* instead of pointers as the first two arguments. Any of the two station *Ids* (or both) can be *NONE*. If the receiver *Id* is *NONE*, it means that the packet is not explicitly addressed to any particular station.

Following a packet reception (Sects. 7.1.6 and 8.2.3), if the received packet should be saved for later use (e.g., retransmission), the program should create a copy of the packet object, rather than saving the pointer to the original object, which will be deallocated when the packet formally disappears from the channel (see Sects. 7.1.1 and 8.1.1). Declaring packet class variables (as opposed to packet pointers) to accommodate packet copies is tricky because a packet class variable can only be declared within a station class (see above) and then it is interpreted as a packet buffer (and receives a special treatment). The most natural and safe way to copy a packet object is to call the following *Packet* method:

*Packet *clone ();*

which creates an exact replica of the packet and returns a pointer to it. This new object can (and should) be deallocated (by the standard C++ operation *delete*) when it is no longer needed. Note that the generic packet pointer presented in *ThePacket* (e.g., after a packet reception) is of the base type *Packet*, while it may point to packets belonging to proper subtypes of *Packet* being objects of various sizes. Thus, the issue of creating accurate replicas of those packets is not completely trivial. The copy constructed by *clone* is always shallow, i.e., the method does not attempt to interpret packet attributes, but accurate, in the sense that it captures the actual, complete packet, as per its proper type.

6.4   Station Groups

Sometimes a subset of stations in the network must be distinguished and identified as a single object. For example, to create a broadcast message, i.e., a message addressed to more than one recipient, one must be able to specify the set of all receivers as a single entity. A station group (an object of type *SGroup*) is just a set of station identifiers (*Id* attributes) representing a (not necessarily proper) subset of all stations in the network.

An *SGroup* is created in the following way:

*SGroup *sg;*

...

sg = create SGroup (... setup arguments ...);

where the "setup arguments" fit one of the following patterns:

( ) (i.e., no arguments)

The communication group created this way contains all stations in the network.

(int ns, int *sl)

If the first argument (*ns*) is greater than zero, it gives the length of the array passed as the second argument. That array contains the *Ids* of the stations to be included in the group. If the first argument is negative, then its negation is interpreted as the length of *sl*, which is assumed to contain the list of exceptions, i.e., the stations that are *not* to be included in the group. In such a case, the group will consist of all stations in the network, except those mentioned in *sl*. The array of station *Ids* pointed to by *sl* can be deallocated after the station group has been created.

(int ns, int n1, ...)

The first argument specifies the number of the remaining arguments, which are station *Ids*. The stations whose *Ids* are explicitly listed as arguments are included in the group.

(int ns, Station *s1, ...)

As above, but pointers to station objects are used instead of the numerical *Ids*.

The internal layout of an *SGroup* is not interesting to the user: there is no need to reference *SGroup* attributes directly. In fact, explicitly created station groups are not used very often.

Note that station groups are not *Objects* (Sect. 3.3.1). They have no identifiers (Sect. 3.3.2) and the nickname variant of *create* (Sect. 3.3.8) is not applicable to them.

The *SGroup* class defines the following two user-accessible methods:

int occurs (Long sid);
int occurs (Station *sp);

which return *YES* (1), if the station indicated by the argument (either as an *Id* or as an object pointer) is a member of the group.

Stations within a station group are ordered (this ordering may be relevant when the station group is used to define a communication group, see Sect. 6.5). If the station group has been defined by excluding exceptions, or with the empty *setup* argument list, the ordering of the included stations is that of the increasing order of their *Ids*. In all other cases, the ordering corresponds to the order in which the included stations have been specified. Notably, a station that has been specified twice counts as two separate elements of the set. It makes no difference from the viewpoint of the station's membership in the group, but is significant when the group is used to define a communication group, as explained in Sect. 6.5.

6.5   Communication Groups

Communication groups (objects of type *CGroup*) are used (explicitly or implicitly) in definitions of refined traffic patterns, i.e., objects of type *Traffic* or its subtypes (Sect. 6.6). A communication group consists of two station groups (Sect. 6.4) and two sets of numerical weights associated with these groups. One group is called the *senders group* and contains the stations that can be potentially used as the senders of a message. Together with its associated set of weights, the senders group constitutes the *senders set* of the communication group. The other group is the *receivers group* and contains the stations that can be used as the receivers. The receivers group together with its associated set of weights is called the *receivers set* of the communication group.

A communication group can be either a *selection group*, in which the station group of receivers specifies individual stations, and a message can be addressed to one of those stations at a time, or a *broadcast group*, in which the receivers group identifies a set of receivers for a broadcast message. A message whose generation has been triggered by such a communication group (see below) is a broadcast message addressed simultaneously to all the stations specified in the receivers set. No weights are associated with the receivers of a broadcast communication group.

To understand how the concept of communication groups works, let us consider a traffic pattern T based on a single communication group G.[6] Let G be a selection group and:

$$S = \langle (s_0, w_0), \ldots, (s_{k-1}, w_{k-1}) \rangle,$$

where s_i denotes a station and w_i its weight, be the set of senders of G. Similarly, by:

$$R = \langle (r_0, v_0), \ldots, (r_{p-1}, v_{p-1}) \rangle,$$

we denote the set of receivers of G. Assume that a message is generated according to T. Among the attributes of this message that must be determined are two station identifiers indicating the sender and the receiver. The sender is chosen at random in such a way that the probability of a station $s_i : (s_i, w_i) \in S$ being selected is equal to:

$$P_S(s_i) = \frac{w_i}{\sum_{j=0}^{k-1} w_j}$$

Thus, the weight of a station in the senders set specifies its relative frequency of being used as the sender of a message generated according to T.

---

[6]Many traffic patterns are in fact defined this way.

Once the sender has been identified (suppose it is station s), the receiver of the message is chosen from the receiver set R in such a way that the probability of station $r_i : (r_i, v_i) \in R$ being selected is equal to:

$$P_R(r_i) = \frac{h_i}{\sum_{j=0}^{p-1} h_j},$$

where:

$$h_i = \begin{cases} v_i & \text{if } r_i \neq s \\ 0 & \text{if } r_i = s \end{cases}$$

In simple words, the receiver is selected at random from R in such a way that the chances of a given station from the receiver set are proportional to its weight. However, the sender station (selected in the previous step) is excluded from the game, i.e., it is guaranteed that the message is addressed to another station. Note that the above formula for P_R makes no sense when the sum in the denominator is zero. In such a case, SMURPH is not able to find the receiver and the execution is aborted with an error message.

If G is a broadcast group, the problem of selecting the receiver is trivial, namely, the entire receiver group is used as a single object and its pointer is stored in the *Receiver* attribute of the message (Sect. 6.2).

The following operation creates a communication group:

$cg = create\ CGroup\ (\dots\ setup\ arguments\ \dots);$

Like for an *SGroup*, the nickname variant of *create* (Sect. 3.3.8) is not available for a *CGroup*. The following configurations of *setup* arguments are possible:

*(SGroup *sg, float *sw, SGroup *rg, float *rw)*
*(SGroup *sg, float *sw, SGroup *rg)*
*(SGroup *sg, SGroup *rg, float *rw)*
*(SGroup *sg, SGroup *rg)*

The first (most general) configuration specifies the senders group (*sg*), the senders weights (array *sw*), the receivers group (*rg*), and the receivers weights (array *rw*). The weights are assigned to individual members of the corresponding station groups according to their ordering within the groups (see Sect. 6.4). Note that if a station appears twice (or more times) within a group, it counts as two (or more) stations and the corresponding number of entries are required for it in the weight array. The effect is as if the station occurred once with the weight equal to the sum of the weights associated with all its occurrences.

With the remaining configurations of *setup* arguments, one set (or both sets) of weights are not specified. In such a case, each station in the corresponding station group is assigned the same weight of $1/n$ where n is the number of stations in the group.

The following two *setup* configurations are used to create a broadcast communication group:

*(SGroup *sg, float *sw, SGroup *rg, int b)*
*(SGroup *sg, SGroup *rg, int b)*

The value of the last argument can only be *GT_broadcast* (-1). Note that no receivers weights are specified for a broadcast communication group. With the second configuration, the default weights (of $1/n$) are assigned to the senders.

The station groups used to build a communication group (passed as arguments of *create*) must not be deallocated for as long as the communication group is in use.

6.6 Traffic Patterns

Traffic patterns are activity interpreters and thus also *Objects* (Sect. 3.3.1). Their union, known as the *Client*, represents the virtual application of the modeled network. The built-in core of a traffic pattern is described by the SMURPH type *Traffic*. That type can be extended to add functionality to the generic, virtual network client.

6.6.1 Defining Traffic Types

The definition of a traffic pattern type obeys the rules introduced in Sects. 3.3.3 and 3.3.4, and has the following format:

traffic ttype : itypename (mtype, ptype) {

    ...

    attributes and methods

        ...

};

where "ttype" is the newly-defined traffic type, "itypename" is the type name of an existing traffic type, and "mtype" and "ptype" are the names of the message type and the packet type (Sect. 6.2) associated with the defined traffic pattern. All messages generated by the traffic pattern will belong to type "mtype," and all packets acquired from this traffic pattern will be of type "ptype." If these types are not specified, i.e., the parentheses together with their contents are absent, they are assumed to be the built-in types *Message* and *Packet*, respectively. If only one identifier appears between the parentheses, it determines the message type; the associated packet type is then assumed to be *Packet*.

The base traffic type *Traffic* can be used directly to create traffic patterns. Messages and packets generated by a traffic pattern of type *Traffic* belong to the basic types *Message* and *Packet*. A new traffic type definition is only required if at least one of the following statements is true:

- the behavior of the standard traffic generator should be changed by replacing or extending some of the standard methods
- the types of messages and/or packets generated by the traffic pattern are proper subtypes of *Message* and/or *Packet*, i.e., at least one of these types needs additional attributes and/or methods on top of those that come with the basic type.

The list of attributes and methods of a user-defined traffic type may contain redeclarations of some virtual methods of *Traffic* and, possibly, declarations of non-standard variables used by these new methods. Essentially, there are two types of *Traffic* methods that can be sensibly overridden by the user's declarations: the methods used in generating messages and transforming them into packets (Sect. 6.6.3), and the methods used for calculating performance measures (Sect. 10.2.3).

6.6.2 Creating Traffic Patterns

Traffic patterns should be built by the *Root* process (Sect. 5.1.8), after the population of stations has been fully defined. A single traffic pattern is created with the *create* operation, in the way described in Sect. 3.3.8. The standard *setup* methods built into type *Traffic* accept the following configurations of arguments:

(CGroup **cgl, int ncg, int flags, …)
(CGroup *cg, int flags, …)
(int flags, …)

The first configuration is the most general one. The first argument points to an array of pointers to communication groups (Sect. 6.5) expected to consist of *ncg* elements. The communication groups describe the distribution of senders and receivers for messages generated according to the new traffic pattern.

The second configuration of arguments specifies only one communication group and provides a shorthand for the popular situation where the array of communication groups contains a single element. With the third configuration, the distribution of senders and receivers is not specified at the time of creation (it can be supplied later, see below), and the definition of the traffic pattern is incomplete. According to the rules for setting up the model, discussed in Sect. 5.1.8, a traffic pattern can be defined incrementally, in several steps. Its definition is completed when the *Root* process finalizes its first state, or when it calls *rootInitDone*, whichever happens first.

The *flags* argument is interpreted as a collection of options represented by the following constants, which can be put together (or'red or added) into a number of useful combinations:

MIT_exp	the message interarrival time is randomized and exponentially distributed (describing a Poisson process)
MIT_unf	the message interarrival time is randomized and uniformly distributed
MIT_fix	the message interarrival time is fixed
MLE_exp	the message length is randomized and exponentially distributed
MLE_unf	the message length is randomized and uniformly distributed
MLE_fix	the message length is fixed
BIT_exp	the traffic pattern is bursty and the burst interarrival time is randomized and exponentially distributed
BIT_unf	the traffic pattern is bursty and the burst interarrival time is randomized and uniformly distributed
BIT_fix	the traffic pattern is bursty and the burst interarrival time is fixed
BSI_exp	if the traffic is bursty (one of the BIT options is set), then the burst size is randomized and exponentially distributed; otherwise, the option is void
BSI_unf	if the traffic is bursty, then the burst size is randomized and uniformly distributed; otherwise, the option is void
BSI_fix	if the traffic is bursty, then the burst size is fixed; otherwise, the option is void
SCL_on	the standard processing for this traffic pattern is switched on, i.e., the traffic pattern will be used automatically to generate messages and queue them at the sending stations (this is the default)
SCL_off	the standard processing is switched off, i.e., the traffic pattern will not automatically generate messages (used if the traffic pattern is to remain permanently inactive, or if the message generation process is to be kept under exclusive user control)
SPF_on	the standard collection of performance measures (Sect. 10.2) will be calculated for the traffic pattern (this is the default); the user can still override or extend the standard methods that take care of this end
SPF_off	no standard performance measures will be collected for this traffic pattern

It is illegal to specify more than one distribution type for a single distribution parameter, i.e., *MIT_exp+MIT_unf*. If no option is selected for a given distribution parameter, this parameter is not defined. For example, if none of *BIT_exp*, *BIT_unf*, or *BIT_fix* is included, the traffic pattern is not bursty. It makes little sense to omit the specification of the message arrival time or the message length distribution, unless *SCL_off* is selected, in which case the user probably defines a special (external) process to take care of message arrival. SMURPH will complain about the incomplete definition of a traffic pattern that is supposed to be automatically run by the simulator.

The number and the interpretation of the remaining setup arguments depend on the contents of *flag*. All these arguments are expected to be *double* numbers.[7] They

---

[7]This is important, because no type conversion is automatically forced for a function/method with a variable-length argument list. Using, say, an *int* argument in place of a *double* one will have the disastrous effect of completely misreading one argument and possibly misaligning (and thus misinterpreting) the remaining ones.

describe the numerical parameters of the arrival process in the following way and order:

- If *MIT_exp* or *MIT_fix* is selected, then the next (single) number specifies the single parameter of the message interarrival time. For *MIT_exp*, this is the mean of the exponential distribution, for *MIT_fix*, it is the fixed interarrival time. In both cases, the time is expressed in *ETUs*. If *MIT_unf* is specified, the next *two* numbers determine the minimum and the maximum message interarrival time (also in *ETUs*).
- If *MLE_exp* or *MLE_fix* is selected, the next number specifies the (mean or fixed) message length (in bits). If *MLE_unf* is selected, the next two numbers determine the minimum and the maximum message length (in bits).
- If *BIT_exp* or *BIT_fix* is selected, the next number specifies the (mean or fixed) burst inter-arrival time in *ETUs* (as for the message arrival process). If *BIT_unf* is specified, the next two numbers determine the minimum and the maximum burst inter-arrival time (also in *ETUs*).
- If *BSI_exp* or *BSI_fix* is selected, the next number specifies the (mean or fixed) burst size as the number of messages contributing to the burst. If *BSI_unf* is selected, the next two numbers determine the minimum and the maximum burst size.

Thus, the maximum number of the numerical distribution parameters is 8. A completely defined traffic pattern describes a message arrival process. For a non-bursty traffic, this process operates according to the following cycle:

1. A time interval t_{mi} is generated at random, according to the message interarrival time distribution. The process sleeps for t_{mi} and then moves to step 2.
2. A message is generated. Its length is determined according to the message length distribution; its sender and receiver are selected using the list of communication groups associated with the traffic pattern (as explained below). The message is queued at the sender, and the generation process continues at step 1.

For a bursty traffic pattern, messages arrive in groups called bursts. Each burst carries a specific number of messages generated according to similar rules as for a regular, non-bursty traffic pattern. The process of burst arrival is run according to this scheme:

1. An integer counter c_{pm} (pending message count) is initialized to 0.
2. A time interval t_{bi} is generated at random according to the burst interarrival time distribution. The process sleeps for t_{bi} and then continues at step 3.
3. A random integer number s_b is generated according to the burst size distribution. This number determines the number of messages to be generated within the burst.
4. The pending message count is incremented by s_b, i.e., $c_{pm} = c_{pm} + s_b$. If the previous value of c_{pm} was zero, a new process is started, which looks similar to the message generation process for a non-bursty traffic pattern (see above) and runs in parallel with the burst arrival process. Each time a new message is generated, the message arrival process decrements c_{pm} by one. When c_{pm} goes down to

zero, the process terminates. Having completed this step, without waiting for the termination of the message arrival process, the burst arrival process continues at step 2.

Typically, the message interarrival time within a burst is much shorter than the burst interarrival time. In particular, it can be 0 in which case all messages of the burst arrive at once, within the same *ITU*. Note that when two bursts overlap, i.e., a new burst arrives before the previous one has dissipated, the new burst adds to the residue of the previous burst.

The procedure of determining the sender and the receiver for a newly-arrived message, when the traffic pattern is based on one communication group, was described in Sect. 6.5. Now we shall discuss the case of multiple communication groups. Suppose that:

$$S = (s_0, w_0), \ldots, (s_{k-1}, w_{k-1}) \text{ and } R = (r_0, v_0), \ldots, (r_{p-1}, v_{p-1})$$

are the sets of senders and receivers of a communication group $G = (S, R)$. If G is a broadcast group, we may assume that all $v_i, i = 0, \ldots, p - 1$ are zeros (they are irrelevant then). The value of:

$$W = \sum_{i=0}^{k-1} w_i$$

is called the sender weight of group G. Suppose that the definition of a traffic pattern T involves communication groups $G_0, \ldots, G_{m-1}$ with their corresponding sender weights $W_0, \ldots, W_{m-1}$. Assume that a message is generated according to T. The sender of this message is chosen in such a way that the probability of a station:

$$s_i^j : \left(s_i^j, w_i^j \right) \in S_j \text{ and } S_j \in G_j$$

being selected is equal to:

$$P_S\left(s_i^j \right) = \frac{w_i^j}{\sum_{l=0}^{m-1} W_l}.$$

Thus, the weight of a station in the senders set specifies its relative frequency of being chosen as a sender, with respect to all potential senders specified in all communication groups belonging to T. Note that the expression for P_S is meaningless when the sum of all W_l's is zero. Thus, at least one sender weight in at least one group must be strictly positive.

Once the sender has been established, the receiver of the message is selected from the receivers set R of the communication group that was used to find the sender. If this communication group happens to be a broadcast group, the entire receivers group is used as the receiver. In consequence, the message will be addressed to all the stations listed in the receivers group, including the sender, if it happens to occur in R. For a non-broadcast message, the receiver is selected from R, as described in Sect. 6.5.

The senders set of a traffic pattern to be handled by the standard *Client* must not be empty: the *Client* must always be able to queue the generated message at a specific sender station. However, the receivers set can be empty in which case the message will not be addressed to any particular station (its *Receiver* attribute will be *NONE*). Packets extracted from such a message and transmitted can still be received (by any process that considers it appropriate), because the absence of a definite receiver attribute only means that the packet will not trigger certain events which are not critical for reception (Sects. 7.1.3 and 8.2.1).[8]

Despite the apparent complexity, the definition of a traffic pattern is quite simple in most cases. For example, the following sequence defines a uniform traffic pattern (in which all stations participate equally) with a Poisson message arrival process and fixed-length messages:

```
...
traffic MyPattType (MyMsgType, MyPktType) { };
MyPattType *tp;
SGroup *sg;
CGroup *cg;
double iart, lngth;
    ...
    readln (iart);
    readln (lngth);
    ...
    sg = create SGroup;
    cg = create CGroup (sg, sg);
    tp = create MyPattType (sg, MIT_exp+MLE_exp, iart, lngth);
    ...
```

MyMsgType and *MyPktType* are assumed to have been defined earlier; *iart* and *lngth* are variables of type *double* specifying the mean message interarrival time and the mean message length, respectively.

To define a simple traffic pattern, like the one above, one does not even have to bother with communication groups. As a matter of fact, any traffic pattern that can be described using explicit communication groups can also be defined without them. With the following methods of *Traffic*, one can construct the sets of senders and receivers one-by-one, without putting them into groups first:

*addSender (Station *s = ALL, double w = 1.0, int gr = 0);*

---

[8]Formally, the reception process can respond to all packets and use arbitrary, programmable criteria to decide whether it should receive a given packet or not. The built-in mechanism of packet addressing merely facilitates a few standard and simple classes of such criteria, thus simplifying the implementation of some standard cases.

addSender (Long sid, double w = 1.0, int gr = 0);
addReceiver (Station *s = ALL, double w = 1.0, int gr = 0);
addReceiver (Long sid, double w = 1.0, int gr = 0);

By calling *addSender* (or *addReceiver*) we add one sender (or receiver) to the traffic pattern, assign an optional weight to it, and (also optionally) assign it to a specific communication group. One exception is a call with the first argument equal *ALL* (an alias for *NULL*) or without any arguments. In such a case all stations are added to the group. The only legal value for *w* is then 1.0; the weights of all stations are the same and equal to 1/*NStations*.

It is possible to use the above methods to supplement the description of a traffic pattern that was created with explicit communication group(s). For *addSender* and *addReceiver*, the explicit communication groups are numbered from 0, in the order of their occurrence on the list that was specified as the first *setup* argument for the traffic pattern.

For *addReceiver*, if *BROADCAST* (−1) is put in place of *w*, it means that the communication group is a broadcast one. This must be consistent with the previous and future additions of receivers to the group.

Using *addSender* and *addReceiver*, the above example can be simplified into the following equivalent form:

```
traffic MyPattType (MyMsgType, MyPktType) { };
MyPattType *tp;
double iart, lngth;
      ...
      readln (iart);
      readln (lngth)
      ...
      tp = create MyPattType (MIT exp+MLE exp, iart, lngth);
      tp->addSender (ALL);
      tp->addReceiver (ALL);
      ...
```

Note that before *addSender* and *addReceiver* are called, the traffic pattern, although formally created, is incompletely specified. If it were left in such a state, SMURPH would detect a problem upon initializing the message generation process for this pattern (when finalizing the system configuration, see Sect. 5.1.8) and complain.

Traffic patterns are *Objects* (Sect. 3.3.1). Their *Id* attributes reflect their creation order. The first-created traffic pattern is assigned the *Id* of 0, the second 1, etc. The global function:

Traffic *idToTraffic (Long id);

converts the traffic *Id* to the *Traffic* object pointer. The global variable *NTraffics* of type *int* contains the number of traffic patterns created so far.

6.6.3 Modifying the Standard Behavior of Traffic Patterns

A traffic pattern created with *SCL_on* (Sect. 6.6.2) is activated automatically, and its behavior is driven by the standard traffic generation process of the *Client* described in the previous section. Traffic patterns created with *SCL_off* can be controlled by user-supplied processes.

A traffic pattern driven by the standard *Client* can also be activated and deactivated dynamically. To deactivate an active traffic pattern, the following *Traffic* method can be used:

void suspend ();

Calling *suspend* has no effect on a traffic pattern already being inactive. An active traffic pattern becomes inactive, in the sense that its generation process is suspended. From then on, the *Client* will not generate any messages described by the deactivated traffic pattern until the traffic pattern is made active again.

The following Traffic method cancels the effect of *suspend* and re-activates the traffic pattern:

void resume ();

Similar to *suspend*, *resume* has no effect on an active traffic pattern. Calling *resume* for an inactive traffic pattern has the effect of resuming the *Client* process associated with the traffic pattern.

The above operations also exist as *Client* methods (with the same names). Calling the corresponding method of the *Client* corresponds to calling this method for all traffic patterns. Thus:

Client->suspend ();

deactivates all (active) traffic patterns and:

Client->resume ();

activates all traffic patterns that happen to be inactive.

The most natural application of the last two methods is in situations when the modeled system should undergo a non-trivial startup phase during which no traffic should be offered to the network. In such a case, all traffic patterns can be suspended before the protocol processes are started (in the first state of *Root*), and then resumed when the startup phase is over.

A traffic pattern driven by a user-supplied process (or a collection of user-supplied processes) is not automatically suspended by *suspend* and resumed by *resume*, but the user process can learn about the status changes of the traffic pattern and respond consistently to these changes. The following two methods of *Traffic*:

int isSuspended ();
int isResumed ();

poll the traffic pattern for its suspended/resumed status. These methods also exist in *Client* variants returning the combined status of all the traffic patterns. The *Client* variant of *isSuspended* returns *YES*, if all traffic patterns have been suspended and *NO* otherwise. The other method is a straightforward negation of *isSuspended*.

The following two events are triggered on a traffic pattern when its suspended/resumed status changes:

SUSPEND	The event occurs whenever the traffic pattern changes its status from "resumed" to "suspended." If a wait request for *SUSPEND* is issued to a traffic pattern that is already suspended, the event occurs immediately, i.e., within the current *ITU*
RESUME	The event is triggered whenever the traffic pattern changes its status from "suspended" to "resumed." If a wait request for *RESUME* is issued to an active (resumed) traffic pattern, the event occurs immediately

A similar pair of events exists for the *Client*. Those events are triggered by the *Client* variants of *suspend* and *resume* operations. Such an operation must be effective to trigger the corresponding event, i.e., the status of at least one traffic pattern must be affected by the operation. Note that by calling the *Client's suspend/resume*, we also trigger the *SUSPEND/RESUME* events for all the traffic patterns that are affected by the operation.

Normally, while a traffic pattern is suspended, the virtual time that passes during that period counts, e.g., in calculating the effective throughput (Sects. 10.2.2 and 10.2.4). Thus, the idleness incurred by the suspension will tend to dilute the standard throughput measures. Generally, it is recommended (or at least makes sense) to (re)initialize the standard performance measures for the traffic pattern (Sect. 10.2.4) whenever its status changes from inactive to active. This suggestion also applies to the entire *Client*. In most cases, all traffic patterns are deactivated and activated simultaneously.

In the course of their action, the *Client* processes governing the behavior of a traffic pattern call a number of methods declared within type *Traffic*. Most of those methods are virtual and accessible by the user; thus, they can be redefined in a user-declared subtype of *Traffic*. Some attributes (variables) of *Traffic* are also made public to facilitate programming non-standard extensions of the *Client*. The following variables store the options specified in the *setup* arguments when the traffic pattern was created:

char DstMIT, DstMLE, DstBIT, DstBSI, FlgSCL, FlgSUS, FlgSPF;

Each of the first four variables can have one of the following values:

UNDEFINED	If the distribution of the corresponding parameter (*MIT* = message interarrival time, *MLE* = message length, *BIT* = burst interarrival time, *BSI* = burst size) was not defined at the time when the traffic pattern was created (see Sect. 6.6.2)
EXPONENTIAL	If the corresponding parameter is exponentially distributed
UNIFORM	If the corresponding parameter has uniform distribution
FIXED	If the corresponding parameter has a fixed value

The remaining two variables can take the values *ON* or *OFF* (1 and 0, respectively). If *FlgSCL* is *OFF*, it means that the standard processing for the traffic pattern has been switched off (*SCL_off* was specified at the creation of the traffic pattern); otherwise, the standard processing is in effect. The value of *FlgSUS* indicates whether the traffic pattern has been suspended (*ON*) or is active (*OFF*). If *FlgSPF* is *OFF* (*SPF_off* was selected when the traffic pattern was created), no standard performance measures are collected for this traffic pattern (see Sect. 10.2).

The following four pairs of attributes contain the numerical values of the distribution parameters describing the arrival process:

double ParMnMIT, ParMxMIT,

    ParMnMLE, ParMxMLE,

    ParMnBIT, ParMxBIT,

    ParMnBSI, ParMxBSI;

If the corresponding parameter (*MIT* = message interarrival time, *MLE* = message length, *BIT* = burst interarrival time, *BSI* = burst size) is uniformly distributed, the two attributes contain the minimum and the maximum values. Otherwise (for the fixed or exponential distribution), the first attribute of the pair contains the fixed or mean value, and the contents of the second attribute are irrelevant (and accidental). The *TIME* values, i.e., *ParMnMIT*, *ParMxMIT*, *ParMnBIT*, *ParMxBIT* are expressed in *ITUs*. They have been converted to *ITUs* from the user *setup* specification in *ETUs*.

The following virtual methods of *Traffic* participate in the message generation process:

virtual TIME genMIT ();

This method is called to generate the time interval until the next message arrival. The standard version generates a random number according to the message interarrival time distribution.

virtual Long genMLE ();

This method is called to determine the length of a message to be generated. The standard version generates a random number according to the message length distribution.

virtual TIME genBIT ();

The method is called to determine the time interval until the next burst arrival. The standard version generates a random number according to the burst interarrival time distribution.

virtual Long genBSI ();

This method is called to generate the size of a burst (the number of messages within a burst). The standard version generates a random number according to the burst size distribution.

virtual int genSND ();

This method is called to determine the sender of a message to be generated. It returns the station *Id* of the sender or *NONE*, if no sender can be generated. The standard version determines the sender according to the configuration of communication groups associated with the traffic pattern, in the manner explained in Sect. 6.6.2.

virtual IPointer genRCV ();

The method is called to find the receiver for a generated message. It returns the station *Id* of the receiver, or a pointer to the station group (for a broadcast receiver), or *NONE* if no receiver can be generated. The standard version generates the receiver from the communication group used to generate the last sender (Sect. 6.6.2). Thus, it should be called after *genSND* (which determines and sets the source group).

virtual CGroup *genCGR (Long sid);

The standard version of this method returns a pointer to the communication group containing the indicated station in the senders set. The station can be specified either via its *Id* or as a pointer to the station object. The latter is possible, because this (alias) method is defined within *Traffic*:

```
CGroup *genCGR (Station *s) {
        return (genCGR (s->Id));
};
```

Note that the alias method is not *virtual*, so when redefining this part of *Traffic* functionality, one only must redefine the first (*virtual*) method.

It may happen that the indicated station (call it s) occurs in the senders sets of two or more communication groups. In such a case, one of these groups is chosen at random with the probability proportional to the sender weight of s in that group taken against its weights in the other groups. Specifically, if $G_1, \ldots, G_k$ are the multiple communication groups that include s as a sender, and $w_s^1, \ldots, w_s^k$ are the respective sender weights of s in them, then G_i is selected with the probability:

$$P^i = \frac{w_s^i}{\sum_{j=1}^{k} w_s^j}.$$

The selected communication group becomes the "current group" and, e.g., the next call to *genRCV* will subsequently select a receiver from that group.

One application of *genCGR* is when the sender is already known, e.g., it has been selected in a non-standard way (behind the scenes), and a matching receiver must be generated in a way that preserves the rules of probability implied by the set of communication groups. More formally, suppose we execute *genSND* multiple times looking only at the cases when a particular station *s* is generated as the sender, and record the frequencies of the various (senders) communication groups from which *s* was chosen. Then, when we run *genCGR* for *s* multiple times, we will see the same frequencies of communication groups returned by the method.

If the specified station does not occur as a sender in any communication group associated with the traffic pattern (i.e., it cannot be a sender for this traffic pattern), *genCGR* returns *NULL*. This way the method can also be used to check whether a station is a legitimate sender for a given traffic pattern.

*virtual Message *genMSG (Long snd, IPointer rcv, Long lgth);*
*Message *genMSG (Long snd, SGroup *rcv, Long lgth);*
*Message *genMSG (Station *snd, Station *rcv, Long lgth);*
*Message *genMSG (Station *snd, SGroup *rcv, Long lgth);*

The method occurs in four guises of which only the first one is *virtual* and provides the replaceable implementation base for the remaining ones. It is invoked to generate a new message and queue it at the sender. The sender station is indicated by the first argument, which can be either a station *Id* or a station object pointer. The second argument can identify a station (in the same way as the first argument) or a station group for a broadcast message. The last argument is the message length in bits. The new message is queued at the sender at the end of the queue corresponding to the given traffic pattern (see Sect. 6.1). The message pointer is also returned as the function value.

6.6.4 Intercepting Packet and Message Deallocation

The allocation and deallocation of storage for messages and packets is mostly handled internally by SMURPH, and we seldom need to concern ourselves with this issue. When a message is queued at the sender station for transmission, a data structure representing the message is created and appended at the end of the respective list. When the last bit of the message is extracted into a packet (Sect. 6.7.1), the message data structure is automatically deallocated. Packets are usually acquired into pre-existing buffers. When the contents of such a buffer are transmitted, SMURPH creates a (flat) copy of the packet structure to represent the packet during its propagation through the medium (be it a wired link or a wireless rfchannel). Normally, when the last bit of the packet has reached the end of the medium (has travelled the longest distance across the medium),[9] the working copy of the packet structure is deallocated. If the packet is expected to live beyond that time, e.g., it has to make another hop, it must be either retransmitted right away (while it is still available) or copied into a buffer (Sect. 6.3).

Sometimes packets and/or messages contain pointers to user-created, dynamically allocated structures that should be freed when the packet (or message) disappears. Unfortunately, using the standard destruction mechanism for this kind of clean-up is not possible in *SMURPH*, because, for the sake of memory efficiency, packets are represented within link/rfchannel activities as variable-length chunks (allocated at the end of the activity object) that formally belong to the standard type *Packet*. Virtual destructors do not help because the function that internally clones a transmitted packet to become part of the corresponding link activity cannot know the packet's actual type: it merely knows the size of the packet's object.

It is possible, however, to intercept packet and message deallocation events in a way that exploits the connection between those objects and the *Client*. To this end, the following two *virtual* methods can be declared within a *Traffic* type:

void pfmPDE (ptype **p*);
void pfmMDE (mtype **m*);

where "ptype" and "mtype" are the types of packets and messages associated with the traffic pattern (Sect. 6.6.1). If declared, *pfmPDE* will be called whenever a packet structure is about to be deallocated, with the argument pointing to the packet. This covers the cases when a packet is explicitly deleted by the protocol program, as well the internal deallocations of *Link* and *RFChannel* activities (Sects. 7.1.1 and 8.1.1). Note that the deallocation of a non-standard packet (i.e., one that has no associated traffic object, Sect. 6.3) cannot be intercepted this way. Link-level cleanup (Sects. 7.4 and 8.3) is recommended for such packets.

Similarly, *pfmMDE* will be called for each message deallocation, which in most cases happens internally, when a queued message has been entirely converted into packets (Sect. 6.7.1).

The fact that the cleaning methods are automatically invoked by the (internal) *Packet* and *Message* destructors makes them potentially dangerous. This is usually more relevant for packets, because it is not uncommon for a packet recipient to clone a copy of the received packet, e.g., with the intention of retransmitting it on its next hop towards the final destination. If the packet includes pointers to dynamically allocated structures that have to be deallocated when the packet is done, then we have a problem, because the original, as well as the copy will be subjected to the same kind of cleaning upon deletion. Then, there are two options:

- The copy must not be shallow, i.e., the dynamically allocated components of the packet must also be copied, because they will be deallocated twice: with the original and with the copy. This option is tricky to apply consistently, because SMURPH automatically makes shallow copies of packets (buffers), e.g., when starting a packet transmission (Sects. 7.1.2 and 8.1.8).

---

[9]See Sects. 4.2.2, 4.4.2 and 7.1.1 for ways of extending the packet's lifetime in the channel.

- The program must use some attributes (or flags) in the packet to let *pfmPDE* differentiate between an intermediate and the final cleaning (and make sure that the dynamically allocated components are only freed in the latter case). This option is recommended for all situations when a copied packet requires non-trivial (deep) cleaning upon deallocation.

The names of the two methods start with *pfm* because they formally belong to the same class as the performance collecting methods (discussed in Sect. 10.2.3), which are called to mark certain stages in packet processing. In this context, the deallocation of a packet/message is clearly a relevant processing stage in the object's life, although its perception is probably useless from the viewpoint of calculating performance measures.

6.6.5 Message Queues

Every station is equipped with two pointers accessible to the protocol program which describe queues of messages awaiting transmission at the station. These pointers are declared within type *Station* in the following way:

*Message **MQHead, **MQTail;*

MQHead points to an array of pointers to messages, each pointer representing the head of the message queue associated with one traffic pattern. The *Id* attributes of traffic patterns (reflecting their creation order, see Sect. 6.6.2) index the message queues in the array pointed to by *MQHead*. Thus,

S->MQHead [tp]

references the head of the message queue (owned by station *S*) corresponding to the traffic pattern number *tp*. Recall that *S* is a standard process attribute identifying the station that owns the process (Sect. 5.1.2). *MQTail* is a similar array of message pointers. Each entry in this array points to the last message in the corresponding message queue, and is used to quickly append messages at its end.

MQHead and *MQTail* are initialized to *NULL* when the station is created. The two pointer arrays are generated and assigned to *MQHead* and *MQTail* when the first message is queued at the station by *genMSG* (Sect. 6.6.3). Thus, a station that never gets any message to transmit will never have a message queue, not even an empty one. An empty message queue is characterized by both its pointers (i.e., *MQHead [tp]* and *MQTail [tp]*) being *NULL*. The way messages are kept in the queues is described in Sect. 6.2.

The *prev* pointer of the first message of a nonempty queue number *tp* (associated with the traffic pattern number *tp*) contains &*MQHead [tp]*, i.e., it points to a fictitious message containing *MQHead [tp]* as its next attribute.

It is possible to impose a limit on the length of message queues, individually at specific stations or globally for the entire network. This possibility may be useful for investigating the network behavior under extreme traffic conditions. By setting a limit on the number of queued messages, the program can avoid overflowing the memory of the simulator. A call to the following global function:

void setQSLimit (Long limit = MAX_Long);

declares a limit on the total number of messages that may be simultaneously queued at all stations combined. When *setQSLimit* is called as a *Traffic* method, it restricts the number of queued messages belonging to the given traffic pattern. The function can also be called as a *Station* method in which case it limits the combined length of all message queues at the indicated station. If the argument is absent, it defaults to *MAX_long* and stands for "no limit" (which is the default setting).

Whenever a message is about to be generated and queued at a station by *genMSG* (Sect. 6.6.3), SMURPH checks whether queuing it would not exceed one of the three possible limits: the global limit on the total number of messages awaiting transmission, the limit associated with the traffic pattern to which the message belongs, or the limit associated with the sender station. If any of the three limits is already reached, no message is generated, and the action of *genMSG* is void. *TheMessage* returns *NULL* in such a case.

Messages belonging to a traffic pattern that was created with the *SPF_off* indicator (Sect. 6.6.2) do not count to any of the three possible limits. For such a traffic pattern, *genMSG* always produces a message and queues it at the corresponding queue of the selected sender.

For efficiency reasons, the queue size limit checking is disabled by default. The program must be created with the *–q* option of *mks* (Sect. B.5) to enable this feature. Otherwise, calls to *setQSLimit* are illegal and trigger execution errors.

The following two *Station* methods tell the number of messages and bits queued at the station:

Long getMQSize (int tp = NONE);
BITCOUNT getMQBits (int tp = NONE);

When the argument is *NONE* (or absent), the methods return the total number of messages/bits queued at the station, over all traffic patterns (queues). Otherwise, the argument must be a legitimate traffic pattern *Id*, and the returned values pertain to the single, indicated traffic pattern.

6.7   Inquiries

A process interested in acquiring a packet for transmission can ask for it. Such an inquiry can be addressed to:

- a specific traffic pattern, if the process is explicitly interested in packets generated according to this specific pattern
- the *Client*, if the process does not care about the traffic pattern and wants to transmit whatever packet is available

- a message, if the process wants to explicitly indicate the message from which the packet is to be extracted.

We shall refer to all these types of inquiries as *Client* inquiries, although only a certain subset of them is implemented as a collection of methods within the *Client* type.

6.7.1 Acquiring Packets for Transmission

The following method of the *Client* can be used to check whether a message awaiting transmission is queued at the station and, if so, to create a packet from that message:

int getPacket (Packet *p, Long min, Long max, Long frm);

The first argument is a pointer to the packet buffer where the acquired packet is to be stored. The remaining three arguments indicate the minimum packet length, the maximum packet length, and the length of the extra information contributing to the packet frame, typically the header and the trailer. All these lengths are in bits.

The method examines all message queues at the current station. If the value returned by the method is *NO*, it means that all the queues are empty and no packet can be acquired. Otherwise, the earliest-arrived message is selected, and a packet is created out of that message. If two or more messages with the same earliest arrival time are queued at the station (possibly in different queues), one of them is chosen nondeterministically.[10]

If *max* is greater than the message length, or zero, which stands for "no limit," the entire message is turned into the packet. Otherwise, only *max* bits are extracted from the message and its remaining portion is left in the queue for further use. If *min* is greater than the length of a packet obtained this way, the packet is expanded (inflated) to the size of *min* bits. The added bits count to the total length of the packet (attribute *TLength*, Sect. 6.2), i.e., they affect the packet's transmission time, but they are not considered payload bits (are not counted in attribute *ILength*), and, e.g., are ignored in calculating the effective throughput (Sects. 10.2.2 and 10.2.4). Note that *min* can be greater than *max*, in which case the packet is always inflated. Finally, *frm* bits are added to the total length (*TLength*) of the packet. This part is used to represent the length of the various headers and trailers required by the protocol, which contribute to the packet's transmission time, but do not count to the payload.

Let *m* denote the message length in bits. The values of the two length attributes *TLength* and *ILength* (Sect. 6.2) of a packet acquired by *getPacket* are determined according to the following algorithm:

ILength = (m->Length > max) ? max : m->Length;
TLength = ILength + frm;

---

[10]If the program has been compiled with the –D (deterministic) option of *mks* (Sect. B.5), the traffic pattern with the lowest *Id* wins in such a case.

if (ILength < min) TLength += min − ILength;
m->Length − = ILength;

Before the message length is decremented, the packet's *setup* method is called with the message pointer passed as the argument (see Sect. 6.2). The standard version of this method is empty. The user may redefine it in proper subtypes of *Packet* to set the packet's non-standard attributes on acquisition. Note that generally, the setting of the non-standard attributes of a packet may depend on the contents of the message from which the packet has been acquired, because that message may belong to a proper subtype of *Message* and may define non-standard attributes of its own, which may have to find their way into the packet.

If, after subtracting the packet length, *m->Length* ends up equal 0, the message is discarded from the queue and deallocated as an object. Note that it is legal for the original message length to be zero. If this is the case, the information length of the acquired packet will be zero and, following the acquisition, the message will be removed from the queue. The standard traffic generator of the *Client* never generates zero-length messages, but they can be created by custom traffic generators programmed by the user (Sect. 6.6.3).

It is illegal to try to acquire a packet into a buffer that is not empty (Sect. 6.3). After *getPacket* puts a packet into a buffer, the buffer status is changed to "full."

Two environment variables (Sect. 5.1.6) are set after a successful return (with value *YES*) of *getPacket*. If the message from which the packet has been acquired remains in the message queue (i.e., the packet's payload length is less than the original message length), the environment variable *TheMessage* (*Info01*) of type *Message** points to that message. Otherwise, *TheMessage* is set to *NULL*. In either case the environment variable *TheTraffic* (*Info02*) of type *int* returns the *Id* of the traffic pattern to which the acquired packet belongs.

Below we list other variants of *getPacket*. All the rules outlined above, unless we explicitly say otherwise, apply to those other variants as well.

The following *Client* method works identically to the one discussed above:

int getPacket (Packet &p, Long min, Long max, Long frm);

the only difference being that the packet buffer is passed by reference rather than a pointer.

Sometimes, the simple selection of the earliest message awaiting transmission is not what the protocol program wants. Here is a variant of the packet acquisition method selecting the packet based on programmable criteria:

int getPacket (Packet *p, int (*f)(Message*), Long min, Long max, Long frm);

where *f* is a pointer to a function which takes a message pointer as an argument and returns an integer, e.g., one declared as:

int qualifier (Message *m);

That function should return nonzero if the message pointed to by *m* satisfies the selection criteria, and zero otherwise. The message from which the packet is to be extracted is determined in a manner similar to the standard version of *getPacket*, with the exception that instead of "the earliest-arrived message," the qualified version looks for "the earliest-arrived message that satisfies *f*." Again, if there are two or more earliest-arrived messages satisfying *f* (i.e., all these messages have arrived at the same time), one of them is chosen at random.[11] There is a variant of qualified *getPacket* in which the buffer is passed by reference rather than through a pointer.

The methods described above are defined within the *Client* and they treat all traffic patterns globally. One possible way of restricting the range of the *Client's getPacket* to a single traffic pattern is to use one of the two qualified variants passing it the following function as the qualifier:

int qualifier (Message *m) { return (m->TP == MyTPId); };

where *MyTPId* is the *Id* of the traffic pattern in question. There is, however, a better (and more efficient) way to achieve the same result. All the *Client* versions of *getPacket* have their counterparts as methods of *Traffic*. Their configurations of arguments and behavior are identical to those of the corresponding *Client* methods with the exception that the search for a message is restricted to the single traffic pattern.

The two unqualified versions of *getPacket* also occur as methods of *Message* (with the same configurations of arguments). They are useful when the message from which a packet is to be extracted is already known. For example, suppose that the *Client* or *Traffic* variant of *getPacket* acquires a packet without using up the entire message (the message was longer than the maximum packet length). When *getPacket* is called again, with the intention to continue extracting data from the same message, it is not guaranteed that the same message will be used to feed the new packet. If there are other messages in the queue(s) with the same (earliest) queueing time, any of them can be chosen at random. If the program wants to make sure that one message is transmitted entirely before another one is started, it should store the pointer to the message used to acquire the previous packet (as returned by *getPacket* in *TheMessage*) and then continue extracting packets from that message until its bits run out, i.e., *TheMessage* becomes *NULL*. With the following call:

Msg->getPacket (buf, min, max, frm);

where *Msg* points to a message (i.e., an object belonging to a subtype of *Message*), the packet is extracted from *Msg* (according to the standard rules) and put into *buf*. If, after this operation, the message pointed to by *Msg* turns out to be empty, its *prev* attribute (Sect. 6.2) is examined to determine whether the message belongs to a queue. If *prev* is *NULL*, it is assumed that the message does not belong to a queue (has been procured outside the *Client's* standard queueing mechanism) and no further action is taken. Otherwise, the empty message is dequeued and deallocated as an object. The program should make sure that the *prev* attribute of a message created in a non-standard way, which should not be automatically dequeued by the *Message* version of *getPacket*, contains *NULL*. As for the other variants of *getPacket*, *TheMessage* is set to *NULL* when the entire message has been exhausted; otherwise it is set to *Msg*.

---

[11] If the program has been compiled with the –*D* (deterministic) option of *mks* (Sect. B.5), the traffic pattern with the lowest *Id* wins in such a case.

There is no way for a *Message* variant of *getPacket* to fail (returning *NO*). For compatibility with the other variants, the method always returns *YES*.

6.7.2 Testing for Packet Availability

Using one of the *getPacket* variants discussed in the previous section, it is possible to acquire a packet for transmission, or learn that no packet (of the required sort) is available. Sometimes one would like to check whether a packet is available without acquiring it (and removing the message from the queue). The following two *Client* methods can be used for this purpose:

int isPacket ();
int isPacket (int (*f)(Message*));

The first version returns *YES*, if there is a message (of any traffic pattern) queued at the current station. Otherwise, the method returns *NO*. The second version returns *YES*, if a message satisfying the qualifier *f* (see Sect. 6.7.1) is queued at the station, and *NO* otherwise.

Similar two versions of *isPacket* are defined within *Traffic*, and they behave exactly as the *Client* methods, except that only the indicated traffic pattern is examined.

Two environment variables are set whenever *isPacket* returns *YES*: *TheMessage* (*Info01*) of type *Message** points to the message located by the method,[12] and *TheTraffic* (*Info02*) of type *int* contains the *Id* of the traffic pattern to which the message belongs.

6.8 Wait Requests

When an attempt to acquire a packet for transmission fails (*getPacket* returns *NO*, Sect. 6.7.1), the process issuing the inquiry may choose to suspend itself until a message becomes queued at the station. By issuing the following wait request:

Client->wait (ARRIVAL, TryAgain);

the process declares that it wants to be awakened (in the indicated state) when a message of any traffic pattern is queued at the station. Thus, the part of the process that takes care of acquiring packets for transmission may look like this:

---

[12]Note that this is not necessarily the same message that will be used by a subsequent call to *getPacket*, unless the program has been generated with the *–D* option of *mks* (Sect. B.5).

...

```
state TryNewPacket:
        if (!Client->getPacket (buf, minl, maxl, frml)) {
                Client->wait (ARRIVAL, TryNewPacket);
                sleep;

        }
```

...

There are two ways of indicating that we are interested in messages that belong to a specific traffic pattern. One way is to put the *Id* of the traffic pattern in place of *ARRIVAL* in the above call to the *Client's wait*. The other (and probably nicer) solution is to use the *Traffic* variant of the *wait* method instead, i.e., to call:

TPat->wait (ARRIVAL, TryAgain);

where *TPat* is a pointer to the traffic pattern in question. Such a call indicates that the process wants to be restarted in the specified state when a message belonging to the given traffic pattern becomes queued at the station.

The keyword *ARRIVAL* can be replaced in the above examples with *INTERCEPT*. The semantics of the two events are almost identical, except that *INTERCEPT* gets a higher priority (a lower order) than *ARRIVAL*, irrespective of the actual values of the order parameters specified with the requests (Sect. 5.1.1). Imagine that a process wants to intercept all messages arriving at the station, e.g., to preprocess them by setting some of their non-standard attributes. At first sight it might seem natural to program this process in the following way:

...

```
state WaitForMessage:
        Client->wait (ARRIVAL, NextMessage);
state NextMessage:
```

...

preprocess the message

...

However, this solution has one serious drawback: if two processes are waiting for the same *ARRIVAL* event, there is no way to tell which one will go first; consequently, we cannot immediately guarantee that one of them will preprocess the message for the other one. One way to make sure that the message preprocessor is run before any process that may use the message would be to assign a low order to the wait request issued by the preprocessor, and make sure that the other processes specify higher values. That requires a special compilation option (–p), which brings about the heavy-duty, global mechanism of three-argument *wait* requests. Another (simpler) option is to replace *ARRIVAL* in the *wait* request issued by the preprocessor with *INTERCEPT*. That event is identical to *ARRIVAL*, but is guaranteed to be presented before *ARRIVAL*, if the two events happen to be triggered within the same *ITU*. Only one process at a given station may be waiting for *INTERCEPT* at any given time, unless each of the multiple *INTERCEPT* requests is addressed to a different traffic pattern, and none of them is addressed to the *Client*. In a nutshell, an *INTER-CEPT* request for a message, regardless where it comes from, is guaranteed to be handled before an *ARRIVAL* request for the same message, The implementation of *INTERCEPT* is similar to the implementation of priority signals (Sect. 5.2.2) and the priority *put* operation for mailboxes (Sect. 9.2.4).

Whenever a process is awakened by a message arrival event (*ARRIVAL* or *INTER-CEPT*), the environment variables *TheMessage* (of type *Message**) and *TheTraffic* (of type *int*) are set as after a successful invocation of *getPacket* (Sect. 6.7.1).

Chapter 7
Links and Ports

In this chapter, we describe the mechanism of wired communication, including the tools available to protocol programs for sending and receiving messages and packets over (wired) links. Following the network creation phase, links become hidden and are (usually) not referenced by the program. This agrees with the fact that networking software running in real-life systems accesses the communication media via interfaces which, in a SMURPH model, are represented by ports.

A protocol process may wish to reference a port for one of the following three reasons:

- to insert into the port (and thus into the underlying link) an activity, e.g., a packet
- to inquire the port about its current or past status, e.g., to check whether an activity is heard on the port
- to issue a wait request to the port, e.g., to await an activity to arrive at the port in the future.

7.1 Activities

Two types of activities can be inserted into ports: packets and jamming signals. *Packets* (Sect. 6.2) represent structured portions of information exchanged between stations. The operation of inserting a packet into a port is called a packet transmission. Jamming signals (or jams for short) are special activities that are clearly distinguishable from packets. Their role is to describe special signals occasionally useful in the models of collision protocols involving the concept of collision consensus [1]. Even ignoring the jamming signals,[1] and assuming that packets are the only activities of interest, it makes sense to use the term "activity" (instead of "packet") to refer to

---

[1]They are not as useful these days as they used to be, and their role is mostly historical.

© Springer Nature Switzerland AG 2019
P. Gburzyński, *Modeling Communication Networks and Protocols*,
Lecture Notes in Networks and Systems 61,
https://doi.org/10.1007/978-3-030-15391-5_7

them, because their representation in links covers more than what can be attributed merely to *Packets* (as introduced in Sect. 6.2).

7.1.1 Activity Processing in Links

Activities are explicitly started and explicitly terminated. When an activity is started, an internal object representing the activity is built and added to the link. One attribute of this object, the starting time, tells the time (in *ITUs*) when the activity was started. Another attribute, the finished time, indicates the time when the activity was terminated. The difference between the finished time and the starting time is called the duration of the activity.

The finished time attribute of an activity that has been started but not terminated yet is undefined, and so is formally the activity's duration. These elements become defined when the activity is terminated.

An activity inserted into one port of a link will arrive at another port connected to the link according to the rules outlined in Sect. 4.2.1. Assume that two ports p_1 and p_2 are connected to the same link l and the propagation distance from p_1 to p_2 is d *ITUs*. [2] If an activity is started on p_1 at time t_s, the beginning of this activity will arrive at p_2 at $t_s + d$. Similarly, if the activity is terminated by p_1 at time t_f, its end will be perceived at p_2 at time $t_f + d$.

A link can be viewed as a dynamic repository for activities. An activity inserted into the link remains in the repository at least for the time needed for the activity to propagate (in its entirety) to all ports connected to the link. If an activity started on port p is terminated at time t, it will reach the end of its official stay in the link at time $t + t_{max}$, where t_{max} is the maximum distance from p to any other port connected to the same link (expressed as the propagation time). Beyond that time, the activity cannot possibly cause any events on any of the link's ports.

To play it safe, SMURPH unconditionally extends the minimum storage time of an activity in the link repository by one *ITU*, which provides a breathing space for the protocol program for a safe examination of the activity in the last moment of its formal existence in the medium. If the activity's removal were scheduled exactly at the last *ITU* of its "physical" presence in the link (at a given port), then, owing to the nondeterminism in processing multiple events occurring within the same grain of time (Sect. 1.2.4) the protocol program could have problems, e.g., properly responding to the end of a packet reception. This is because the wake-up event caused by the reception of the last bit of the packet (events *EOT* and *EMP*, Sect. 7.1.3) would compete (within the same *ITU*) with the internal event scheduled to remove the packet activity from the link.

One can think of situations when it might be useful to extend the "breathing space" even more. While formally such extensions can be deemed unphysical, they occa-

---

[2]Recall that the distances are specified in *DUs* when the link is built and internally stored in *ITUs* (Sect. 4.3.3).

sionally become handy as shortcuts, without necessarily compromising the model's realism. In some cases, in response to an event or an inquiry, the protocol process will receive a pointer to a related activity, e.g., representing the packet being received or perceived. The object representing the activity is only available for as long as the activity remains within the link repository. If the process wants to save it for later, it will have to create a copy, which operation may be cumbersome and inefficient. If the said "later" happens to be a well-bounded, short while into the simulated future, then it may make better sense to ask the link to extend the activity's stay in the repository for some short time after its formal disappearance. This will not affect the interpretation of an activity as far as triggering events and/or interfering with other activities.

For another illustration, consider protocols that are (informally) specified with references to the past. For example, the standard CSMA/CD access scheme of the original Ethernet says something like this: "having acquired a packet for transmission, make sure that the medium has been quiet for the time corresponding to the prescribed packet space interval, and then transmit the packet." While the above specification is hardly a recipe for implementation, one may be tempted to program a model operating according to it, if only for illustration, presentation, or convenience. Formally, that would call for inquiries of the kind: "when was the last activity perceived on the port?" Such an inquiry (see Sect. 7.2.2) only makes sense, if the port (or rather the link) can "remember" past activities, say, up to a certain time limit.

Generally, in a situation when the activities that have "physically" disappeared from the link must be kept around for a while longer, the link can define a nonzero *linger* attribute, which can be set at the time of the link's creation, or redefined dynamically by calling the link's *setPurgeDelay* method (Sect. 4.2.2). The values of 0 and 1 (*TIME_0* and *TIME_1*) amount to the same default of one *ITU*; any value above 1 replaces the standard margin added to the time of the definitive and ultimate removal of the activity from the link's repository.

As an activity that has overstayed its "physical" presence in the link can contribute nothing to future events, SMURPH moves it to a special "secondary" repository dubbed the link's archive. This is just a technical matter aimed at improving the efficiency of this (optional) feature. If the linger time is defined as *TIME_0* or *TIME_1*, the archive is not maintained.

7.1.2 Starting and Terminating Activities

The following *Port* method starts a packet transmission:

*void startTransmit (Packet *buf);*

where *buf* points to the buffer containing the packet to be transmitted (Sect. 6.3). A shallow copy of this buffer is made and inserted into a data structure representing a new link activity. A pointer to that copy is returned via the environment variable *ThePacket* (*Info01*) of type *Packet**. Another environment variable, *TheTraffic*

(*Info02*) of type *int*, returns the *Id* attribute of the traffic pattern that was used to generate the packet. Recall that the *Id* attribute of a packet created by hand, behind the *Client's* back (Sect. 6.2), is *NONE* (−1).

A jamming signal is started with this *Port* method:

void startJam ();

Jamming signals have no structure that would call for an argument to *startJam*, they all look the same and are solely characterized by their timing attributes.

There are two ways to terminate a packet transmission. The following port method:

int stop ();

terminates the transmission of a packet inserted by a previous call to *startTransmit*, or a jamming signal started by a previous call to *startJam*. If *stop* is applied to a packet, then the terminated packet is marked as complete, i.e., formally equipped with a pertinent trailer (end marker) that will allow the receiving party to recognize the activity as a complete packet. Another way to terminate an activity is to invoke this port method:

int abort ();

which aborts the transmission. A packet terminated this way is interrupted in the middle and will not be perceived as a complete (receivable) packet by its prospective recipients. For a jamming signal, both methods effect the same action.

Each of the two methods returns an integer value that tells the type of the terminated activity. This value can be either *TRANSMISSION* or *JAM*. It is illegal to attempt to interrupt a non-existent activity, i.e., to execute *stop* or *abort* on an idle port.

If the terminated activity is a packet transmission, two environment variables *ThePacket* and *TheTraffic* (see above) are set (as for *startTransmit*). The first one points to the packet in question, the second one stores the *Id* of the traffic pattern to which the packet belongs.

Formally, the amount of time needed to transmit a packet pointed to by *buf* on port *prt* is equal to:

*buf->TLength * prt->getXRate ()*

i.e., to the product of the total length of the packet (Sect. 6.2) and the port's transmission rate (Sect. 4.3.1). Thus, a sample process code for transmitting a packet might look as follows:

...

```
state Transmit:
       MyPort->startTransmit (buffer);
       Timer->wait (buffer->TLength*MyPort->getXRate ( ), Done);
state Done:
       MyPort->stop ( );

       ...
```

Note that generally the time spent on transmitting a packet need not have much to do with the contents of the packet's *TLength* attribute. That time is determined solely by the difference between the moments when the transmission of the packet was stopped and when it was started. For example, if a process starts transmitting a packet and then forgets to terminate it (by *stop* or *abort*), the packet will be transmitted forever.

In most cases, the protocol program would like to transmit the packet for the time determined by the product of the packet's length and the transmission rate of the port, possibly aborting the transmission earlier, if a special condition occurs. The port method:

*void transmit (Packet *buf, int done);*

starts transmitting the packet pointed to by *buf* and automatically sets up the *Timer* to wake up the process in state *done* when the transmission is complete. The transmission still must be terminated explicitly. Thus, the code fragment listed above is equivalent to:

...

state Transmit:

    MyPort->transmit (buffer, Done);

state Done:

    MyPort->stop ( );

    ...

The following port method converts a number of bits to the number of *ITUs* required to transmit these bits on a given port:

TIME bitsToTime (int n = 1);

Calling:

t = p->bitsToTime (n);

has the same effect as executing:

*t = p->getXRate () * n;*

When called without an argument, or with the argument equal 1, the method returns the amount of time needed to transmit a single bit, i.e., the port's transmission rate.

The transmission rate of a port can be changed at any time with *setXRate* (see Sect. 4.3.1). If *setXRate* is invoked after *startTransmit* or *transmit*, but before *stop*, it has no impact on the packet being transmitted. It is also possible to change the transmission rates of all ports connected to a given link by executing the *setXRate* method of the link. The link method accepts the same argument as the port method, but returns no value (its type is *void*). When called with *RATE_inf* (or with no argument at all), the link method uses the (current) default rate of the link. Otherwise, the argument redefines the link's default rate and becomes the current transmission rate of all ports connected to the link.

Similar to a packet transmission, it is also possible to start emitting a jamming signal and simultaneously set the *Timer* for a specific amount of time. A call to this port method:

void sendJam (TIME d, int done);

is equivalent to the sequence:

MyPort->startJam ();
Timer->wait (d, done);

It is illegal to insert more than one activity into a single port at a time. However, it is perfectly legitimate to have two ports connected to the same link and separated by the distance of 0 *ITUs*. Thus, two or more activities can be inserted simultaneously into the same place of a link (albeit from different ports).

7.1.3   Wait Requests and Events

The first argument of a *wait* request addressed to a port is the event identifier, usually represented by a symbolic constant. The symbolic event identifiers for the *Port AI* are listed below.

SILENCE	The event occurs at the beginning of the nearest silence period perceived on the port. If no activity is heard on the port when the request is issued, the event occurs immediately, i.e., within the current *ITU*
ACTIVITY	The event occurs at the beginning of the nearest activity (a packet or a jamming signal) heard on the port. The activity must be preceded by a silence period to trigger the event, i.e., two or more activities that overlap (from the viewpoint of the sensing port) are treated as a single continuous activity. If an activity is heard on the port when the request is issued, the event occurs within the current *ITU*
BOT	(Beginning of Transmission). The event occurs as soon as the port perceives the nearest beginning of a packet transmission. The event is not affected by interference or collisions, i.e., the possible presence of other activities on the port at the same time. It is up to the protocol program to detect such conditions and interpret them
EOT	(End of Transmission). The event occurs when the nearest end of a packet terminated by *stop* (Sect. 7.1.2) is perceived by the port. Aborted transmission attempts, i.e., packets terminated by *abort*, do not trigger this event. Like *BOT*, the event is not affected by the possible presence of other activities on the port at the same time

(continued)

(continued)

BMP	(Beginning of My Packet). The event occurs when the port perceives the beginning of a packet for which the current station (the one whose process issues the wait request) is the receiver (or one of the receivers, for a broadcast packet). Like *BOT* and *EOT*, the event is not affected by the possible presence of other activities on the port at the same time. *BMP* implies *BOT*, but not the other way around
EMP	(End of My Packet). The event occurs when the port perceives the end of a packet for which the current station is the receiver (or one of the receivers). The event is only triggered if the packet has been terminated by *stop*, but (like *BOT*, *EOT* and *BMP*) it is not affected by the possible presence of other activities on the port at the same time. *EMP* implies *EOT*, but no the other way around
BOJ	(Beginning of a Jamming signal). The event occurs when the port hears the nearest beginning of a jamming signal preceded by anything not being a jamming signal. If multiple jams overlap (according to the port's perception), only the first of those jams triggers the *BOJ* event, i.e., overlapping jams are heard as one continuous jamming signal. If a jam is being heard on the port when the request is issued, the event occurs within the current *ITU*
EOJ	(End of a Jamming signal). The event occurs when the port perceives the nearest end of a jamming signal (followed by anything not being a jamming signal). Two (or more) overlapping jams are heard as one, so only the end of the last of those activities will trigger the event. If no jam is heard when the request is issued, the event is *not* triggered immediately
ANYEVENT	This event occurs whenever the port begins to sense a new activity or stops to sense some activity. Overlapping activities are separated, e.g., two overlapping activities generate four separate events, unless some of those events occur within the same *ITU* (see below)
COLLISION	The event occurs when the earliest collision is sensed on the port (see Sect. 7.1.4). If a collision is present on the port when the request is issued, the event occurs immediately

When a process is awakened by one of the port events listed above, except for *ANYEVENT* (see below), the environment variable *ThePort* (*Info02*) of type *Port** stores the pointer to the port on which the waking event has occurred. If the waking event is related to a packet (i.e., it is one of *BOT*, *EOT*, *BMP*, *EMP*, and also *ANYEVENT*, as explained below), the environment variable *ThePacket* (*Info01*) of type *Packet** returns the pointer to the packet object. That object is a copy of the packet buffer given to *startTransmit* or *transmit* (Sect. 7.1.2) when the transmission of the packet was started.

The object pointed to by *ThePacket* is deallocated when the activity carrying the packet is removed from the link's repository. If the *linger* attribute is left at its default setting (Sect. 7.1.1), the activity (and the packet) is deallocated as soon as its last bit disappears from the link. According to what we said in Sect. 7.1.1, "as soon" includes a grace period of 1 *ITU*. This way, regardless of the priorities of the requisite events, the most remote recipient can always properly recognize and safely receive all perceived packets.

With *ANYEVENT*, if two (or more) events occur at the same time (within the same *ITU*), only one of them (chosen at random) will be presented. The restarted process can learn about all events that occur within the current *ITU* by calling *events* (see Sect. 7.2.1).

When a process awaiting *ANYEVENT* is awakened, the environment variable *TheEvent* (*Info02*) of type *int* contains a value identifying the event type. The possible values returned in *TheEvent* are represented by these symbolic constants:

BOT	beginning of packet
EOT	end of a complete packet (one that was terminated by *stop*)
ABTPACKET	end of an aborted packet, i.e., one terminated by *abort*
BOJ	beginning of a jamming signal
EOJ	end of a jamming signal

If the event type is *BOT*, *EOT*, or *ABTPACKET*, *ThePacket* (*Info01*) points to the packet responsible for the event.

The purpose of *ANYEVENT* is to cover all the circumstances when anything (anything at all) changes in the port's perception of the link. Thus, the event provides a hook for implementing custom functionality that is not available through a combination of other (specific events). One potential problem with *ANYEVENT* (used as a Swiss-knife handler of all possible port state changes) is that the persistent nature of port events may make the processing a bit tricky. Consider the following generic code fragment:

...

```
state MonitorEvents:
        MyPort->wait (ANYEVENT, CheckItOut);
state CheckItOut:
        ... determine and handle the case ...
        proceed MonitorEvents;
```

...

which probably isn't going to work, because whatever condition has triggered the event in state *MonitorEvents*, it remains pending when, having completed handling the event in state *CheckItOut*, the process proceeds back to the initial state. The problem is that port events require a time advance to disappear (Sect. 2.2.3). One possible way out is to replace *proceed* with *skipto* (Sect. 5.3.1), but then it becomes difficult to intercept absolutely all conditions, like multiple events falling within the same *ITU*.

To this end, it is possible to declare a semantics of *ANYEVENT* whereby the triggering event never remains pending and requires an actual condition change that follows the *wait* request to be triggered. This means that a once-perceived condition will not trigger another wakeup until something new happens on the port, or, put differently, that the triggers are the condition changes rather than the conditions themselves. Here is the port method to switch between the two interpretations of *ANYEVENT*:

Boolean setAevMode (Boolean mode = YESNO);

where the argument can be *NO* (equivalent to *AEV_MODE_TRANSITION*), or *YES* (equivalent to *AEV_MODE_PERSISTENT*). In the first case, the non-persistent mode of triggering *ANYEVENT* is assumed. The second value declares the "standard" (and default) handling of *ANYEVENT*, whereby events persist until time is advanced. The method returns the previous setting of the option.

When *setAevMode* is invoked without an argument (or with *YESNO*, which is interpreted as an undefined *Boolean* value, Sect. 3.2.5), the method will set the port's *AEV* mode to the current default mode of the link, which is initialized to *AEV_MODE_PERSISTENT*. The link's default mode can be changed by calling the *setAevMode* method of the link, which will redefine the link's default, as well as set the mode of all ports connected to the link to the new default. When this is done at the initialization, when no ports are yet connected to the link, the ports will automatically copy the new default as they become configured into the link.

The link variant of the method is declared with the following header:

void setAevMode (Boolean mode = YESNO);

i.e., the method returns no value. When called without arguments (or with *YESNO*), is sets the *AEV* mode of all ports connected to the link to the current default for the link.

The recommended way of receiving *ANYEVENT* in the non-persistent mode is this:

```
...

state MonitorEvents:

    ... check for pending events and handle them ...

    MyPort->wait (ANYEVENT, MonitorEvents);

    sleep;

    ...
```

To avoid missing any events, before going to sleep, the process should account for all the events that are available (present) at the time of its wakeup. This can be accomplished through port inquiries (Sect. 7.2.1). Note that, with multiple events falling within the same *ITU*, the process still should take measures to avoid interpreting the same event more than once. However, it does not have to worry that the same event will wake it up repeatedly, in an infinite loop. In many practical cases, the interpretation of events is idempotent, i.e., processing them multiple times is not an issue. Otherwise, the process can mark them as processed, e.g., using a spare packet flag (Sect. 6.2).

7.1.4 *Collisions*

By a collision, we understand a situation when two or more transmission attempts perceived by a port overlap, and the information carried by each of the packets involved is (at least partially) destroyed. SMURPH provides for automatic collision detection (built into the link model), because its predecessor was devised to model collision protocols (CSMA/CD) in shared, broadcast, wired channels [2].

We say that a port perceives a collision, if at least one of the following two statements is true:

- a jamming signal is heard on the port
- two or more packets are heard simultaneously.

The purpose of jamming signals is to represent special link activities that are always different from correctly transmitted packets. The explicit occurrence of a jamming signal is equivalent to a collision. This interpretation is assumed for historical compatibility with CSMA/CD protocols which sometimes used special jamming signals to enforce the so-called collision consensus.

7.1.5 *Event Priorities*

While nondeterminism in processing multiple events scheduled for the same time slot is generally a good thing (Sect. 1.2.4), there are situation when it is natural, or even necessary, to assign different priorities to different events that may occur simultaneously on the same port. We mean here formal events, i.e., different SMURPH "presentations" of the same "actual" event. For illustration, consider the following fragment of a process code method:

```
...

state SomeState:
        MyPort->wait (BOT, NewPacket);
state NewPacket:
        MyPort->wait (EMP, MyPacket);
        MyPort->wait (EOT, OtherPacket);
        MyPort->wait (SILENCE, Garbage);

        ...
```

Most likely, the intended interpretation of the three (simultaneous) wait requests issued in state *NewPacket* is as follows. Upon detection of the end of a packet addressed to the current station we want to get to state *MyPacket*; otherwise, if the packet whose end we are just perceiving happens to be addressed to some other station, we want to continue at *OtherPacket*; finally, if the packet does not terminate properly (it has been aborted), we want to resume in state *Garbage*. The problem is that the three awaited events will all occur at the same time (within the same *ITU*). Therefore, special measures must be taken to enforce the intended order of their processing.

If the program has been created with −*p* (Sect. B.5), it is possible to use the third argument of the wait request to specify the ordering of the awaited events, e.g.,

...

state SomeState:

    MyPort->wait (BOT, NewPacket);

state NewPacket:

    MyPort->wait (EMP, MyPacket, 0);

    MyPort->wait (EOT, OtherPacket, 1);

    MyPort->wait (SILENCE, Garbage, 2);

    ...

However, on a closer scrutiny, the issue looks different than an occasional need to compensate for the crudeness of the *ITU*, which the −*p* option is intended for, and which practically never surfaces in authoritative network models. The problem is not caused by multiple, independent events accidentally scheduled to the same *ITU* because of the discrete nature of time, but by the same event whose different interpretations appear like different SMURPH events. Subjecting those interpretations to the same kind of nondeterminism as completely unrelated events makes it difficult to process them in a systematic way.

As the problem is rather common, SMURPH automatically prioritizes multiple events triggered by the same packet in the intuitively natural way, without resorting to the (heavy-duty) *wait order* mechanism. In fact, the default ordering is *not* automatically imposed, if the program has been compiled with −*p* (the order option), as it is then assumed that the program wants to exercise full explicit control over event priorities.[3]

The default ordering applies to the multiple events triggered by the same packet which can be *BMP*, *BOT*, and *ACTIVITY* (for the packet's beginning) and *EMP*, *EOT*, *SILENCE* (for its end). In a nutshell, the most specific event has the highest priority, i.e., *BMP* wins over *BOT*, which in turn wins over *ACTIVITY*. Similarly, *EMP* is presented before *EOT*, which takes precedence over *SILENCE*, i.e., the above code is going to work as desired. However, the rule only applies to events triggered by the same packet. If, for example, the beginnings of two packets are perceived simultaneously, and one of these packets is addressed to the current station while the other is not, the *BOT* event triggered by the second packet can be presented before

---

[3]This feature is disputable. The reader should assume that −*p* should *never* be used in network models. I have never needed it in mine. If it looks like the order argument of *wait* is needed, it means that there is a problem with the model design.

the *BMP* triggered by the first one. This is to say, the nondeterminism always holds by default for *truly* independent events scheduled for the same *ITU*.

7.1.6 Receiving Packets

By sensing *BMP* and *EMP* events, a process can recognize packets that are addressed to its station. A packet triggering one of those events must be either a non-broadcast packet whose *Receiver* attribute contains the *Id* of the current station, or a broadcast packet with *Receiver* pointing to a station group that includes the current station (Sect. 6.5).

Given a packet, the program can check whether it is addressed to a specific station by calling this method of the packet:

int isMine (Station *s = NULL);

which returns nonzero, if and only if the indicated station is the packet's receiver (or one of the receivers in the broadcast case). If the argument is *NULL* (or absent), the inquiry refers to the station pointed to by *TheStation*, i.e., normally to the station that runs the inquiring process.

The receiving process usually waits until the end of the packet arrives at the port (event *EMP*), before assuming that the packet has been received completely. Then the process can access the packet via the environment variable *ThePacket* (Sect. 7.1.3) and examine its contents. If *ThePacket* has been set in response to *EMP*, there is no need to check whether it is addressed to the current station (by definition of the *EMP* event, the argument-less variant of *isMine* returns nonzero on such a packet). The most often performed operation after the complete reception of a standard packet (i.e., one that originated in the *Client*) is:

Client->receive (ThePacket, ThePort);

The purpose of this operation is to update the performance measures associated with the traffic pattern to which the packet belongs and with the link on which the packet has been received. The semantics of *receive* are discussed in detail in Sects. 7.1.6 and 8.2.3. From the viewpoint of the present section suffices it to say that *receive* should always be executed for a standard packet (for which performance measures are to be collected) as soon as, according to the protocol rules, the packet has been completely and successfully received at its final destination (has completed its trip within the network).

The method exists in the following variants:

void receive (Packet *pk, Port *p);
void receive (Packet *pk, Link *lk = NULL);
void receive (Packet &pk, Port &pr);
void receive (Packet &pk, Link &lk);

The second variant accepts a link pointer rather than a port pointer, and the last two variants are equivalent to the first two, except that instead of pointers they take object references as arguments. If the second argument is not specified, it is assumed to be *NULL*, which means the link performance measures will not be affected by the packet's reception.

An identical collection of *receive* methods is declared within type *Traffic*. Those methods operate exactly as the corresponding *Client* methods, with the additional check whether the received packet belongs to the indicated traffic pattern. Thus, they double as simple assertions.

All versions of *receive* verify whether the station whose process is executing the operation is authorized to receive the packet, i.e., if the packet is addressed to the station. A packet whose *Receiver* attribute is *NONE* can be received by any station. Note that such a packet never triggers a *BMP* or *EMP* event.

Below is the simplest code method of a process responsible for receiving packets.

```
perform {
        state WaitForPacket:
                RcvPort->wait (EMP, NewPacket);
        state NewPacket:
                Client->receive (ThePacket, ThePort);
                skipto WaitForPacket;
};
```

Note that *RcvPort* could be used, instead of *ThePort*, as the second argument of receive with the same result (this is the only port on which a packet is ever received by the process). After returning from *receive*, the process uses *skipto* (Sect. 5.3.1) to skip the pending *EMP* event and resume waiting for a new packet from the next *ITU*.

The *Receiver* attribute of a packet can be interpreted in two ways: as the *Id* of a single station or as a station group pointer (Sects. 6.2 and 6.4). The latter interpretation is assumed for a broadcast packet. To tell whether a packet is a broadcast one, the recipient can examine its flag number 3 (*PF_broadcast*, see Sect. 6.2) which is set internally for broadcast packets. The following two packet methods:

```
int isBroadcast ();
int isNonbroadcast ();
```

return the broadcast status of the packet without referencing this flag explicitly. Another flag (number 30 or *PF_last*) is used to mark the last packet of a message. It is possible (see Sect. 6.7.1) that a single message is split into multiple packets which are transmitted and received independently. If the *PF_last* flag of a received packet is set, it means that the packet is the last (or the only) packet acquired from its message. This flag is used internally by SMURPH in calculating the message delay (Sect. 10.2.1). The following two packet methods:

int isLast ();
int isNonlast ();

tell whether the packet is the last packet of a message without referencing the *PF_last*
flag directly.

7.2 Port Inquiries

By issuing a port wait request, a process declares that it would like to learn when
something happens on the port in the future. It is possible to ask (poll) a port about
its present status, i.e., to determine the configuration of activities currently perceived
by the port, about its past status, within the window allowed by the link's linger time
(Sect. 7.1.1), or (in a way) about the nearest future, without necessarily waiting for
events.

7.2.1 Inquiries About the Present

The following port method:

Boolean busy ();

returns *YES*, if the port is currently perceiving any activity (a packet transmission,
a jamming signal, or a number of overlapping activities), and *NO* otherwise. The
complementary method:

Boolean idle ();

is a simple negation of *busy*.

Here is a way to check if there is an (own) activity in progress, one that has
originated on the port. This port method:

Boolean transmitting ();

returns *YES*, if there is an ongoing packet transmission or a jamming signal that was
started on the port and not yet completed (by *stop* or *abort*, Sect. 7.1.2).

In the case when multiple overlapping activities are perceived at the port at the
same time, it is possible to count them by calling one of the following port methods:

int activities (int &t, int &j);
int activities (int &t);
int activities ();

With the first method, the two reference arguments return the number of packet
transmissions (t) and the number of jamming signals (j) sensed by the port within
the current *ITU*. The method's value tells the total number of all those activities,

i.e., the sum of t and j. The second method does not explicitly return the number of jams; however, the method's value still gives the total number of all activities (so the amount of information returned by the first two methods is the same). The last method has no arguments and it merely returns the total number of all activities perceived by the port.[4]

In the case when exactly one packet transmission is sensed by the port and no other activity is present at the same time, each of the above methods sets the environment variable *ThePacket* to point to the packet object. *ThePort* is always set to point to the port to which the inquiry was addressed. Note that when the condition is not met (there is more than just a single packet), the value of *ThePacket* is undefined (the environment variable is not touched by the method and retains its previous setting).

When a process is awakened by *ANYEVENT* (Sect. 7.1.3), it may happen that two or more simultaneous events have triggered the waking event. It is possible to learn how many events of a given type occur on the port by calling this port method:

int events (int etype);

where the argument identifies the event type in the following way:

BOT	beginning of packet
EOT	end of a complete packet (one that was terminated by *stop*)
BOJ	beginning of a jamming signal
EOJ	end of a jamming signal

The method returns the number of events of the type indicated by the argument. By calling *activities* (see above) the program can learn the number of packets that are simultaneously heard on the port. If the process subsequently wishes to examine all these packets, the following port method is available:

Packet *anotherPacket ();

The method can be invoked after receiving one of the events *BOT*, *EOT*, *BMP*, *EMP*, or after calling *events* with the second argument equal *BOT* or *EOT*, or after calling *activities*. When the method is invoked after the process has been awakened by one of the four packet events, its subsequent calls return pointers to all packets triggering the event at the same time. These pointers are returned both via the method's value and in *ThePacket*. If *NULL* is returned, it means that there are no more packets, i.e., all of them have been examined. If n packets have triggerred a given event simultaneously (within the same *ITU*), n calls to *anotherPacket* are required to scan them all.

---

[4]Many protocol programs do not use jamming signals at all.

When invoked after *events*, the method behaves exactly as if the process were awakened by *BOT* or *EOT*, depending on the argument of *events*. After a call to *activities*, the method returns (in its subsequent calls) pointers to all packets currently perceived by the port.

7.2.2 Inquiries About the Past

In a real network, it is generally impossible to ask a port about its past status, unless some memory feature is implemented. In SMURPH this possibility has been provided to simplify the implementations of some protocols. Of course, it is always possible to re-program a protocol in such a way that no port inquiries about the past are made: processes can maintain their private data bases of the interesting past events, as they would have to in real life.

The history of past activities in a link is kept in the link's archive for the time determined by the current setting of the link's *linger* attribute (Sects. 4.2.2 and 7.1.1). If no inquiries about the past are ever issued to a link (i.e., to a port connected to the link), the linger time should be set to 0, which will tend to reduce the simulation time. Otherwise, the linger interval should be big enough to make sure that the protocol works correctly, but not too big, to avoid overloading the simulator with a huge number of unnecessarily archived activities.

Below, we list the port methods that perform inquiries about the past. All of them return values of type *TIME* and take no arguments, so we just list their names.

lastBOA	The method returns the time when the last beginning of (any) activity (the end of the last silence period) was perceived on the port. *TIME_0* is returned, if no activity has been heard on the port within the linger time. The method can also be referenced as *lastEOS* (for "last End Of Silence")
lastEOA	The method returns the time of the beginning of the current silence period on the port (the last End Of Activity). *TIME_inf* (meaning "undefined") is returned, if an activity is currently heard on the port. If no activity has been noticed on the port within the linger time, the method returns *TIME_0*. The method can also be called as *lastBOS* (for "last Beginning Of Silence")
lastBOT	The method returns the time when the last beginning of a packet was heard on the port. *TIME_0* is returned, if no beginning of packet has been seen within the link's linger time. The event is unaffected by other activities that may overlap with the packet. For example, if collisions are possible, and they affect the reception status of packets, they must be recognized explicitly, because *lastBOT* is unaffected by them
lastEOT	The method returns the time when the last end of a complete packet transmission (i.e., terminated by *stop*, see Sect. 7.1.2) was perceived on the port. *TIME_0* is returned, if no end of packet has been noticed on the port within the linger time. Similar to *lastBOT*, other activities occurring at the same time have no impact on the event, as long as the packet has been terminated by *stop* (rather than *abort*)

(continued)

(continued)

lastBOJ	The method returns the time when the last beginning of a jamming signal was heard on the port. Overlapping jams are recognized as a single continuous jamming signal, so only the beginning of the earliest of them is detected. *TIME_0* is returned, if no jamming signal has been perceived on the port within the linger time
lastEOJ	The method returns the time when the last end of a jamming signal was heard on the port. *TIME_inf* is returned, if a jamming signal is currently present. If no jam has been observed on the port within the linger time, the method returns *TIME_0*
lastCOL	The method returns the time of the beginning of the last collision that has occurred on the port. By the beginning of a collision we mean the first moment when the collision was recognized (Sect. 7.1.4). *TIME_0* is returned, if no collision has been perceived on the port within the linger time

Note that the link's linger time need not be nonzero for the above methods to operate sensibly, assuming that we understand their limitations. For example, the two port methods *busy* and *idle* (Sect. 7.2.1) are implemented as macros that, respectively, expand into:

undef (lastEOA ())
def (lastEOA ())

If a port is perceiving an activity, e.g., receiving a packet, then it knows for a fact that the entire activity (its complete description) is present and available in the link and is going to be so at least until the last bit of the activity is gone from the port. Before that time, a query about the time of the activity's commencement (as seen by the port) is going to produce the correct value, regardless of the setting of linger and the activity's duration.

7.2.3   Inquiries About the Future

The essence of event-driven simulation is that inquiries about the future are best addressed by waiting for events, which in SMURPH means issuing *wait* requests to ports. Ports also offer a collection of methods for asking about the timing of future events, to the extent that they can be predicted based on the present configuration of activities in the link. These methods are seldom useful in serious models, but they may prove handy as shortcuts in some situations. They have been made available, because the link model uses them internally in timing the events awaited with the "proper" *wait* requests. All the methods return values of type *TIME* and take no arguments, so we just list their names.

activityTime	The method returns the nearest time when the port will appear busy (in the sense of the *busy* method from Sect. 7.2.1). If the port is busy at the time when the method is invoked, the method will return *Time* (i.e., now). If the port is idle, but there exist activities in the link that will reach the port at some known time in the future, the method will return that time. Otherwise, the method returns *TIME_inf* (Sect. 3.1.5) interpreted as "undefined"
silenceTime	The method returns the nearest time when the port will appear idle (in the sense of the *idle* method from Sect. 7.2.1). If the port is idle at the time when the method is invoked, the method will return *Time* (i.e., now). If the port is perceiving an activity, or a set of overlapping activities, and the time when they will disappear is definite (known), the method will return that time. Otherwise, the method returns *TIME_inf* interpreted as "undefined"
bojTime	The method returns the nearest time when the port will perceive a jamming signal. If the port is perceiving a jamming signal at the time when the method is invoked, the method will return *Time* (i.e., now). If the port is not perceiving a jamming signal, but there exist jamming signals in the link that will reach the port at some known time in the future, the method will return that time. Otherwise, the method returns *TIME_inf* interpreted as "undefined"
eojTime	The method returns the nearest time when the port will stop perceiving a jamming signal. If the port is not perceiving a jamming signal at the time when the method is invoked, the method will return *Time* (i.e., now). If the port is perceiving a jamming signal, but the time when the signal will disappear from the port is known, the method will return that time. Otherwise, the method returns *TIME_inf* interpreted as "undefined"
collisionTime	The method returns the nearest time when the port will perceive a collision (interpreted as the event described in Sect. 7.1.4). If the port is perceiving a collision at the time when the method is invoked, the method will return *Time* (i.e., now). If the port is not perceiving a collision, but the time when a collision will occur on the port is known, the method will return that time. Otherwise, the method returns *TIME_inf* interpreted as "undefined"
botTime	The method returns the nearest time when the port will perceive the beginning of a packet, i.e., a *BOT* event (according to Sect. 7.1.3). If the port is perceiving a *BOT* event at the time when the method is invoked, the method will return *Time* (i.e., now). If the time when a *BOT* event will occur on the port is known, the method will return the earliest time when that will happen. Otherwise, the method returns *TIME_inf* interpreted as "undefined"
eotTime	The method returns the nearest time when the port will perceive the end of a packet, i.e., an *EOT* event (according to Sect. 7.1.3). If the port is perceiving an *EOT* event at the time when the method is invoked, the method will return *Time* (i.e., now). If the time when an *EOT* event will occur on the port is known, the method will return the earliest time when that will happen. Otherwise, the method returns *TIME_inf* interpreted as "undefined"

(continued)

(continued)

bmpTime *empTime*	The methods are like *botTime* and *eotTime*, but they additionally make sure that the packet(s) involved are addressed to the current station (see the *BMP* and *EMP* events in Sect. 7.1.3)
aevTime	The method returns the nearest time when anything of interest will happen on the port. This effectively means the time when *ANYEVENT* would be triggered (Sect. 7.1.3), if the present configuration of link activities were taken into account. Note that the method can return *Time* (now)

Needless to say, the future, as "predicted" by the above methods can only be known based on the configuration of activities present in the link at the moment when the method is invoked. For example, if an activity has been started by some other port, it has reached the current port, but it has not yet been terminated on the originating port, *silenceTime* will return *TIME_inf* (and *activityTime* will return *Time*).

The methods are used to compute (and recompute) waking times for wait requests addressed to ports. Suppose that a process issues a wait request for a *BOT* event. The simulator would like to create a description of the event and insert it into the event queue with the proper time stamp (Sect. 1.2.1). The time stamp of that event is determined by invoking *botTime* on the port. It can be "undefined" (as in "don't know"), when there is no activity in the link that the port can perceive (now or in the future). In that case the event will be added at the end of the queue with the time stamp of *TIME_inf*, marking it as dormant. The time stamp may have to revised (recalculated) when a new activity is inserted into the link (the event may then change its position in the queue). A dormant event may receive a definite time stamp, or an event scheduled at some definite time may be rescheduled earlier. One example of the former scenario is when a process is waiting for an *EOT* (or *EMP*) event, and the activity that (according to the present configuration of activities in the link) may trigger the event has not been terminated yet (the packet is still being transmitted by the sender). When the sender eventually stops the transmission, the pending *EOT/EMP* events will be rescheduled from *TIME_inf* to a definite time.

7.3 Faulty Links

It is possible to implant into a link a faulty behavior, whereby packets transmitted over the link may be damaged with a certain probability. A damaged packet is marked in a special way, and this marking can affect the interpretation of certain events and the outcome of certain inquiries. The following link method declares the fault rate for the link:

setFaultRate (double rate, int ftype);

288 7 Links and Ports

The first argument specifies the probability of error for a single packet bit, i.e., the bit error rate (BER). Its value must be between 0 and 1; usually it is much less than 1. The second argument selects the processing type for invalid packets, as explained below. By default, i.e., before *setFaultRate* is called, the link is error-free, and packets inserted into it are never damaged.

A faulty link can be reverted to its error-free status by calling *setFaultRate* with the second argument equal *NONE* (the value of the first argument is then irrelevant).

The valid/invalid status of a packet is stored in the packet's *Flags* attribute Sect. 6.2). Two flags are used for this purpose, and a packet can be damaged in one of the following two ways:

- The packet's header is unrecognizable and the packet's destination cannot be (formally) determined. This status is indicated by two flags: *PF_hdamaged* (flag 28) and *PF_damaged* (flag 27) being set simultaneously.
- The packet's header is correct, but the packet is otherwise damaged (e.g., its trailer checksum is invalid).

Note that a packet that is header-damaged is also damaged, but not vice-versa. The damage status of a packet can always be determined directly, by calling any of the following packet methods:

int isDamaged ();
int isValid ();
int isHDamaged ();
int isHValid ();

Each of the above methods returns either *YES* or *NO*, in the obvious way. Note that *isHDamaged* implies *isDamaged*; similarly, *isValid* implies is *isHValid*.

Depending on the value of the second argument of *setFaultRate*, the damaged status of a packet can be interpreted automatically, by suppressing certain events that are normally triggered by a valid packet. In addition to *NONE* (which makes the link error-free), the following three values of *ftype* are legal:

FT_LEVEL0	No automatic processing of damaged packets is in effect. Damaged packets trigger the same events and respond to the same inquiries as valid packets. The only way to tell a damaged packet from a valid one is to examine the packet's *Flags* attribute, e.g., with one of the above methods
FT_LEVEL1	A header-damaged packet does not trigger the *BMP* event. A damaged packet does not trigger the *EMP* event (Sect. 7.1.3). This also applies to *anotherPacket* (Sect. 7.2.1) called after *BMP* or *EMP*), which ignores header-damaged (damaged) packets. The packet method *isMine* (Sect. 7.1.6) returns *NO* for a header-damaged packet (but works as usual if the packet is damaged without being header-damaged)
FT_LEVEL2	The interpretation of damaged packets is as for *FT_LEVEL1*, but *BOT* and *EOT* events are also affected. Thus, a header-damaged packet does not trigger the *BOT* event and a damaged packet does not trigger the *EOT* event. This also applies to *anotherPacket* and to the link inquiries about past *BOT/EOT* events

Regardless of the processing type for damaged packets, the events *SILENCE*, *ACTIVITY*, *COLLISION*, *BOJ*, *EOJ*, and *ANYEVENT* (Sect. 7.1.3) are not affected by the damage status of a packet. Note that jamming signals are never damaged.

The standard link method used by SMURPH to generate damaged packets is declared as *virtual*; consequently, it can be overridden by the user in a *Link* subtype declaration. The header of this method looks like this:

virtual void packetDamage (Packet *p);

The method is called for each packet given as an argument to *transmit* (Sect. 7.1.2) to decide whether it should be damaged or not. First, based on the length of the packet header (p->$TLength$ –p->$ILength$), *packetDamage* determines the probability that the packet will be header-damaged. This probability is equal to:

$$1-(1-r)^h,$$

where r is the fault rate and h is the header length in bits. A uniformly distributed random number between 0 and 1 is then generated, and if its value is less than the damage probability, the packet is marked as both header-damaged and damaged (the flags *PF_hdamaged* and *PF_damaged* are both set). Otherwise, the probability of the packet being damaged (without being header-damaged) is determined based on the packet's *ILength* attribute and another random number is generated, in the same way as before. If this number turns out to be less than the damage probability for the packet's payload, the packet is marked as damaged (*PF_damaged* is set, but *PF_hdamaged* is cleared). If the packet is not to be damaged at all, both damage flags are cleared.

The possibility of declaring links as faulty is switched off by default, to avoid unnecessary overhead. To enable faulty links, the program must be compiled with the $-z$ option of mks (Sect. B.5).

7.4 Cleaning After Packets

Sometimes a packet needs to carry a pointer to a dynamically allocatable data structure that must be deallocated when the packet disappears from the network. This issue was discussed in Sect. 6.6.4, where tools for cleaning up after standard (i.e., *Client*) packets (and messages) were presented. It is possible to perform a similar kind of operation at the link level, in a way that makes it applicable to all packets (including non-standard ones) reaching the end of their stay within the link (or its archive, Sect. 7.2.2). With this *Link* method:

void setPacketCleaner (void (*)(Packet*));

one can plug into the link a function to be called for every packet disappearing from the link. By examining the packet's *TP* attribute (Sect. 6.2) and casting the packet to the proper type, the function can then selectively perform the required cleaning duties, e.g.:

```
void cleanPackets (Packet *p) {
        if (p->TP == −1)
                delete ((HelloPacket*)p)->NeighborList;
}
...

EtherLink->setPacketCleaner (cleanPackets);
...
```

Note that the *cleaner* function is global. One must be aware that the data structure deallocated by the cleaner is volatile. Thus, if the packet's recipient needs to preserve it beyond the life time of the packet's activity in the link, it must make itself a copy. Also, if traffic-level cleaning is applied together with the link-level cleaning, one should make sure that the same packet is not cleaned twice.

If *setPacketCleaner* is called with a *NULL* argument, the packet cleaner function is removed from the link.

For a packet belonging to a traffic pattern (i.e., one generated and absorbed by the *Client*), packet deallocation is most naturally intercepted with a *virtual* method defined within the traffic type to which the packet belongs (Sect. 6.6.4). Thus, a link-level interceptor is only recommended for those situations when either all or most packets must be processed that way, or special packets (i.e., ones that do not originate in the *Client*, and have no associated traffic objects) must be subjected to the procedure. Beware that, unlike the cleaning methods associated with traffic patterns, which are invoked when the packets are destructed (i.e., on *delete*), the link-level cleaners are only invoked when a packet disappears from the link or its archive. The packet is then automatically deleted by the link, so cleaning implies deletion, but if the packet is explicitly deleted by the program (beyond the link purge mechanism), the link-level cleaner will not be called.

As link types are practically never extended by the user, link-level cleaning has been implemented as a "plug-in" function rather than a *virtual* method (like for traffic patterns). Extending a link type to accommodate this single feature into a *virtual* method would be (arguably) an overkill.

References

1. P. Gburzyński, P. Rudnicki, Modeling low-level communication protocols: a modular approach, in *Proceedings of the 4th International Conference on Modeling Techniques and Tools for Computer Performance Evaluation*, Palma de Mallorca, Spain, 1988
2. P. Gburzyński, P. Rudnicki, LANSF—a modular system for modelling low-level communication protocols, in *Modeling Techniques and Tools for Computer Performance Evaluation*, ed. by R. Puigjaner, D. Potier (Plenum Publishing Company, 1989), pp. 77–93

Chapter 8
Radio Channels and Transceivers

The role of transceivers is similar to that of ports; thus, transceivers share many methods with ports and trigger similar events. Operations on transceivers cover:

- starting and terminating packet transmissions
- inquiring about the present status of the transceiver, i.e., its perception of the radio channel
- wait requests, e.g., to await an event related to a packet arrival.

One difference with respect to ports is that transceivers implement no methods for inquiring about the past (Sect. 7.2.2) and about the future (Sect. 7.2.3).

8.1 Interpreting Activities in Radio Channels

There is only one type of activity that can be inserted into a transceiver, i.e., a packet transmission.[1] A packet is always preceded by a preamble, which consists of a specific number of "physical" bits. These bits do not count to the total length of the packet (attribute *TLength*, Sect. 6.2), but they do take bandwidth, i.e., need time to transmit. A preamble can be as short as one bit, but its length cannot be zero. The preamble length can be defined when the transceiver is created (Sect. 4.5.1) or re-defined at any time with *setPreamble* (Sect. 4.5.2). Different transceivers can use preambles of different length and can even set that length dynamically, e.g., before every packet transmission.

---

[1] In contrast to ports, there are no jamming signals.

© Springer Nature Switzerland AG 2019
P. Gburzyński, *Modeling Communication Networks and Protocols*,
Lecture Notes in Networks and Systems 61,
https://doi.org/10.1007/978-3-030-15391-5_8

8.1.1   The Stages of Packet Transmission and Perception

A packet is transmitted, and also perceived[2] at those transceivers that it will ever reach, in three stages, which mark the timing of the possible events that the packet may ever trigger on a perceiving transceiver:

1. *Stage P*: the beginning of preamble. This stage marks the moment when the first bit of the preamble is transmitted at the source or arrives at a perceiving transceiver.
2. *Stage T*: the end of preamble, which usually coincides with the beginning of packet. This is the moment when the last bit of the preamble has been transmitted (or perceived) and the first bit of the actual packet begins.
3. *Stage E*: the end of packet representing the moment when the last bit of the packet has been transmitted (or has passed through a perceiving transceiver).

The timing of the three stages at the transmitter is determined by the transceiver's transmission rate (*XRate*) and the lengths of the corresponding packet components. If stage *P* occurs at time *t*, then stage *T* will take place at $t + $ *tcv->XRate* * *tcv->Preamble*, where *tcv* is the transmitting transceiver and *Preamble* is the preamble length. Then, stage *E* will occur *RFC_xmt (tcv->XRate, pkt) ITUs* after stage *T*, where *pkt* points to the packet being transmitted. Recall that *XRate* is a transceiver attribute representing the number of *ITUs* required to transmit a single bit (Sect. 4.5.1), and *RFC_xmt* is one of the so-called assessment methods of the radio channel to which the transceiver is interfaced (Sect. 4.4.3). Note that the default version of *RFC_xmt* calculates the packet's transmission time as *tcv->XRate* * *pkt->TLength*, assuming a straightforward interpretation of the packet's bits as the actual bits to be sent over the channel at the specified rate. This need not always be the case. In many encoding schemes, the "logical" packet bits (represented by *pkt->TLength*) are transformed into symbols expressed as sequences of "physical" bits. For example, those symbols may balance the number of physical zeros and ones and/or provide for forward error correction (FEC). It is the responsibility of *RFC_xmt* to account for such features and determine the correct timing for sending the packet contents over the channel.

The time separation between the three stages at any perceiving transceiver reached by a packet is identical to their separation at the packet's transmission. The delay of their appearance at the receiver, with respect to the transmission time, is equal to the distance in *ITUs* separating the receiver from the transmitter. This distance, in turn, is determined by the Cartesian coordinates of the two transceivers (Sect. 4.5.2). For illustration, suppose that a packet is transmitted by transceiver *v* with coordinates (x_v, y_v).[3] Let *w*, with coordinates (x_w, y_w), be a neighbor of *v* perceiving the packet. If

---

[2] We use the word "perception" instead of "reception" in reference to the fact that the packet is noticed at all by the transceiver. This means that the packet is audible and its signal may be recognized and, for example, may interfere with other packets, but need not mean that the packet can be received. The word "reception" will be used to refer to the latter circumstance.

[3] This example assumes a 2d deployment of nodes, but it all looks the same in a 3d deployment, with the difference of one extra coordinate.

stage P of the packet occurs at time t at v (i.e., the first bit of the preamble is transmitted by v at time t), then it will occur at w at time $t + \sqrt{(x_w - x_v)^2 + (y_w - y_v)^2}$, assuming that the transceiver coordinates are expressed in *ITUs*.[4] The same transformation concerns the remaining two stages of the packet, i.e., their occurrence at w is offset by $\sqrt{(x_w - x_v)^2 + (y_w - y_v)^2}$ with respect to v.

8.1.2 Criteria of Event Assessment

A packet being transmitted on one transceiver v and perceived by another transceiver w may trigger on w events very similar to those for ports and a wired link (Sect. 7.1.3). For one difference, there is no explicit *COLLISION* event for a transceiver. A packet transmitted on a transceiver t may contribute to the overall signal level perceived by t (its receiver), but it never triggers on t the standard packet events (*BOP*, *BOT*, *EOT*, and so on). This means that a packet transmitted from a given transceiver cannot be (automatically and inadvertently) received by the same transceiver.

The extent to which a transmitted packet is going to affect the activities perceived by other transceivers (and whether it is going to be received as a valid packet) is described by the collection of user-redefinable (*virtual*) assessment methods of the underlying radio channel (*RFChannel*). Those methods (see Sect. 4.4.3) can base their decisions on the following criteria:

The transmission power	This is a *double* attribute of the transmitting transceiver, which can be set with *setXPower* (see Sect. 4.5.2)
The receiver gain	This is a *double* attribute of the perceiving transceiver, which can be set with *setRPower* (see Sect. 4.5.2)
The transmission tag	This is an *IPointer*-type (Sect. 3.1.1) attribute of the transmitting transceiver, which can be set with *setXTag* (see Sect. 4.5.2)
The receiver tag	This is an *IPointer*-type attribute of the receiving transceiver, which can be set with *setRTag* (see Sect. 4.5.2)
The distance	This is the Euclidean distance in *DUs* separating the transmitting transceiver from the receiving transceiver (see Sect. 4.5.2)
The interference histogram	This is a representation of the complete interference history of the packet. It describes how much other signals perceived by the transceiver at the same time as the packet have interfered with the packet at various moments of its perception

Two of the assessment methods, *RFC_erb* and *RFC_erd*, facilitate converting user-defined distributions of bit-error rates (typically dependent on the signal to interference ratio) into conditions and events determining the success or failure of a packet reception. These methods are optional: they provide convenient shortcuts, e.g., potentially useful to other assessment methods like *RFC_bot* and *RFC_eot*,

---

[4]They are internally; however, the user specifies them in *DUs* (see Sect. 4.5.2).

and help implement tricky events whose timing depends on the distribution of bit errors in received packets. Note that the bits referred to above are physical bits (see Sect. 8.1.1).

Since the assessment methods are (easily) exchangeable by the user, the radio channel model provided by SMURPH is truly open-ended. The standard versions of those methods (Sect. 4.4.3) implement a very naive and practically useless channel model. In contrast to wired channels, the large population of various propagation models and transmission/reception techniques makes it impossible to satisfy every-one with a few complete built-in models. Consequently, SMURPH does not attempt to close this end. Instead, it offers a simple and orthogonal collection of tools for describing different channel models with a minimum effort.

8.1.3   Neighborhoods

For every transceiver t, SMURPH keeps track of its neighborhood, i.e., the population of transceivers (interfaced to the same *RFChannel*) that may perceive (not necessarily receive) packets transmitted from t. The configuration of neighborhoods is dynamic and may change in the following circumstances:

1. A transceiver changes its location. In such a case, the neighborhood of the moving transceiver may change. Also, the neighborhoods of other transceivers may have to be modified by either including or excluding the moved transceiver.
2. A transceiver changes its transmission power (attribute *XPower*). If the power increases, the neighborhood of the transceiver may include more transceivers; if it decreases, the neighborhood may have to be trimmed down.
3. A transceiver changes its reception gain (attribute *RPower*). If the gain increases, the transceiver may have to be included in some new neighborhoods; if it decreases, the transceiver may have to be removed from some neighborhoods.

The actual decisions as to which transceivers should be included in the neigh-borhood of a given transceiver are made by invoking *RFC_cut* (Sect. 4.4.3). The method returns the maximum distance in *DUs* on which a signal transmitted at a given power (the first argument) will be perceived by a receiver operating at a given gain (the second argument). Note that the method can be conservative, i.e., the model should remain correct, if the neighborhoods are larger than they absolutely must be. In particular, one may be able to get away with a "global village," i.e., a single neighborhood covering all transceivers (*RFC_cut* returning *Distance_inf*). Practi-cally, the only advantage of trimmed neighborhoods is the reduced simulation time and, possibly, reduced memory requirements of the simulator. This advantage is gen-erally quite relevant, especially for networks consisting of hundreds and thousands of transceivers.

Note that one more circumstance (ignored in the above list) potentially affecting the neighborhood status of a pair of transceivers is a *tag* change (in one or both). This, however, has been intentionally eliminated from the scope of *RFC_cut*, which method deals with the pure signal level (plus pure receiver gain) ignoring the tags of the assessed parties. This is because otherwise, due to the free interpretation of the tags, any change in a tag would require a complete reconstruction of the neighborhoods, which would be a rather drastic and time-consuming action. Consequently, if tags are used by the channel model, *RFC_cut* should assume that their configuration is always most favorable from the viewpoint of signal propagation, i.e., the returned neighborhood distance should be the maximum over all possible values of the tags involved.

8.1.4 Event Assessment

At any moment, a given transceiver may perceive several packets arriving from different neighbors and seen as being at different stages. The role of the assessment procedure at a receiving transceiver is to determine whether any of those packets should be received or, more specifically, whether it should trigger the events that will effect its reception. The decisions are arrived at by a collective interaction of the (user-exchangeable) assessment methods listed in Sect. 4.4.3.

First, *RFC_att* is called to determine the signal level of the packet at the perceiving port. Its first argument describes the transmission power at which the packet was transmitted at the sending transceiver, as well as the tag of that transmission, the second argument gives the distance between the sender and the perceiving transceiver, and the last argument points to the sending transceiver. The receiving (destination) transceiver can be accessed through the environment variable *TheTransceiver* (*Info02*).

Many simple channel models ignore tags, assuming that all transmissions and receptions are homogeneous. In a simple distance-based propagation model, the method may use a (possibly randomized) function of the original power level and the distance to produce the attenuated signal level. For one example of a scenario involving tags, consider multiple channels with nonzero crosstalk. With the tags representing the different channels, *RFC_att* may apply an additional factor to the signal depending on the difference between the channel numbers (Sect. A.1.2).

Note that, although the sender transceiver is directly available to *RFC_att* (via the third argument), the method should avoid consulting its attributes, if only possible. For example, instead of using the *Tag* value from the first (*SLEntry**) argument, the method may be tempted to invoke *getXTag* for the sender's transceiver. However, these two values can be legitimately different, as the *tag* of the perceived signal was associated with it in the past (when the packet was transmitted), and the current tag setting of the sender's transceiver need not be the same. On the other hand, the attributes of the receiving transceiver (whose pointer is available via the environment variable *TheTransceiver*) are up to date and they directly apply to the situation under assessment.

The signal level returned by *RFC_att* is associated with the packet at its perceiving transceiver. Suppose that some transceiver v perceives n packets denoted

$p_0, \ldots, p_{n-1}$, with $r_0, \ldots, r_{n-1}$ standing for their received signal levels. The interference level experienced by one of those packets, say p_k is determined by a combination of all signals of the remaining packets. The assessment method *RFC_add* is responsible for calculating this combination. Generally, the method takes a collection of signal levels and returns their combined signal level. One should remember that the signals presented to *RFC_add* (in the array passed in the third argument) have been already attenuated, i.e., they are outputs of *RFC_att*. Consequently, although the tags of those signals are available to *RFC_add*, the method should not apply to them the same kind of "factoring" as that prescribed for *RFC_att*, although, of course, it may implement whatever subtle ways of adding those signals are required by the model.

If the second argument of *RFC_add* is nonnegative, it should be interpreted as the index of one entry in the signal array that must be ignored. This is exactly what happens when the method is called to calculate the interference experienced by one packet. The collection of signals passed to the method in the third argument always covers the complete population of signals perceived by the transceiver. The indicated exception refers then to the one signal for which the interference produced by the other signals is to be calculated.

Sometimes *RFC_add* is invoked to compute the global signal level caused by all perceived packets with no exception. In such a case, the second argument is *NONE*, i.e., -1. In all cases, the *Tag* element of the last argument of *RFC_add* refers to the current value of *XTag* at the perceiving transceiver. The *Level* component of that argument is zero, unless the transceiver is currently transmitting a packet of its own, in which case it contains the transceiver's transmission power setting.

The actual decision regarding a packet's reception is made by *RFC_bot* and *RFC_eot*. Their roles are similar, but they are called at different stages of the packet's perception. *RFC_bot* is invoked at stage T, i.e., at the end of the preamble and before the first bit of the proper packet. It takes the following arguments:

- the transmission rate r at which the packet was transmitted
- the perceived signal level of the packet along with the tag passed in the *SLEntry** argument *sl*
- the reception gain of the perceiving transceiver combined with the transceiver's tag passed in another *SLEntry** argument *rs*
- the interference histogram of the preamble *ih*.

The method returns a *Boolean* value: *YES*, if the perceived quality of the preamble makes it possible to formally start the packet's reception, and *NO*, if the packet cannot be received. In real-life terms, this decision determines whether the receiver has been able to recognize that a packet is arriving and, based on the quality of the preamble, clock itself to the packet. If the decision is *NO*, the packet will not be received. In particular, its next stage (E) will not be subjected to another assessment, and the packet will trigger no reception events (Sect. 8.2.1). If the method returns *YES*, the packet will undergo another assessment at stage E regarding the successful reception of its last bit. This second assessment will be handled by *RFC_eot*. In any case (also following a negative assessment by *RFC_bot*) the packet will continue contributing its signal to the population of activities perceived by the transceiver until stage E, i.e., until the packet is no more heard by the transceiver.

Note that *RFC_eot* has the same header as *RFC_bot*. The method is called in stage E to decide whether the packet has been successfully received, or rather is receivable, as the actual reception must be carried out explicitly by the respective process of the protocol program. The idea is the same as for the preamble, except that this time the interference histogram covers the proper packet, i.e., the part following the preamble until and including the packet's last bit.

8.1.5   Interference Histograms

The interference histogram passed as the last argument to *RFC_bot* refers solely to the preamble component of the packet. On the other hand, the histogram passed to *RFC_eot* applies to the packet proper, excluding the preamble. In both cases, the histogram describes the complete interference history of the respective component as a list of different interference levels experienced by it, along with the intervals during which the interference has remained constant.

The interference histogram is a class comprising several useful methods and two public attributes:

int NEntries;
*IHEntry *History;*

where *History* is an array of size *NEntries*. Each entry in the *History* array is a structure looking like this:

```
typedef struct {
        TIME Interval;
        double Value;
} IHEntry;
```

The sum of all *Intervals* in *History* is equal to the total duration (in *ITUs*) of the activity being assessed. The corresponding *Value* attribute of the *History* entry gives the interference level (as calculated by *RFC_add*) experienced by the activity within that interval.

The complete *History* consists of as many intervals as many different interference levels have happened during the transceiver's perception of the assessed activity. If the preamble or packet has experienced absolutely no interference, its *History* consists of a single entry whose Interval spans its entire duration and whose *Value* is 0.0.[5]

---

[5]Recall that signals are represented internally on the linear scale.

Below we list the public methods of class *IHist*, which may be useful for examining the contents of the interference history.

void entry (int i, TIME &iv, double &v);
void entry (int i, double &iv, double &v);
void entry (int i, RATE r, Long &iv, double &v);

These are three different versions of essentially the same method that produces the contents of a single entry (number *i*) from the interference histogram. In all three cases, the last argument returns the *Value* attribute from the *i*-th entry. The first variant returns the *Interval* in *ITUs* (via the second argument passed by reference), while the second variant converts the *Interval* to *ETUs* and returns it as a *double* number. The third version takes four arguments and converts the *Interval* to the number of bits based on the transmission rate specified as the second argument. If the *Interval* happens to be divisible by *r*, then the value returned in *iv* is simply *Interval/r*. Otherwise, it is either *Interval/r* or *Interval/r* + 1, with the probability of the latter equal to the fractional part of the exact (*double*) division. In other words, the average result over an infinite number of calls to *entry* converges on the exact ratio *(double)Interval/r*.

TIME duration ();

This method returns the total duration of the activity represented by the histogram, i.e., the sum of all *Intervals*, in *ITUs*.

Long bits (RATE r);

This method returns the total duration of the activity represented by the histogram, i.e., the sum of all *Intervals*, expressed in bits, based on the specified transmission rate *r*. The result is obtained by dividing the value returned by *duration* by *r* and, if the fractional part is nonzero, applying the same randomized adjustment as for *entry* (see above).

double avg ();
double avg (TIME ts, TIME du = TIME_inf);
double avg (RATE r, Long sb, Long nb = −1);

These methods return the average interference level within the specified interval of the histogram:

1. Over the entire histogram. The average is taken over time, not over entries, i.e., longer intervals contribute proportionally to their duration. In other words, the returned average is the integral of the interference over time divided by the total time.

2. Over the specified time: *ts* is the starting time in *ITUs* counting from zero (the beginning of the history), and *du* is the duration (also in *ITUs*). If the interval exceeds the duration of the histogram, the excessive component is ignored, i.e., it does not count to the time over which the average is taken. If the interval falls entirely beyond the histogram's duration, the returned average interference is zero.

If the second argument is not specified (or is *TIME_inf*), it stands for the end of the histogram's duration. If the first argument is *TIME_inf*, the second argument is interpreted as the interval from the end of the histogram's duration, i.e., the last *du ITUs* of the activity are examined.

3. Starting at bit *sb* and extending for *nb* bits, based on the specified rate *r*. The first bit of the histogram is numbered 0. Intervals extending beyond the histogram are treated as in the previous case. If the last argument is not specified (or is negative), all the bits of the histogram starting at bit number *sb* are considered. If *sb* is negative, *nb* bits are scanned back from the end of the activity represented by the histogram.

The following methods accept exactly the same configurations of arguments as the *avg* methods above:

double max ();
double max (TIME ts, TIME du = TIME_inf);
double max (RATE r, Long sb, Long nb = −1);

They return the maximum interference over the specified segment of the histogram.

8.1.6 Event Reassessment

In some reception models, the conditions determining the calculation of signal levels, interference, the bit error rate and, consequently, the event assessment for *BOT* and *EOT* may change half way through a packet reception. We do not consider here mobility, whose impact is only re-evaluated at activity (packet) boundaries. In all realistic mobility scenarios, the amount of position change during the reception of a single packet is completely negligible. For an example of what we have in mind, suppose that the receiver uses a directional antenna that can be reset instantaneously at will [1]. The right way of implementing this operation in the channel model is to use a dynamic gain factor in *RFC_att* (Sects. 4.4.3 and 8.1.4) depending on the antenna angle and the location of the transmitter and the receiver. When the antenna setting changes, the transceiver must re-evaluate all signals reaching it at the time, based on the new setting. This is accomplished with the following *Transceiver* method:

void reassess ();

which is to be called in all circumstances where the perception of signal levels by the receiver may have changed "behind the scenes." Note that all standard situations, like, for example, changing the receiver gain with *setRPower* (Sect. 4.5.2), are handled automatically by the simulator. By "behind the scenes," we mean "in a way not naturally perceptible by the simulator," which practically always means "affecting the results produced by *RFC_att*."

Note that straightforward directional transmission is (usually) different (and safe) in this respect. This is because (at least in all sane schemes), the transmitter sets the antenna before sending out a packet, and the entire packet is sent out with the same antenna setting. Also note that an obsessive experimenter may use this feature to implement "extreme fidelity" mobility models, whereby signals at the receiver are reassessed after a minuscule distance change occurring half-way through a packet reception. Needless to say, reassess is rather costly, and its frivolous usage is discouraged.

8.1.7   The Context of Assessment Methods

The assessment methods, whose role is described in Sects. 4.4.3 and 8.1.6, constitute the bulk part of the user-specified channel model. One should be aware that they execute in a context that is somewhat different from that of a protocol process. In this respect, they are unique examples of user-supplied code that does not belong to any particular process, but instead constitutes a kind of plug-into the simulator kernel.

Consequently, an assessment method should not assume, e.g., that, while it is run, *TheStation* points to any particular station, or *TheProcess* points to any particular process of the protocol program. On the other hand, the idea of environment variables (Sect. 5.1.6) is extrapolated over (some of) the assessment methods to implicitly extend their lists of arguments. Two standard environment variables, *TheTransceiver* and *ThePacket* are made available to some assessment methods, to provide them with extra context information that may facilitate the implementation of tricky models. Specifically, for these methods: *RFC_att*, *RFC_add*, *RFC_act*, *RFC_bot*, and *RFC_eot*, *TheTransceiver* points to the transceiver carrying out the assessment. Additionally, for *RFC_att*, *RFC_bot*, and *RFC_eot*, i.e., those assessment methods that are clearly concerned with a single activity, *ThePacket* points to the assessed packet. For *RFC_xmt*, *TheTransceiver* identifies the sending transceiver.

Sometimes, an assessment method may want to know the identity of the station for which the assessment is carried out, not only the identity of the transceiver. That may be useful for debugging or in those cases where the decision is based on exotic, dynamic criteria described in data structures stored at the receiving station. Given a transceiver, its owning station can be accessed by this *Transceiver* method:

Station *getOwner ();

which returns the pointer to the *Station* that owns the transceiver. Thus:

TheTransceiver->getOwner ()

is the correct way to reference "the station" from an assessment method, and then it means "the station for which the assessment is being carried out." Note that *TheStation* (Sect. 4.1.3), pointing to "the current station," will refer to the station whose activity has immediately triggered the current assessment, which is often the *System* station (Sect. 4.1.2).

8.1.8   Starting and Terminating Packet Transmission

Similar to a port (see Sect. 7.1.2), a packet transmission on a transceiver can be started with this method:

void startTransmit (Packet *buf);

where *buf* points to the buffer containing the packet to be transmitted. A shallow copy of this buffer is made and inserted into a data structure representing a new activity in the underlying radio channel, to be propagated to the neighbors of the transmitting transceiver. A pointer to this packet copy is returned via the environment variable *ThePacket* (*Info01*) of type *Packet**. Another environment variable, *TheTransceiver* (*Info02*) of type *Transceiver**, returns a pointer to the transceiver executing the operation. Note that this is different from the *Port* variant of *startTransmit*.

A packet transmission in progress can be terminated in two ways. The following transceiver method:

int stop ();

terminates the transmission in a way that renders the packet complete, i.e., formally outfitted with the pertinent trailer that allows the receiving party to recognize the complete packet. Instead of *stop* the program can use this method:

int abort ();

to abort the transmission. This means that the packet has been interrupted in the middle, and it will not be perceived as a complete (receivable) packet. For compatibility with the similar methods available for *Ports* (Sect. 7.1.2), both the above methods return *TRANSMISSION*.[6] Note that both *stop* and *abort* will trigger an error if no packet is being transmitted at the time of their invocation.

Both methods set two environment variables *ThePacket* (to point to the packet whose transmission has been terminated) and *TheTraffic* (to the *Id* of the traffic pattern to which the packet belongs).

The following transceiver method:

void transmit (Packet *buf, int done);

starts transmitting the packet contained in *buf* (by invoking *startTransmit*) and automatically sets up the *Timer* to wake up the process in state *done* when the transmission is complete. As explained in Sect. 8.1.1, each packet transmitted over a radio channel is preceded by a preamble, whose length cannot be zero. The total amount of time needed to transmit the packet pointed to by *buf* via a transceiver pointed to by *tcv* is equal to:

RFC_xmt (tcv->XRate, buf) + (tcv->Preamble * tcv->XRate)

and this is the amount of time after which the process will wake up in state *done*.

---

[6]There are no jamming signals in radio channels.

A started transmission must be terminated explicitly. Thus, the following sample code illustrates the full incantation of a packet transmission:

```
...
state Transmit:
        MyTransceiver->transmit (buffer, Done);
state Done:
        MyTransceiver->stop ( );

        ...
```

which is identical to the transmission sequence for a port (Sect. 7.1.2).

The transmission rate of a transceiver can be changed at any time by calling *setXRate* (see Sect. 4.5.2). If *setXRate* is invoked after *startTransmit* or *transmit* but before *stop*, the new setting has no impact on the packet being currently transmitted.

It is illegal to transmit more than one packet on the same transceiver at the same time. The transmission of a packet can be aborted prematurely (by *abort*) within the preamble (in stage *P*, Sect. 8.1.1), i.e., before the first bit of the proper packet has been transmitted. Such a packet has a degenerate stage *T* (the preamble does end but the proper packet never begins), and has no stage *E*. No assessment will ever be carried out for such a packet: *RFC_bot* will not be called after the preamble has been received, and the packet will trigger no packet events (Sect. 8.2.1), although it will trigger *EOP* (the end-of-preamble event). Note that formally *stop* can also be called before the end of preamble. This makes little sense from the viewpoint of the model, but does not cause an error. In such a case, *stop* is interpreted as *abort*.

A packet aborted past the preamble stage will never trigger an end-of-packet event (and no *RFC_eot* assessment will be carried out for it), although it may trigger beginning-of-packet events (and *RFC_bot* will be called for the packet in stage *T*).

8.2   Packet Perception and Reception

The receiver part of a transceiver can be switched off and on. A switched off receiver is incapable of sensing any activities, and will not trigger any events related to packet reception. This feature is essential from the viewpoint of modeling the behavior of real-life transceivers, whose receivers may often remain in the off state for some periods, e.g., when transiting from transmit to receive mode in a half-duplex model. For example, if the receiver is switched on after the arrival of a packet preamble, but before the arrival of the first bit of the actual packet, it may still be able (at the appropriately reduced opportunities) to recognize a sufficiently long fragment of the preamble to receive the packet. The preamble interference histogram passed to *RFC_bot* (Sects. 4.4.3 and 8.1.4) accounts only for the portion of the preamble that was *actually* perceived by the transceiver while it was switched on. Thus, the protocol program can realistically model the deaf periods of the transceiver by appropriately switching its receiver on and off. This is handled by the following methods of *Transceiver*:

void rcvOn ();
void rcvOff ();

whose meaning is obvious. Note that if the receiver is switched on after the arrival of the first bit of the proper packet, the packet will not be received (its end-of-packet event will not even be assessed by *RFC_eot*). By default, e.g., immediately after the transceiver has been created, the receiver is in the "on" state.

Note that a similar handling of the transmitter would be redundant. If the transmitter is to remain off for some time, the protocol program should simply refrain from transmitting packets during that period. A sane program would not try to transmit while the transmitter is switched off, so maintaining this status internally would make little sense.

The following two transceiver methods:

Boolean busy ();
Boolean idle ();

one being the simple negation of the other, tell the presence or absence of a (any) signal perceived by the transceiver. If the receiver is off, its status is always "idle." Otherwise, it is determined by *RFC_act*, one of the user-exchangeable assessment methods (Sect. 4.4.3), whose sole responsibility is to tell whether the channel is sensed busy or idle. In addition to computing the output the above two methods, *RFC_act* is also used in timing two transceiver events: *SILENCE* and *ACTIVITY* (Sect. 8.2.1). The assessment method takes two arguments. The first one gives the combined level of all signals currently perceived on the transceiver, as determined by *RFC_add* with the second argument equal *NONE* (Sect. 8.1.4). The second argument of *RFC_act* provides the reception gain of the transceiver (attribute *RPower* combined with the receive *Tag*). Intentionally, the method implements a threshold on the perceived signal level based on the current setting of the receiver gain.

Similar to a port, this transceiver method:

Boolean transmitting ();

returns *YES*, if the transceiver is currently transmitting is own packet.

8.2.1 Wait Requests and Events

The first argument of a *wait* request addressed to a transceiver is a symbolic constant identifying the event type. Below we list those constants and explain the conditions for the occurrence of the respective events.

SILENCE	The event occurs at the beginning of the nearest silence period perceived by the transceiver. This will happen as soon as a) the receiver is switched off, or b) the total signal level perceived by the transceiver makes *idle* return *YES* (or *busy* return *NO*, see Sect. 8.2). If the transceiver is idle when the request is issued, the event occurs within the current *ITU*
ACTIVITY	The event occurs at the nearest moment when a) the receiver is on, and b) the total signal level perceived by the transceiver makes *busy* return *YES* (or *idle* return *NO*). If the transceiver is busy when the request is issued, the event occurs within the current *ITU*
BOP	The event occurs at the nearest moment when the beginning of a preamble is heard on the transceiver. If the beginning of a preamble occurs exactly at the time when the request is issued (within the current *ITU*), the event occurs within the same *ITU* Note that this event is not assessed (Sect. 8.1.4) and is not a realistic event to be signalled by any physical hardware. For example, the beginning of any preamble, no matter how much drowned in other activities will trigger the event. Its role is to enable exotic assessment schemes and other tricks that the user may want to play with the simulator If the receiver is switched on after the first bit of the preamble has passed through the transceiver, *BOP* will no longer be triggered: the event only occurs at the "true" beginning of a preamble, as issued by the sender
EOP	The event occurs at the nearest moment when the end of a preamble is heard on the transceiver, i.e., immediately after the last bit of a preamble. If the end of a preamble occurs exactly at the time (within the same *ITU*) when the request is issued, the event occurs within the same *ITU* Like *BOP*, this event is not assessed, and its role is to enable tricks rather than provide for modeling realistic phenomena. It will be triggered by the end of an aborted preamble, even if it is not followed by a packet (Sect. 8.1.8) If the preamble is followed by a packet, then *EOP* will coincide with *BOT* (see below), except that the latter event is assessed
BOT	The event (beginning of transmission) occurs as soon as the transceiver perceives the beginning of a packet following a preamble (stage T), and *RFC_bot* returns *YES* for that packet (Sect. 8.1.4). The interpretation of this event is that the preamble has been recognized by the transceiver which can now begin to try to receive the packet. If the event condition is present at the time when the wait request is issued, the event is triggered within the same *ITU*
EOT	The event (end of transmission) occurs as soon as the following three conditions are met simultaneously: (1) the transceiver perceives the end of a packet (stage E), (2) *RFC_bot* returned *YES* for that packet in stage T, (3) *RFC_eot* returns *YES* for the packet in the present stage (Sect. 8.1.4).[a] The interpretation of this event is that the packet has been successfully received (can be deemed by the process responding to the event to have been successfully received). If the event condition is present at the time when the wait request is issued, the event is triggered within the same *ITU*
BMP	The event (beginning of my packet) occurs under the same conditions as *BOT* with one extra necessary condition: the packet must be addressed to the station running the process that has issued the wait request. Formally, it means that *isMine* (Sect. 7.1.6) returns *YES* for the packet (this also covers broadcast scenarios in which the current station is one of the possibly multiple recipients)

(continued)

(continued)

EMP	The event (end of my packet) occurs under the same conditions as *EOT* with one extra necessary condition: the packet must be addressed to the station running the process that has issued the wait request (see above). This is the most natural way of receiving packets at their target destinations
ANYEVENT	This event occurs whenever the transceiver perceives anything of a potential interest (any change in the configuration of sensed activities), e.g., when any of the preceding events would be triggered, if it were awaited. Beginnings and ends of all packets (stages *T* and *E*), not necessarily positively assessed by *RFC_bot* or *RFC_eot*, also trigger *ANYEVENT*. If any activity boundary or stage occurs at the time when the wait request for *ANYEVENT* is issued, the event is triggered within the current *ITU*

[a]If the assessment by *RFC_bot* has failed, *RFC_eot* is never invoked for the packet

The next four event types, listed below, deal with signal thresholds in two kinds of scenarios. First, the protocol program may want to perceive changes in the received signal level, e.g., to implement various "listen before talk" collision avoidance schemes. It is possible to define two separate signal thresholds, low and high, and receive events when the perceived signal level crosses them. This is accomplished with the following transceiver methods:

double setSigThresholdLow (double stl);
double setSigThresholdHigh (double sth);
double setSigThreshold (double stb);

The first method sets the low signal threshold, the second method sets the high signal threshold, and the last one sets both thresholds to the same value. The threshold signal levels are interpreted in the linear domain. Each method, in addition to setting the signal threshold(s) to a new value returns the previous setting of the threshold. The last method returns the previous setting of the high threshold. By default, i.e., immediately after a transceiver is created, both thresholds are set to the same value 0.0.

The two threshold events are:

SIGLOW	This event occurs whenever the received signal level perceived by the transceiver (as returned by *RFC_add* with the second argument equal *NONE*, Sect. 8.1.4) becomes less than or equal to the current low threshold setting. If the condition holds when the wait request is issued, the event occurs within the current *ITU*
SIGHIGH	This event occurs whenever the received signal level perceived by the transceiver becomes greater than the current high threshold setting. If the condition holds when the *wait* request is issued, the event occurs within the current *ITU*

It is also possible to directly monitor the level of interference suffered by a selected activity, typically a packet being in some partial stage of reception. This transceiver method:

Boolean follow (Packet *p = NULL);

allows the program to declare the activity to be monitored (followed). If a non-*NULL* argument is present, it should point to a packet being carried by one of the activities currently perceived by the transceiver. Such a packet pointer can be obtained, e.g., from *ThePacket* by receiving a packet-related event (*BOT* or *BMP*), or through one of the inquiries addressed to the transceiver, as explained in Sect. 8.2.5. If the argument is absent (or *NULL*), it implies the activity last examined by a transceiver inquiry. The method returns *OK*, if the argument (or the last transceiver inquiry) identifies an activity currently perceived by the transceiver (in which case the activity has been successfully marked as being followed), or *ERROR* otherwise, in which case the method's action has been void.

When a *BOT/BMP* event is triggered and received by a process, the packet triggering the event is automatically marked as followed (which mostly obviates the need to call *follow* directly). This comes handy, e.g., if the subsequent invocation of *RFC_eot* wants to make sure that the *EOT* event is caused by the packet whose *BOT* event was in fact previously received (i.e., responded to by the transceiver), not only positively assessed. This kind of reception scheme is implemented in the ALOHA model in Sect. 2.4.3.

All packets that pass through a powered-on receiver (Sect. 8.2) are subjected to automatic assessment by *RFC_bot*, and a positive assessment at that stage is a prerequisite for a subsequent (positive) assessment by *RFC_eot*. A packet that hasn't passed the assessment by *RFC_bot* will not even be assessed by *RFC_eot*: it will never trigger an *EOT/EMP* event. A packet whose *BOT/BMP* event has caused a process to respond (i.e., wake up on the event) is marked as followed. At most one packet can be followed by a given transceiver at any given time, so the next interception of BOT/BMP on the transceiver will switch the followed packet. Assuming *p* points to one of the packets currently being perceived by a transceiver, this *Transceiver* method:

Boolean isFollowed (Packet *p);

returns *YES*, if and only if the packet pointed to by *p* is the one being followed.

The threshold interference levels are set by *setSigThresholdLow* and *setSigThresholdHigh*, in the same way as for *SIGLOW* and *SIGHIGH*. The following two events are available:

INTLOW	This event occurs whenever the interference level of the followed activity becomes less than or equal to the threshold. If the condition holds when the wait request is issued, the event occurs within the current *ITU*
INTHIGH	This event occurs whenever the interference level of the followed activity becomes greater than the threshold. If the condition holds when the wait request is issued, the event occurs within the current *ITU*

Note that the transceiver must be in the "on" state for all packet perception/reception events to be ever triggered. Only *SILENCE* and *SIGLOW* are triggered when the receiver is (or goes) off. The perceived signal level of a switched-off receiver is zero. Neither *INTLOW* nor *INTHIGH* is triggered when the receiver is switched off. The followed activity must be perceptible (at least in principle) to trigger any of these two events. This means that its signal level must be above the cut-off threshold (Sect. 4.5.2).

When a process is awakened by any of the above events (except for *ANYEVENT*), the environment variable *TheTransceiver* (*Info02*) of type *Transceiver** stores the pointer to the transceiver on which the waking event has occurred. If the waking event is related to a specific (proper) packet (this concerns *BOT*, *BMP*, *EOT*, *EMP*, and also *ACTIVITY*, as explained below), the environment variable *ThePacket* (*Info01*) of type *Packet** returns the packet pointer. The object pointed to by *ThePacket* is the copy of the packet buffer created by *startTransmit* or *transmit* (Sect. 8.1.8) when the packet's transmission was initiated.

Before triggering an *ACTIVITY* event, SMURPH determines whether the transceiver is perceiving a "receivable" packet, and if this is the case, its pointer is returned via *ThePacket*. Otherwise, *ThePacket* contains *NULL*. A packet is deemed receivable if (a) it was positively assessed by *RFC_bot*, (b) its portion received so far is positively assessed by *RFC_eot*. Thus, if the protocol program wants to take advantage of this feature, *RFC_eot* must be prepared to meaningfully assess incomplete packets.[7]

One should be aware that the object pointed to by *ThePacket* is deallocated when the activity carrying the packet is removed from the radio channel. By default, this happens one *ITU* after the last bit of the packet disappears from the most distant transceiver of the sender's current neighborhood (Sect. 4.5.2). If needed, that delay can be increased with the *linger* argument of the *RFChannel setup* method, or by invoking *setPurgeDelay* on the channel (Sect. 4.4.2).

The philosophy of the event handling mechanism of SMURPH enforces the view that one event conveys information about one thing. While generally this makes perfect sense, there may be situations where the protocol program wants to discern several conditions possibly present within the same *ITU*. Similar to the port case (Sect. 7.1.3), *ANYEVENT* directly returns information about a single event, and if two (or more) of them occur within the same *ITU*, the one actually presented is chosen at random. The awakened process can learn about all of them by calling *events* (Sect. 8.2.5).

When a process awaiting *ANYEVENT* is awakened, the environment variable *TheEvent* (*Info02*) of type *int* contains a value identifying the event type. The possible values returned in *TheEvent* are represented by the following constants (coinciding with the identifiers of the respective events):

---

[7]The meaning of that assessment is up to the channel model and protocol program to interpret, e.g., it can be always positive, with the actual (authoritative) assessment reserved for the true end-of-packet event. The method can tell that the packet is incomplete, if the number of bits in the interference histogram is less than the packet's total length (attribute *TLength*, Sect. 6.2).

BOP	beginning of preamble
EOP	end of preamble
BOT	beginning of packet (stage *T*)
EOT	end of packet (stage *E*)

If the event type is *BOT* or *EOT*, *ThePacket* points to the related packet object. Like for the port variant of the event, it is possible to declare a special mode of *ANYEVENT*, whereby the event is never pending (see Sect. 7.1.3). This is accomplished by invoking the *setAevMode* method of *Transceiver*, whose role is identical to its *Port* counterpart.

Similar to links and ports, when *setAevMode* is invoked without an argument (or, equivalently, with *YESNO*, Sect. 3.2.5), the method will set the transceiver's *AEV* mode to the default mode of the channel, which is *AEV_MODE_PERSISTENT* by default. The channel's default mode can be changed by calling the *setAevMode* method of *RFChannel*, which will redefine the channel's default as well as set the mode of all transceivers connected to the channel. When this is done at the initialization, when no transceivers are yet connected to the channel, the transceivers will automatically copy the default as they are being configured.

8.2.2 Event Priorities

Section 7.1.5 (referring to ports) also applies to transceivers. If the explicit event ordering mechanism is switched off, i.e., the third argument of *wait* is unavailable (Sect. 5.1.1), SMURPH imposes implicit ordering on multiple events awaited by the same process that may refer to the same actual condition on the transceiver. Thus, *BMP* is triggered before *BOT*, and *EMP* is triggered before *EOT*, which in turn takes precedence over *SILENCE*. Technically, *ACTIVITY*, if occurring together with *BMP* or *BOT*, has the lowest priority; however, for a transceiver, *ACTIVITY* is unlikely to occur exactly on the packet boundary (stage *T*), as the packet is preceded by a nonzero-length preamble, which (in all sane circumstances) should trigger *ACTIVITY* earlier. Even if the receiver is switched on exactly on the packet boundary, in which case *ACTIVITY* will formally coincide with *BOT*, the total lack of preamble will fail the assessment of this event without ever consulting *RFC_bot*.

The events *BOP* and *EOP* are not covered by the implicit priority scheme.

8.2.3 Receiving Packets

Refer to Sect. 7.1.6 for the explanation of how packets should be extracted from ports and formally received. Everything we say in that section applies also to transceivers: just mentally replace all the occurrences of "port" with "transceiver." The *receive* methods of the *Client* and traffic patterns exist in the following additional variants:

void receive (Packet *pk, Transceiver *tr);
void receive (Packet *pk, RFChannel *rfc = NULL);
void receive (Packet &pk, Transceiver &pr);
void receive (Packet &pk, RFChannel &rfc);

which perform the necessary bureaucracy of packet reception, including updating various performance-related counters associated with the radio channel.

A sample code of a process responsible for receiving packets from a transceiver may look like this:

```
perform {
        state WaitForPacket:
                RcvXcv->wait (EMP, NewPacket);
        state NewPacket:
                Client->receive (ThePacket, TheTransceiver);
                skipto WaitForPacket;
};
```

where *RcvXcv* is a transceiver pointer. This scheme assumes that the *EMP* event is assessed in such a way as to hide all the intricacies involved in the mechanism of packet reception, e.g., accounting for the bit error rate implied by the signal-to-interference ratio in the history of the packet's signal. Note that the packet can ever successfully get to stage *E* at the transceiver (which makes it possible to trigger the *EMP* event), if its beginning (stage *T*) has been previously assessed as well. This implicitly involves *RFC_bot* in the assessment of the *EMP* event, even though the *BOT/BMP* event is never mentioned in the reception code.

In the wireless environment, where interference affects packet reception in a much subtler way than in a wired link, one can think of more elaborate reception schemes that make the model closer to the way the reception is handled by real-life transceivers. In Sect. 2.4.3, we saw a scheme with an explicit *BMP* event whose purpose was to mark the packet as "followed" from the moment when its beginning was recognized by the transceiver until the *EMP* assessment. In Sect. 8.2.4, we shall see yet another realistic reception scheme.

8.2.4 Hooks for Handling Bit Errors

The simulation model of a communication channel can be viewed as consisting of two main components: the part modeling signal attenuation, and the part transforming the received signal level (possibly affected by interference and background noise) into bit errors. The results from the first component are pipelined into the second one, whose ultimate purpose is to tell which packets have been correctly received.

For example, suppose that we know how to calculate the bit error rate based on the received signal strength (sl), receiver gain (rs), and the interference level (il). This part of our channel model can be captured by a function, say:

double ber (double sl, double rs, double il);

which returns the probability of a single-bit error.[8] One way to arrive at such a function is to calculate the signal-to-noise ratio, e.g., in the linear domain:

$SNR = (sl \times rs)/(il \times rs + bg);$

where *bg* is the fixed (mean) level of background noise. Then, *SNR* can be transformed into the probability that a single bit is received in error, e.g., by using an interpolated table obtained experimentally (Sect. A.1.1), or resorting to some closed formula from a formal model. With the *ber* function in place, *RFC_eot* may look like this:

```
Boolean RFC_eot (RATE r, const SLEntry *sl, const SLEntry *sn, const IHist *h) {
        int i;
        Long nb;
        double intf;
        for (i = 0; i < h->NEntries; i++) {
                h->entry (i, r, nb, intf);
                if (rnd (SEED_toss) > pow (1.0 —ber (sl->Level, sn->Level, intf), nb))
                        return NO;
        }
        return YES;
}
```

For clarity, we assume homogeneous signals (tags are not used). The method goes through all chunks of the packet that have experienced a fixed level of interference. For each chunk of length *nb* bits that has suffered interference *intf*, it assumes that the chunk contains no error with probability $(1 - E)^{nb}$, where E is the error rate at the given interference level, as prescribed by *ber*. As soon as a single bit error is discovered, the method stops and returns *NO*. Having examined all chunks and found no error, the method returns *YES*, which translates into a positive assessment of the entire packet.

Note that the number of bits, *nb*, returned by every invocation of *entry* in the above code, is an integer number (see Sect. 8.1.5), even though the duration of the corresponding interference level need not fall on precise bit boundaries. When it does not (as is most oftent the case), the value of *nb* returned by *entry* will be randomized in such a way that its average over a large number of samples will approach the precise fractional number of bits covered by the histogram entry. We did mention this in Sect. 8.1.5 (while discussing interference histograms), but now we can see how this randomness nicely fits a larger scheme of generating randomized error bits.

The above example is merely an illustration. An actual scheme for transforming the reception parameters of a packet into its assessment decision need not be based on the notion of bit error rate or signal-to-noise ratio. However, this view is inherent in the vast majority of all practical channel models. Thus, SMURPH goes one step

---

[8]For simplicity, we ignore the impact of the transmission rate (which would translate into one more argument to *ber*): in many models the rate is fixed and can thus be hardwired into the function.

further towards accommodating it as an integral (albeit optional) part of the overall assessment framework. Two special assessment methods, *RFC_erb* and *RFC_erd*, are provided to capture a complete description of the bit-error distribution as a function of the reception parameters of an incoming packet. By defining those methods, the program:

- localizes the bit-error model in one place, separate from any other model components
- activates certain tools to make it easier to carry out packet assessment, and enables events resulting from possibly complicated, programmable configurations of bit errors.

The two methods have identical signatures:

Long RFC_erb (RATE r, const SLEntry *sl, const SLEntry *rs, double il, Long nb);
Long RFC_erd (RATE r, const SLEntry *sl, const SLEntry *rs, double il, Long nb);

RFC_erb is expected to return a randomized number of error bits within a sequence of *nb* bits received at rate *r*, signal parameters *sl* (*Level* and *Tag*), receiver parameters (gain *Level* and *Tag*) *rs*, and interference level *il*. For example, supposing that the *ber* function from the previous example transforms the reception parameters into the probability of a bit error, and that bit errors (under steady interference) are independent, here is one possible variant of the method:

```
Long RFC_erb (RATE r, const SLEntry *sl, const SLEntry *rs, double il, Long nb) {
        return IRndBinomial (ber (sl->Level, rs->Level, il), nb);
};
```

which returns the outcome of a random experiment consisting in tossing *nb* times a biased coin with the probability of "success" equal to the bit error rate. Once *RFC_erb* is defined, the following methods of *RFChannel* become automatically available[9]:

```
Long errors (RATE r, const SLEntry *sl, const SLEntry *rs, const IHist *h, Long sb = NONE,
        Long nb = NONE);
Boolean error (RATE r, const SLEntry *sl, const SLEntry *rs, const IHist *h, Long sb = NONE,
        Long nb = NONE);
```

Suppose first that the last two arguments are not used (retaining their default settings). Then, the first method calculates a randomized number of error bits in the histogram *h*. The second method returns *YES*, if the histogram contains one or more bit errors, and *NO* otherwise. Formally, it is equivalent to *errors (r, sl, rs, h) != 0*, but executes considerably faster.

The last two arguments allow us to narrow down the error lookup to a fragment of the histogram. If both *sb* and *nb* are nonnegative,[10] then *sb* specifies the starting bit position (recall that bits in a histogram are numbered from 0, Sect. 8.1.5) and *nb*

---

[9]Or rather usable, as they are always formally available, but calling them only makes sense when *RFC_erb* is present.

[10]Recall that *NONE* is defined as −1.

gives the number of consecutive bits to be looked at. If that number falls behind the activity represented by the histogram, or is negative (or simply skipped), the lookup extends to the end of the histogram. If *sb* is negative and *nb* is not, then *nb* refers to the trailing number of bits in the histogram, i.e., the errors will be looked up in the last *nb* bits of the activity.

With these tools in place, the *RFC_eot* method from the above example can be simplified to:

```
Boolean RFC_eot (RATE r, const SLEntry *sl, const SLEntry *sn, const IHist *h) {
        return !error (r, sl, sn, h);
}
```

The two *RFChannel* methods are also available as transceiver methods, in the following variants:

Long errors (SLEntry *sl, const IHist *h);
Boolean error (SLEntry *sl, const IHist *h);

The missing arguments, i.e., the reception rate, receiver gain, and receiver tag, are taken from the current settings of *this* transceiver. Note that the packet's reception rate is assumed to be equal to the transceiver's current transmission rate, which need not be always the case.[11]

Two more variants of *error/errors* are mentioned in Sect. 8.2.5. Note that the availability of these methods (six of them altogether) is the only net contribution of *RFC_erb*. Thus, the program need not define *RFC_erb*, if it does not care about the *error/errors* methods.

The role of *RFC_erd* is to generate randomized time intervals (in bits) preceding events related to bit errors. To illustrate its possible applications, consider a receiver that extracts symbols (i.e., sequences of bits) from the channel and stops (aborts the reception) upon the first occurrence of an illegal symbol. To model the behavior of such a receiver with a satisfying fidelity, one may want to await simultaneously several events, e.g.,

1. an *EOT* event indicating that the packet has been successfully received (because no symbol error was encountered during its reception)
2. a signal (*SIGLOW*) event indicating that the packet signal has been lost (say, the packet has been aborted by the transmitter)
3. an event representing the occurrence of an illegal symbol in the received sequence.

Generally, events of the third type require a custom description capturing the various idiosyncrasies of the encoding scheme. While their timing is strongly coupled to the bit error rate, the exact nature of that coupling must be specified by the user as part of the error model. This is the role of *RFC_erd*, whose first four arguments describe the reception parameters (exactly as for *RFC_erb*), while the last argument

---

[11] If it isn't, then a sensible assessment (by *RFC_bot* or *RFC_eot*) should fail before the method gets a chance to be called.

identifies a custom function (one of possibly several options) indicating what exactly the function is expected to determine. This way, *RFC_erb* provides a single custom generator of time intervals for (potentially multiple types of) events triggered by interesting configurations of bit errors.

The second part of this mechanism is a dedicated event named *BERROR*, which is triggered when the delay generated by *RFC_erd* expires. The complete incantation includes a call to the following transceiver method:

Long setErrorRun (Long em);

which declares the so-called error run, i.e., the effective value of the last argument to *RFC_erd* selecting a specific action of the method. Having recognized the beginning of a potentially receivable packet (via the *BOT* event), the program will mark the packet as "followed" (Sect. 8.2.1). Then, *BERROR* events will be timed by *RFC_erd* applied to the reception parameters of the followed packet.

As most "set" methods, *setErrorRun* returns the previous setting of the error run option. Also, an *RFChannel* variant of the method is available, which returns no value and selects the same value of error run for all transceivers interfaced to the channel, also resetting the channel's default, e.g., as for the channel variant of *setXRate* (Sect. 4.5.2). The channel's default for the error run, which is assumed by transceivers as their default, is 1. There is a matching *getErrorRun* method of *Transceiver*, which returns the current setting of the option.

For illustration, the following method:

```
Long RFC_erd (RATE r, const SLEntry *sl, const SLEntry *rs, double il, Long nb) {
        return (Long) dRndPoisson (1.0/pow (ber (sl->Level, rs->Level, il), nb));
};
```

returns the randomized number of bits preceding the first occurrence of *nb* consecutive error bits. In this case, the last argument actually gives the length of an error run.[12] Now consider this code fragment:

```
...
state RCV_WAIT:
        Tcv->wait (BOT, RCV_START);
state RCV_START:
        Tcv->follow (ThePacket);
        Tcv->setErrorRun (3);
        Tcv->setSigThresholdLow (dBToLin (-40.0));
        Tcv->wait (EOT, RCV_GOTIT);
        Tcv->wait (BERROR, RCV_IGNORE);
        Tcv->wait (SIGLOW, RCV_GONE);

        ...
```

---

[12]The name of the method selecting the value of that argument (*setErrorRun*) was inspired by this illustration.

The invocation of *follow* (the first statement in state *RCV_START*) is redundant, because the packet becomes "followed" by the virtue of its having triggered the *BOT* event, but we want to emphasize that fact. Having recognized the beginning of a packet, the process awaits one of three events: its successful reception, three consecutive error bits, or a low signal indicating that the packet has been lost. The second event may be synonymous with an incorrect symbol (within the framework of a forward error correction encoding), which will abort the reception before the last bit of the packet arrives at the transceiver.

The presence of *RFC_erd* among the assessment methods of a radio channel enables *BERROR* as one more event type that can be awaited on a transceiver interfaced to the channel. If the program does not care about this event, *RFC_erd* is not needed. Note that the assessment method enables a class of reception schemes represented by the above code fragment. With this approach, *RFC_eot* is trivial and, e.g., can boil down to:

```
Boolean RFC_eot (RATE r, const SLEntry *sl, const SLEntry *sn, const IHist *h) {
        return TheTransceiver->isFollowed (ThePacket);
}
```

because when the packet reaches stage *E*, and none of the other events has been triggered in the meantime, it means that the packet has made it without an error.

8.2.5   Transceiver Inquiries

In contrast to ports, transceivers implement no inquiries about the past (and no inquiries about the future). Three inquiries about the present, the methods *busy*, *idle*, and *transmitting*, were already covered in Sect. 8.2. It is possible to examine all activities (packets) being perceived by a transceiver at the current moment. These *Transceiver* methods:

int activities (int &p);
int activities ();

return the total number of activities (packets possibly at different stages) currently sensed by the transceiver. The argument in the first variant returns the number of those activities that are past stage *T*, i.e., within their proper packets, following the preamble. If the population of such activities is nonempty, *ThePacket* returns a pointer to one of them (selected at random).

When a process is awakened by *ANYEVENT* (Sect. 8.2.1), only one of the possibly multiple conditions being present at the same *ITU* is returned to the process in *TheEvent*. It is possible to learn how many events of a given type occur at the same time by invoking this transceiver method:

int events (int etype);

where the argument identifies the event type in the following way:

BOP	beginning of a preamble
EOP	end of a preamble
BOT	beginning of a packet (stage T)
EOT	end of a packet (stage E)

The method returns the number of simultaneous events of the type indicated by the argument.

The following *Transceiver* method scans through all activities currently perceived by the transceiver:

int anotherActivity ( );

The exact behavior of *anotherActivity* depends on the circumstances of the call. If the method is invoked after one of the events *BOP, EOP, BOT, EOT, BMP, EMP*, then its subsequent invocations will scan all the activities that have triggered the event simultaneously. The same action is performed after a call to *events* (whose argument specifies an event type). For illustration consider the following code fragment:

...

```
state WaitPacket:
        Tcv->wait (EOT, LookAtThem);
state LookAtThem:
        while (Tcv->anotherActivity ( ) != NONE)
                handlePacket (ThePacket);
        skipto WaitPacket;
```

...

which traverses all packets sensed by the transceiver and found to be in stage E. Note *skipto* used to transit back to state *WaitPacket*, to remove the *EOT* event from the transceiver (Sect. 5.3.1).

The method returns *NONE* when it runs out of activities to present. Otherwise, its value is *PREAMBLE* or *TRANSMISSION*, depending on whether the activity is still in the preamble stage or within the proper packet. In the above example, all activities traversed by anotherActivity are bound to be within the proper packet (stage E), so there is no need to worry about this detail.

If *anotherActivity* is called after *BOP* or *EOP*, then its subsequent invocations will return *PREAMBLE* as many times as many activities are found in the respective stage. The value of *ThePacket* will be set to *NULL* on each call.

If the method is called after *ANYEVENT, ACTIVITY*, or after a call to *activities*, then it will scan through all the activities perceived by the transceiver at the current *ITU*, regardless of their stage. Whenever it returns *TRANSMISSION* (as opposed to *PREAMBLE*), *ThePacket* will be set to point to the respective packet.

Further inquiries about the activity spotted by the method are possible. In particular, the following *Transceiver* method:

Packet *thisPacket ( );

returns a pointer to the packet carried by the last activity processed by *anotherActivity*, even if the activity is before stage T (i.e., *anotherActivity* has returned *PREAMBLE*). This may look like cheating, but may also be useful in tricky situations, e.g., when the packet structure carries some extra information pertaining to the properties of the preamble.

For scanning packets, i.e., activities past stage T, this method of *Transceiver* can be used as a shortcut, instead of *anotherActivity*:

Packet *anotherPacket ();

It behaves in a manner similar to *anotherActivity*, except that it ignores all activities before stage T. If *anotherPacket* is called after one of the events *BOT, EOT, BMP, EMP*, then its subsequent calls will scan through all the packets that have triggered the event simultaneously. The same action is performed after a call to *events* whose argument specifies the respective event type. In all other circumstances, *another-Packet* will scan through all packets (activities past stage T) currently perceived by the transceiver. For each packet, the method sets *ThePacket* to the packet pointer and returns the same pointer as its value. If the pointer is *NULL*, it means that the list of packets has been exhausted.

Several types of inquiries deal with signal levels, including various elements of the interference affecting packet reception. This method of *Transceiver*:

double sigLevel ();

returns the total signal level perceived by the transceiver, as calculated by *RFC_add* with the second argument equal *NONE* (Sect. 4.4.3).

Following a call to *anotherActivity* or *anotherPacket*, this method:

double sigLevel (int which);

will return the signal level related to the last-scanned activity. The value returned by *sigLevel* depends on the argument whose value can be one of the following constants:

SIGL_OWN	The perceived signal level of this activity
SIGL_IFM	The maximum interference experienced by the activity so far. This maximum refers to the activity's current stage, i.e., if the packet is past stage T, the interference experienced by the preamble is no longer relevant
SIGL_IFA	The average interference experienced by the activity so far. This average refers to the current stage, i.e., if the packet is past stage T, the interference suffered by the preamble is no longer relevant
SIGL_IFC	The current interference level experienced by the activity

One more variant of the method:

double sigLevel (const Packet *pkt, int which);

applies to the activity carrying the indicated packet, if the activity is currently perceived by the transceiver. The packet pointer passed to *sigLevel* must match the address of the packet as it appears within the activity object; in other words, the packet is located by its address rather than contents, so the pointer should be obtained via a (previous) inquiry or an event related to the activity. If the method cannot locate the current activity (this applies to the last two variants of *sigLevel*), it returns a negative value (−1.0). Note that a packet triggering *EOT* is eligible for this kind of inquiry: its activity is still formally available to the transceiver.

The following code fragment illustrates the usage of *anotherPacket* and *sigLevel*:

```
...
state StartWaiting:
        Tcv->wait (EOT, EndPacket);
state EndPacket:
        Packet *p;
        double sig = 0.0;
        while ((p = Tcv->anotherPacket ( )) != NULL)
                sig += Tcv->sigLevel (p, SIGL_OWN);
        ...
```

The loop in state *EndPacket* executes for all packets simultaneously found in stage *E* calculating the sum of their signal levels. Note that *anotherPacket*, being executed in response to the *EOT* event, scans through all packets that might trigger the event for the current *ITU*. Following a call to *anotherPacket*, current activity is properly set, so the loop can be simplified to:

```
...
state EndPacket:
        double sig = 0.0;
        while (Tcv->anotherPacket ( ) != NULL)
                sig += Tcv->sigLevel (SIGL_OWN);
        ...
```

If the above example appears too far-fetched (it rather is), here is a modified version:

```
...
state EndPacket:
        double sig = 0.0;
        Packet *p = ThePacket;
        activities ( );
        while (Tcv->anotherPacket ( ) != NULL)
                if (ThePacket != p)
                        sig += Tcv->sigLevel (SIGL_OWN);
        ...
```

The call to *activities* before the loop is entered sets the stage for *anotherPacket*, which will now scan through all packets currently perceived by the transceiver.

The packet pointer identifying the activity that has triggered the *EOT* event is stored in *p* and compared against *ThePacket* after every call to *anotherPacket*. Thus, the loop calculates the sum of the signal levels of all packets audible at the time, except for the one that has triggered the *EOT* event.

As it turns out, the *if* statement within the loop is not even necessary, because the triggering packet will not show up on the list of activities examined by *anotherPacket*. This is because a packet that triggers *EOT* is no longer audible: its period of perception has just ended! The situation would change, if the event awaited in state *StartWaiting* were replaced by *BOT*. In that case, the loop would also account for the packet that has triggered the event.

Two more *Transceiver* methods accept a packet pointer with the same interpretation as the argument of *sigLevel*, i.e., to identify an activity. One of them is:

IHist iHist (const Packet *pkt);

which returns a pointer to the packet's interference histogram (Sect. 8.1.5). This histogram can be used in tricky scenarios (see Sect. 8.2.4 for an illustration), e.g., to affect the processing of a received (or perceived) packet in a manner depending on its experienced interference.

The protocol program may issue inquiries or wait requests that formally call for multiple assessments of effectively the same event related to the same packet. To save on execution time and enforce the consistency of multiple assessments, all activities carry a flag whose role is to kill the packet when it fails any of the necessary tests for receivability in a way that makes its future positive assessment impossible. Here is a transceiver method making that flag explicitly available to the program:

Boolean dead (const Packet *pkt = NULL);

The method returns *YES*, if the packet has been found non-receivable. This will happen in any of the following circumstances:

1. The packet is past the preamble (stage *T*) and the preamble assessment by *RFC_bot* was negative.
2. The packet has been aborted by the sender and this event has made it to the perceiving transceiver (the packet will fail the *EOT* assessment without even bothering the assessment method).
3. The packet is in stage *E* and the assessment by *RFC_eot* has been negative.
4. The packet is not perceived any more, i.e., its last bit has disappeared from the transceiver. This is mostly a precaution, because such a packet will not trigger any events and should normally be ignored by the receiving process on other grounds.
5. The packet is past stage *T* and before stage *E*, its preamble assessment (by *RFC_bot*) was positive, but its partial assessment by *RFC_eot* has been negative.

In all other cases, *dead* returns *NO*.

One added value of the method is that, if it is invoked while the packet is past stage *P* and before stage *E*, it will carry out the *EOT* assessment (invoke *RFC_eot*) on the partial packet. Thus, if this inquiry is used, *RFC_eot* must be prepared to sensibly assess packets before stage E. This is like the assessment carried out for an *ACTIVITY* event (see Sect. 8.2.1).

If *dead* is called without an argument (or with the argument equal *NULL*), then it applies to the activity last scanned by *anotherActivity* or *anotherPacket*. Any activity found to be before stage T is never dead unless it is an aborted preamble whose last bit has just arrived at the transceiver (such a preamble is never followed by a packet).

If the receiver is off (Sect. 8.2.3), no activities can be perceived on it, and no signals can be measured. In particular, *activities* and *events* return zero, *anotherActivity* immediately returns *NONE*, *anotherPacket* immediately returns *NULL*, *sigLevel* returns 0.0, and *dead* returns *YES* regardless of the argument.

IF *RFC_erb* is present among the assessment methods (Sect. 8.2.4), these two *Transceiver* methods become available:

Long errors (Packet *p = NULL);
Boolean error (Packet *p = NULL);

If the argument is not *NULL*, it must point to a packet currently perceived by the transceiver. The first method returns the randomized number of bit errors found in the packet so far, the second returns *YES*, if and only if the packet contains one or more bit errors, as explained in Sect. 8.2.4. If the argument is *NULL*, the inquiry applies to the activity last scanned by *anotherActivity* or *anotherPacket*. If p does not point to a packet currently perceived by the transceiver, or p is *NULL* and no meaningful activity is available for the inquiry, *errors* returns *ERROR* (a negative value) and *error* returns *NO*.

8.3 Cleaning After Packets

The same feature for postprocessing packets at the end of their lifetime in the radio channel is available for rfchannels as for links (Sect. 7.4). Even though, in contrast to links, type *RFChannel* is extensible by the user (and the cleaning function could be naturally accommodated as a *virtual* method), the idea of a plug-in global function has been retained, for (questionable) compatibility with links.

Reference

1. R.J. Mailloux, *Phased Array Antenna Handbook* (Artech House, Boston, 2005)

Chapter 9
Mailboxes

One way to synchronize collaborating processes, which usually concerns processes running at the same station, is to take advantage of the fact that each process is an autonomous *AI* capable of triggering events that can be perceived by other processes (Sect. 5.1.7). A more systematic tool for inter-process communication is the *Mailbox AI*, which can also serve as an interface of the SMURPH program to the outside world (in the real or visualization mode).

9.1 General Concepts

The *Mailbox* class provides a frame for defining mailbox types and creating objects of these types. While there seems to be a unifying theme in all the different mailbox variants described in this chapter, one can see three distinct types of functionality, which, in combination with some other aspects, call for a bit of taxonomy.

A mailbox is primarily a repository for "messages"[1] that can be sent and retrieved by processes. Viewed from the highest level of its role in the program, a mailbox can be *internal* or *external* (also called *bound*). An internal mailbox is used as a synchronization and communication vehicle for SMURPH processes, while a bound mailbox is attached to an external device or a socket, thus providing the program with a reactive interface to the outside world. Bound mailboxes are only available in the real or visualization mode, while internal mailboxes can be used in both modes.

An internal mailbox, in turn, can be a *barrier* or a *fifo*. Both kinds implement different variants of events (signal) passing, with the option of queuing pending events (possibly accompanied with some extra information) in the second case. A barrier mailbox is a place where any process can block awaiting the occurrence of an identifiable event, which can be delivered by another process. An arbitrary number of processes can wait on a barrier mailbox at the same time, for the same or different

---

[1] Not to be confused with the *Client* messages described in Sect. 6.2.

© Springer Nature Switzerland AG 2019
P. Gburzyński, *Modeling Communication Networks and Protocols*,
Lecture Notes in Networks and Systems 61,
https://doi.org/10.1007/978-3-030-15391-5_9

event identifiers. When an event with a given identifier is delivered, all the processes awaiting that event are immediately awakened.

A fifo mailbox acts as a queued repository for items. One attribute of such a mailbox is its capacity, i.e., the maximum number of items that can be stored in the mailbox awaiting acceptance. This capacity can be zero, which gives the mailbox a special (somewhat degenerate but useful) flavor. The taxonomy is further confounded by the fact that a fifo mailbox can be either *counting* or *typed*. In the first case, the items deposited in the mailbox are illusory: all that matters is their count, and the entire fifo is fully represented by a single counter. In the second case, the items are actual. Their type must be simple (meaning convertible to *void**), but it can be a pointer representing members of structures or classes.

9.2   Fifo Mailboxes

A fifo mailbox has a capacity, which is usually determined at the time of its creation, but can also be redefined dynamically. A mailbox with capacity $n \geq 0$ can accommodate up to n pending items. Such items are queued inside the mailbox, if submitted and not immediately awaited, and can be extracted later. If the capacity is huge (e.g., *MAX_long*, see Sect. 3.1.5), the mailbox becomes an essentially unlimited FIFO-type storage.

A fifo mailbox whose capacity is zero (which is the default) is unable to store even a single item. This makes sense because a process can still receive a deposited item, if it happens to be already waiting on the mailbox when the item arrives.

The type of elements handled by a fifo mailbox can be declared with the mailbox type extension. Fifo mailboxes belonging to the standard type *Mailbox* are counting, i.e., the items are structure-less tokens that carry no information other than their presence. A typed mailbox may specify a simple type for the actual items, which can be a pointer type.

At any moment, a fifo mailbox (typed or counting) can be either empty or nonempty. A non-empty counting mailbox holds one or more pending tokens that await acceptance. As these tokens carry no information, they are all represented by a single counter. Whenever a token is put into a counting mailbox, the mailbox element counter is incremented by one. Whenever a token is retrieved from a nonempty counting mailbox, the counter is decremented by one. A non-empty typed mailbox is a FIFO queue of some tangible objects.

9.2.1   Declaring and Creating Fifo Mailboxes

The standard type *Mailbox* is usable directly and it describes (by default) a capacity-0, counting (i.e., untyped), fifo mailbox. The following sequence:

Mailbox *mb;

...

mb = create Mailbox;

illustrates how to create such a mailbox.

The mailbox capacity can be specified as the *setup* argument, e.g.,

mb = create Mailbox (1);

creates a capacity-1 mailbox.

A user-provided extension of type *Mailbox* is required to define a typed mailbox capable of storing objects of a definite type. The type extension must specify the type of elements to be stored in the mailbox queue. The declaration has the following format:

mailbox mtype *:* itypename *(* etype *) {*

    ...

    attributes and methods

    ...

};

where "mtype" is the new mailbox type. The optional argument in parentheses, if present, must identify a simple C++ type. Note that although the element type must be simple (it is internally cast to *void**), it can be a pointer type. If the type argument is absent, the new mailbox class describes a counting mailbox.

For example, the following declaration describes a fifo mailbox capable of storing integer values:

mailbox IntMType (int);

Note the empty body of the declaration. As the type argument is present, the above declaration will not be mistaken for an announcement (see Sect. 3.3.6).

It is possible to intercept the standard operations on a fifo mailbox (e.g., to carry out some extra action) by declaring within the mailbox class the following two virtual methods:

void inItem (etype *par);*
void outItem (etype *par);*

Each of the two methods has one argument whose type should be the same as the mailbox element type. The first method (*inItem*) will be called whenever a new item is added to the mailbox. The argument will then contain the value of the item. Similarly, *outItem* will be called whenever an item (indicated by the argument) is removed from the mailbox. The two methods are independent, which means that it is legal to declare only one of them. For example, if only *inItem* is declared, no extra action will be performed upon element removal.

Note that when a fifo mailbox has reached its capacity (is full), an attempt to add a new item to it will fail (Sect. 9.2.3). In such a case, *inItem* will not be called.

Similarly, *outItem* will not be invoked upon a failed attempt to retrieve an item from an empty mailbox.

A mailbox can be owned either by a station or by a process. The most significant difference[2] is that a mailbox owned by a station cannot be destroyed, while that owned by a process can. This applies to all kinds of mailboxes, not only fifos. Those mailboxes that have been created during the initialization stage of the protocol program (i.e., when the *Root* process was in its first state, Sect. 5.1.8) have ownership properties similar to those of ports (Sect. 4.3.1), transceivers (Sect. 4.5.1), and packet buffers (Sect. 6.3), i.e., they belong to the stations at which they have been created. The *Id* attribute of such a mailbox combines the serial number of the station owning it and the creation number of the mailbox within the context of its station. The naming rules for a station mailbox are basically identical to those for ports. The standard name of a station mailbox (Sect. 3.3.2) has the form:

mtypename mid *at* stname

where "mtypename" is the type name of the mailbox, "mid" is the station-relative mailbox number, and "stname" is the standard name of the station owning the mailbox. The base standard name of a mailbox has the same format as the standard name, except that "mtypename" is always *Mailbox*.

The following *Station* method:

*Mailbox *idToMailbox (Long id);*

converts the numerical station-relative mailbox identifier into a pointer to the mailbox object. An alternative way to accomplish the same goal is to call a global function declared exactly as the above method which assumes that the station in question is pointed to by *TheStation* (Sect. 4.1.3).

A mailbox can also be declared as a class variable within the station class, e.g.,

```
mailbox MyMType (ItemType) {
    ...
};
...
station MySType {
    ...
    MyMType Mbx;
    ...
};
```

The capacity of a mailbox, including one that has been declared "statically" as a class variable (as in the above example) can be set after its creation by calling the *setLimit* method of *Mailbox*, e.g.,

---

[2]Another difference will come into play when the mailbox is exposed (Sect. 12.5.2).

```
station MyStation {

      ...

      Mailbox MyMType;

      ...

      void setup ( ) {

            ...

            MyMType.setLimit (4);
            MyMType.setNName ("My Mailbox");

            ...

      };

      ...

};
```

When the protocol program is in the initialization stage (Sect. 5.1.8), a dynamically created mailbox (with *create*) is assigned to the current station. Note that this station can be *System* (Sect. 4.1.2). This kind of ownership makes sense, e.g., for a "global" mailbox interfacing processes belonging to different regular stations.

A mailbox created dynamically after the initialization stage is owned by the creating process, and its *Id* attribute is a simple serial number incremented globally, e.g., as for a process (Sect. 5.1.3). As we said before, a process mailbox is more dynamic than one owned by a station, because it can be destroyed (by *delete*).

A mailbox type declaration can specify private *setup* methods. These methods are only accessible when the mailbox is created dynamically, with the *create* operation. As a private setup method subsumes the standard method that defines the mailbox capacity, it should define this capacity explicitly by invoking *setLimit* (see Sect. 9.2).

9.2.2   Wait Requests

An event identifier for a fifo mailbox is either a symbolic constant (always representing a negative value) or a nonnegative number corresponding to the item count. For example, by calling:

mb->wait (NONEMPTY, GrabIt);

the process declares that it wants to be awakened in state *GrabIt* at the nearest moment when the mailbox pointed to by *mb* becomes nonempty. If the mailbox happens to be nonempty at the time when the request is issued, the event occurs immediately, i.e., within the current *ITU*.

Another event that can be awaited on a fifo mailbox is *NEWITEM* or *UPDATE*, which are two different names for the same event. A process that calls:

mb->wait (NEWITEM, GetIt);

will be restarted (in state *GetIt*) at the nearest moment when an item is put into the mailbox (Sect. 9.2.3). Note the difference between *NONEMPTY* and *NEWITEM*. First, *NEWITEM* (aka *UPDATE*) is only triggered when an element is *being* stored in the mailbox. Thus, unlike *NONEMPTY*, it does not occur immediately when the mailbox already contains some item(s). Moreover, no *NONEMPTY* event ever occurs on a capacity-0 mailbox. By definition, a capacity-0 mailbox is never *NONEMPTY*. On the other hand, a process can sensibly await the *NEWITEM* event on a capacity-0 mailbox. The event will be triggered at the nearest moment when some other process executes *put* (Sect. 9.2.3) on the mailbox.

When a process is restarted by *NEWITEM* on a nonzero-capacity mailbox, or by *NONEMPTY*, it is not guaranteed that the mailbox is in fact nonempty when the process is run. It can happen that two or more processes have been waiting for *NONEMPTY* (or *NEWITEM*) on the same mailbox. When the event is triggered, they will be all restarted within the same *ITU*. Then one of the processes may empty the mailbox before the others are given their chance to look at it.

Here is a way to await a guaranteed delivery of an item. A process calling

mb->wait (RECEIVE, GotIt);

will be awakened in state *GotIt* as soon as the mailbox becomes nonempty and the process is able to retrieve something from it. Like *NONEMPTY*, the event occurs immediately if the mailbox is nonempty when the wait request is issued. Before the process is run, the first element is removed from the mailbox and stored in the environment variable *TheItem* (see below).[3]

Suppose that two or more processes are waiting for *RECEIVE* on the same mailbox. When an item is put into the mailbox, only one of those processes will be awakend,[4] and the item will be automatically removed from the mailbox. If some other processes are waiting for *NONEMPTY* and/or *NEWITEM* at the same time, all these processes will be restarted even though they may find the mailbox empty when they eventually get to run.

If a *RECEIVE* event is triggered on a typed mailbox that defines the *outItem* method (Sect. 9.2.1), the method is called with the removed item passed as the argument. This is because triggering the event results in the automatic removal of an item.

If the first argument of a mailbox wait request is a nonnegative integer number, it represents a count event which is triggered when the mailbox contains precisely the specified number of items. In particular, if the specified count is 0 (symbolic constant *EMPTY*), the event will occur when the mailbox becomes empty. If the mailbox contains exactly the specified number of items when the wait request is issued, the event occurs within the current *ITU*. Note that when multiple processes access the mailbox at the same time, the number of items when the process is restarted may not match the awaited count.

---

[3]This does not happen for a counting mailbox, in which case *TheItem* returns *NULL*.

[4]Nondeterministically, or according to the order attributes of the wait requests (Sect. 5.1.1).

When a process is awakened due to a mailbox event, the environment variable *TheMailbox* (*Info02*) of type *Mailbox*, points to the mailbox on which the event has occurred. For a typed mailbox, and when it makes sense, the environment variable *TheItem* (*Info01*) of type *void** contains the value of the item that has triggered the event. Note that for some events this value may not be up to date. For a counting mailbox, *TheItem* is always *NULL*.

If a process is waiting on a mailbox, and that mailbox becomes deleted (deallocated as an object),[5] the process will be awakened regardless of the event awaited by it. This is potentially dangerous. If the program admits such situations at all, the process should make sure that the mailbox is still around when it wakes up. This can be easily found out by looking at *TheMailbox*, which will be *NULL*, if the mailbox has been destroyed. This feature applies to all mailbox types and all events that can be awaited on them.

Generally, the interpretation of *TheItem* for *NONEMPTY*, *NEWITEM*, and a count event makes sense under the assumption that only one process at a time is allowed to await events on the mailbox and remove elements from it. This is because when multiple processes are awakened by the same event, they will all see the same contents of *TheItem*, so whichever one acts on it first will deprive the others of that opportunity (without making them aware of that fact). For example, the process may deallocate the object pointed to by *TheItem*. This will not automatically nullify *TheItem*, as presented to the other processes, so they may end up referencing (and corrupting) deallocated memory. On the other hand, the interpretation of *TheItem* for *RECEIVE* is always safe. When a process is restarted by this event occurring on a typed mailbox, *TheItem* contains the value of the *actually received* item. Note that the item has been removed from the mailbox by the time the process is run, so *TheItem* is in fact the only way to get hold of it.

Of course, it is possible for multiple processes to safely contend for an item deposited into a mailbox. The first process that consumes the item can notify the other contenders that the item is gone, but the obvious idea of nullifying *TheItem* will not work, because the environment variables are local to the process wakeup scenario, i.e., the value of *TheItem* presented to the next process awakened by the same event will come from elsewhere (that value is effectively preset when the event is triggered).

Yet another event type that can be awaited on a fifo mailbox is *GET*, which is triggered whenever any process executes a successful *get* operation on the mailbox. This also applies to *erase* (see Sect. 9.2.3). For symmetry, there exists a *PUT* event, which is yet another (in addition to *UPDATE*) alias for *NEWITEM*.

---

[5]This can only happen for a mailbox owned by a process, as opposed to one owned by a station.

9.2.3   Operations on Fifo Mailboxes

The following *Mailbox* method adds a token to a counting mailbox:

int put ();

For a typed mailbox, the method accepts one argument whose type coincides with the item type specified with the mailbox type declaration (Sect. 9.2.1). The method returns one of the following three values:

ACCEPTED	The new element triggers an event that is currently awaited on the mailbox, i.e., a process awaiting a mailbox event will be awakened within the current *ITU*. Note, however, that this event need not be the one that will actually restart the waiting process[a]
QUEUED	The new element has been accepted and stored in the mailbox, but it does not trigger any events immediately
REJECTED	The mailbox is full (it has reached its capacity) and there is no room to accept the new item

[a]Which may happen when the process is expecting multiple events scheduled for the same *ITU*

The second value (*QUEUED*) is never returned for a capacity-0 mailbox. If a *put* operation is issued for a typed mailbox that defines the *inItem* method (Sect. 9.2.1), the method is called with the argument equal to the argument of *put*, but only when:

- the mailbox capacity is greater than 0 and the item is *ACCEPTED* or *QUEUED*, or
- the mailbox capacity is zero.

For a typed mailbox, this method checks whether the mailbox contains a specific item:

*Boolean queued (*etype *item);*

where "etype" stands for the item type. The method returns *YES* is the specified item is present in the mailbox, and *NO* otherwise. The check involves linearly scanning the mailbox queue for the specified value (its first occurrence) and may be costly if the queue is large.

One of the following four methods can be invoked to determine the full/empty status of a mailbox:

int empty ();
int nonempty ();
int full ();
int notfull ();

The first method returns *YES* (1), if the mailbox is empty and *NO* (0) otherwise. The second method is a simple negation of the first. The third method returns *YES*, if the mailbox is completely filled (its capacity has been reached). The fourth method is a straightforward negation of the third.

The following method removes the first item from a mailbox:

rtype *get ();*

For a counting mailbox, "rtype" is *int* and the method returns *YES*, if the mailbox was nonempty (in which case one pending token has been removed from the mailbox) and *NO* otherwise. For a typed mailbox, "rtype" coincides with the mailbox element type. The method returns the value of the removed item, if the mailbox was nonempty (the first queued item has been removed from the mailbox), or 0 (*NULL*) otherwise. Note that 0 may be a legitimate value for an item. If this is possible, *get* can be preceded by a call to *empty* (or *nonempty*), to make sure that the mailbox status has been determined correctly.

If a successful *get* operation is executed for a typed mailbox that defines the *outItem* method (Sect. 9.2.1), the method is called with the removed item passed as the argument.

One may want to peek at the first item of a typed mailbox without removing it. The method:

rtype *first ();*

behaves like *get*, except that the first item, if present, is not removed from the mailbox. For a counting mailbox, the method type is *int* and the returned value is the same as for *nonempty*.

Methods *put, get,* and *first* set the environment variables *TheMailbox* and *TheItem* (Sect. 9.2.2). *TheMailbox* is set to point to the mailbox whose method has been invoked and *TheItem* is set to the value of the inserted/removed/peeked at item. For a counting mailbox, *TheItem* is always *NULL*.

Here is the operation to clear a mailbox:

int erase ();

The method removes all elements stored in the mailbox, so that immediately after a call to *erase* the mailbox appears empty. The value returned by the method gives the number of removed elements. In particular, if erase returns 0, it means that the mailbox was empty when the method was called. If erase is executed for a typed mailbox that defines the *outItem* method (Sect. 9.2.1), the method is called individually for each removed item, in the order in which the items would be extracted by *get*, with the removed item passed as the argument on each turn.

Operations on mailboxes may trigger events awaited by processes (Sect. 9.2.2). Thus, a *put* operation triggers *NEWITEM* (aka *PUT* and *UPDATE*), and may trigger *NONEMPTY*, *RECEIVE*, and a count event. Operations *get* and *erase* trigger *GET* and may trigger count events (the latter also triggers *EMPTY*). Also, remember that deleting the mailbox itself triggers all possible events (see Sect. 9.2.2).

Operations *get, first,* and *nonempty* performed on a capacity-0 mailbox always return 0. Note, however, that *full* returns *YES* for a capacity-0 mailbox: the mailbox is both full and empty at the same time. The only way for a *put* operation on a

capacity-0 mailbox to formally succeed, is to match a wait request for *NEWITEM*
that must be already pending when the *put* is issued. Even then, there is no absolute
certainty that the element will not be lost. This is because the process to be awakened
may be awaiting a number of events (on different activity interpreters) and some of
them may be scheduled at the same *ITU* as the *NEWITEM* event. In such a case, the
actual waking event will be chosen nondeterministically, and it does not have to be
the mailbox event. One possible way of addressing this problem (if it is a problem)
is discussed in Sect. 9.2.4.

Even if the *NEWITEM* event is the only event awaited by the process, synchronization based on capacity-0 mailboxes may be a bit tricky. For illustration, consider
the following process:

```
process One (Node) {
        Mailbox *Mb;
        int Sem;
        void setup (int sem) { Sem = sem; Mb = &S->Mb; };
        states { Start, Stop };
        perform {
                state Start:
                        if (Sem) {
                                Mb->put ( );
                                proceed Stop;
                        } else
                                Mb->wait (NEWITEM, Stop);
                state Stop:
                        terminate;
        };
};
```

and assume that two copies of the process are started at the same station *s* in the
following way:

```
create (s) One (0);
create (s) One (1);
```

At first sight, it seems that both copies should terminate (in state *Stop*) within the
ITU of their creation. However, if *Mb* is a capacity-0 mailbox, it need not be the
case. Although the order of the *create* operations suggests that the process with *Sem*
equal 0 is created first, all we know is that the two processes will be started within
the same (current) *ITU*. If the second copy (the one with *Sem* equal 1) is run first, it
will execute *put* before the second copy is given a chance to issue the *wait* request to
the mailbox. Thus, the token will be ignored, and the second copy will be suspended
forever (assuming that the awaited token cannot arrive from some other process).

Of course, it is possible to force the right startup order for the two processes, e.g.,
by rewriting the *setup* method in the following way:

```
void setup (int sem) {
        setSP (sem);
        Sem = sem;
        Mb = &S->Mb;
};
```

i.e., by assigning a lower order to the startup event for the first copy of the process (Sect. 5.1.4). One may notice, however, that if we know the order in which things are going to happen, there is no need to synchronize them. Another way to solve the problem is to replace the *put* statement with a condition:

if (Mb->put () == REJECTED) proceed Start;

This solution works, but, to the people versed in operating systems and synchronization, it has the unpleasant flavor of indefinite postponement. A nicer way to make sure that both processes terminate is to create *Mb* as a capacity-1 mailbox, and replace *NEWITEM* with *NONEMPTY* or, even better, *RECEIVE*. Note that with *NEWITEM* the solution still wouldn't work. With *NONEMPTY*, the two processes would terminate properly, but the mailbox would end up containing a pending token.

Now, for a change, let us have a look at a safe application of a capacity-0 mailbox. The following two processes:

```
process Server (MyStation) {
        Mailbox *Request, *Reply;
        int Rc;
        void setup ( ) {
                Rc = 0;
                Request = &S->Request;
                Reply = &S->Reply;
        };
        states { Start, Stop };
        perform {
                state Start:
                        Request->wait (RECEIVE, Stop);
                state Stop:
                        Reply->put (Rc++);
                        proceed Start;
        };
};
```

```
process Customer (MyStation) {
        Mailbox *Request, *Reply;
        void setup ( ) {
                Request = &S->Request;
                Reply = &S->Reply;
        };
        states { ..., GetNumber,... };
        perform {

                ...

                state GetNumber:
                        Request->put ( );
                        Reply->wait (NEWITEM, GetIt);
                state GetIt:
                        Num = TheItem;

                        ...

        };
};
```

communicate via two mailboxes *Request* and *Reply*. Note that when the *Server* process issues the *put* operation for *Reply*, the other process is already waiting for *NEWITEM* on that mailbox (and is not waiting for anything else). Thus, *Reply* can be a capacity-0 mailbox. We assume that *Request* is a capacity-1 counting mailbox.

The following method:

Long getCount ();

returns the number of elements in the mailbox.

The capacity of a fifo mailbox can be modified and checked with these three methods:

Long setLimit (Long lim = 0);
Long getLimit ();
Long free ();

For example, a call to *setLimit* can be put into a user-supplied *setup* method for a mailbox subtype (Sect. 9.2.1). The argument of *setLimit* must not be negative. The method returns the previous capacity of the mailbox. The new capacity cannot be less than the number of elements currently present in the mailbox. The last method provides a shorthand for *getLimit () − getCount ()*.

9.2.4 The Priority Put Operation

Similar to the case of signal passing (Sect. 5.2.2), it is possible to give a mailbox event a higher priority, to restart the process waiting for the event before anything else can happen. This mechanism is independent of event ranking (Sect. 5.1.4), which provides another means to achieve similar effects. As an alternative to the obvious solution of assigning the lowest order to the wait request, the present approach consists in triggering a priority event on the mailbox, and assumes no cooperation from the event recipient.

A priority event on a mailbox is issued by the priority *put* operation implemented by the following *Mailbox* method:

int putP ();

For a typed mailbox, the method accepts one argument whose type coincides with the mailbox item type. The semantics of *putP* is similar to that of regular *put* (Sect. 9.2.3), i.e., the operation inserts a new item into the mailbox. The following additional rules apply exclusively to *putP*:

- At most one operation triggering a priority event (*putP* or *signalP*, see Sect. 5.2.2) can be issued by a process from the moment it wakes up until it goes to sleep. This is equivalent to the requirement that at most one priority event can remain pending at any given time.
- At the time when *putP* is executed, the mailbox must be empty and there must be exactly one process waiting on the mailbox for the new item. In other words: the exact deterministic fate of the new element must be known when *putP* is invoked.

If no process is waiting on the mailbox when *putP* is called, the method fails and returns *REJECTED*. Otherwise, the method returns *ACCEPTED*.

As soon as the process that executes *putP* is suspended, the waiting recipient of the new mailbox item is restarted. No other process is run in the meantime, even if there are other events scheduled for the current *ITU*.

The receiving process of a priority *put* operation need no special measures to implement its end of the handshake. It just executes a regular wait request for *NEWITEM*, *NONEMPTY*, or a count-1 event. The priority property of the handshake is entirely in the semantics of *putP*.

For illustration, let us return to the second example from Sect. 9.2.3 and assume that the *Customer* process communicates with two copies of the server in the following way:

...

```
state GetNumber:
        NRcv = 0;
        Request1->put ( );
        Request2->put ( );
        Reply1->wait (NEWITEM, GetIt);
        Reply2->wait (NEWITEM, GetIt);
state GetIt:
        Num = TheItem;
        if (!NRcv++) {
                Reply1->wait (NEWITEM, GetIt);
                Reply2->wait (NEWITEM, GetIt);
                sleep;

        }
```

...

The process wants to make sure that it receives both items; however, unless the servers use *putP* instead of *put*, one item can be lost, because the following scenario is possible:

- Server 1 executes *put*, and the *Customer* is scheduled to be restarted in the current *ITU*. The order of this request is irrelevant.
- Before the customer is awakened, server 2 executes *put*. Thus, there are two events that want to restart the *Customer* at the same time.
- The *Customer* is restarted and it perceives only one of the two events. Thus, one of the items is lost.

One may try to eliminate the problem by assigning a very low order to the wait requests issued by the *Customer* (Sect. 5.1.4). Another solution is to make the servers use *putP* instead *put*. Then, each *put* operation will be immediately responded to by the *Customer*, irrespective of the order of its wait requests. Yet another solution is to create mailboxes *Reply1* and *Reply2* as capacity-1 and replace *NEWITEM* with *RECEIVE*.

It might seem that the above piece of code could be rewritten in the following (apparently equivalent) way:

. . .

```
state GetNumber:
       NRcv = 0;
       Request1->put ( );
       Request2->put ( );
transient Loop:
       Reply1->wait (NEWITEM, GetIt);
       Reply2->wait (NEWITEM, GetIt);
state GetIt:
       Num = TheItem;
       if (!NRcv++) proceed Loop;
```

. . .

One should be careful with such simplifications. Note that *proceed* (Sect. 5.3.1) is in fact a *Timer* wait request (for 0 *ITUs*). Thus, it is possible that when the second server executes *putP*, the *Customer* is not ready to receive the item (although it will become ready within the current *ITU*). By replacing *proceed* with *sameas*, we retain the exact functionality of the original code, as *sameas* effectively "replicates" the body of state *Loop* at its occurrence (no state transition via the event queue is involved).

9.3   Barrier Mailboxes

A barrier mailbox is an internal mailbox whose formal capacity is negative. The exact value of that capacity does not matter: it is used as a flag to tell a barrier mailbox from a fifo.

9.3.1   *Declaring and Creating Barrier Mailboxes*

A barrier mailbox is usually untyped. The simplest way to set up a barrier mailbox is to use the default mailbox type and set its capacity to something negative, e.g.:

*Mailbox *bm;*

...

bm = create Mailbox (–1);

It may be convenient to introduce a non-standard type, to improve code clarity, e.g.:

```
mailbox Barrier {
        void setup ( ) {
                setLimit (–1);
        };
};
```

Here, the *setup* method will automatically render the mailbox a barrier upon creation. It may make some sense to use a typed mailbox for this purpose, with the item type being *int*, which coincides with the type of signals awaited and triggered on the mailbox. Thus, we can modify the above declaration into:

```
mailbox Barrier (int) {
        void setup ( ) {
                setLimit (–1);
        };
};
```

The primary advantage of a typed barrier mailbox is that it may sensibly define an *inItem* method (Sect. 9.2.1) to be called when a signal is delivered to the mailbox. There is no use for *outItem* for a barrier mailbox.

9.3.2   Wait Requests

The event argument of a wait request addressed to a barrier mailbox identifies the signal that the process wants to await. The event will occur as soon as some other process executes *signal* (Sect. 9.3.3) whose argument identifies the same event.

At any moment, there may be an unlimited number of pending wait request issued to a barrier mailbox, identifying the same or different events. Whenever an event is signaled, all processes waiting for this particular event are awakened. Other processes, waiting for other events, remain waiting until their respective events are delivered.

Note that events triggered on a barrier mailbox are never queued or pending. To be affected, a process must already be waiting on the mailbox at the time of the event's arrival. A signaled event not being awaited by any process is ignored.

When a process is awakened by an event on a barrier mailbox, the environment variable *TheMailbox* (*Info02*) of type *Mailbox**, points to the mailbox on which the event has occurred. *TheBarrier* (*Info01*) of type *int* contains the signal number that has triggered the event. Similar to other mailbox types, when a barrier mailbox is deleted (Sect. 9.2.2), all processes waiting on it will be awakened. If a situation like this is possible, the awakened process should check the value of *TheMailbox*, which will be *NULL*, if the mailbox has been deleted.

9.3.3 Operations on Barrier Mailboxes

The practically single relevant and characteristic operation on a barrier mailbox is provided by this method:

int signal (int sig);

which triggers the indicated signal (dubbed the barrier). All processes awaiting the barrier are awakened. Similar to *put* (Sect. 9.2.3), *signal* returns *ACCEPTED*, if there was at least one process awakened by the operation, and *REJECTED* otherwise. If the mailbox is typed (and its item type is *int*), *signal* can be replaced by *put* (note that *put* in that case accepts an *int* argument).

A priority variant of *signal* (Sects. 5.2.2 and 9.2.4) is also available. Thus, by calling:

mb->signalP (sig);

the caller sends the signal as a priority event. This is similar to *putP*. The semantics is exactly as described in Sect. 9.2.4, except that the event in question concerns waking up a process waiting on a barrier.

A barrier mailbox formally appears as empty (and not full), i.e., *empty* returns *YES* (and *nonempty* returns *NO*), while *full* returns *NO* (and *notfull* returns *YES*). Its item count is always zero; thus, *erase* performed on a barrier mailbox is void and returns 0.

If the capacity of a barrier mailbox is changed to a nonnegative value (with *setLimit*, Sect. 9.2.3), the mailbox becomes a fifo. This is seldom useful, if at all, but possible, similar to changing a fifo mailbox to a barrier (by resetting a nonnegative capacity to negative). A necessary condition for the status change of a mailbox to succeed is that there are no outstanding wait requests on the mailbox. If that is not the case, the operation is treated as an error, i.e., *setLimit* aborts the program with an error message.

9.4 Bound Mailboxes

A mailbox that has been bound to a device or a TCP/IP port becomes a buffer for messages to be exchanged between the SMURPH program and the other, external party. One way to carry out this communication is to use the same standard operations *put* and *get*, which are available for an internal mailbox (Sect. 9.2.3). For a bound mailbox, those operations handle only one byte at a time. This may be sufficient for exchanging simple status information, but is rather cumbersome for longer messages. Thus, bound mailboxes provide other, more efficient tools for operating on larger chunks of data.

9.4.1 Binding Mailboxes

Any fifo mailbox (created as explained in Sect. 9.2.1) can be bound to a device or a socket port. This operation is performed by the following method defined in class *Mailbox*:

int connect (int tp, const char *h, int port, int bsize = 0, int speed = 0,
  int parity = 0, int stopbits = 0);

If it succeeds, the method returns *OK* (zero). The first argument indicates the intended type of binding. It is a sum (Boolean or arithmetic) of components/flags whose values are represented by the following constants:

INTERNET, LOCAL, or DEVICE	This component indicates whether the mailbox is to be bound to a TCP/IP socket, to a local (UNIX-domain) socket, or to a device
CLIENT or *SERVER*	This flag is only applicable to a socket mailbox. It selects between the client and server modes of our end of the connection. In the client mode, *connect* will try to set up a connection to an already-existing (server) port of the other party. In the server mode, *connect* will set up a server port (without actually connecting it anywhere) expecting a connection from a client One additional flag that can be specified with *SERVER* is *MASTER* indicating a "master" mailbox that will be solely used for accepting incoming connections (rather than sustaining them directly)
WAIT or *NOWAIT*	This flag only makes sense together with *SERVER* (i.e., if the connection is to be established in the server mode). With *WAIT*, the *connect* operation will block until there is a connection from a client. Otherwise, which is the default action if neither *WAIT* nor *NOWAIT* has been specified, the method will return immediately
RAW or *COOKED*	This flag is used together with *DEVICE*. If the device is a tty (representing a serial port), *RAW* selects the raw interface without any tty-specific preprocessing (line buffering, special characters). With *COOKED*, which is the default, the stream is preprocessed as for standard terminal I/O
READ and/or *WRITE*	These flags tell what kind of operations will be performed on the device to which the mailbox is bound. For a socket mailbox, this specification is irrelevant because both operations are always legitimate. For a device mailbox, at least one of the two options must be chosen. It is also legal to specify them both

Upon a successful completion, *connect* returns *OK*. The operation may fail in a fatal way, in which case it will abort the program. This will happen if some hard rules are violated, i.e., the program tries to accomplish something formally impossible. The method may also fail in a soft way, returning *REJECTED* or *ERROR*. Here are the possible reasons for a fatal failure:

1. the mailbox is already bound, i.e., an attempt is made to bind the same mailbox twice
2. the mailbox is a barrier; it is illegal to bound barrier mailboxes
3. the mailbox is nonempty, i.e., it contains some items that were deposited while the mailbox wasn't bound and have not been retrieved (or erased) yet
4. there are processes waiting for some events on the mailbox
5. there is a formal error in the arguments, e.g., conflicting flags.

A bound mailbox (except for a master mailbox, Sect. 9.4.4) is equipped with two buffers: one for input and the other for output. The *bsize* (buffer size) argument of *connect* specifies the common size of the two buffers (i.e., each of them is capable of storing *bsize* bytes). The argument redefines the capacity of the mailbox for the duration of the connection. If the specified buffer size is zero, the (common) buffer size is set to the current capacity of the mailbox. Unless the mailbox is a master mailbox (Sect. 9.4.4), the resulting buffer size must be greater than zero.

Information arriving in the mailbox from the other party will be stored in the input buffer, from where it can be retrieved by the operations discussed in Sect. 9.4.7. If the buffer fills up, the arriving bytes will be blocked until some buffer space is reclaimed. Similarly, outgoing bytes are first stored in the output buffer which is emptied asynchronously by the SMURPH kernel. If the output buffer fills up, the mailbox will refuse to accept more messages until some space in the buffer becomes free. Both buffers are circular: each of them can be filled and emptied independently from the two ends.

The contents of a bound mailbox, as perceived by a SMURPH program, are equivalenced with the contents of its input buffer. For example, when we say that the mailbox is empty, we usually mean that the input buffer of the mailbox is empty. All operations of acquiring data from the mailbox, inquiring the mailbox about its status, etc., refer to the contents of the input buffer. The bytes deposited/written into the mailbox (the ones that pass through the output buffer) are destined for the other party and, conceptually, do not belong to the mailbox.

An attempt to bind a barrier mailbox, i.e., one whose capacity is negative, is treated as an error and aborts the program. Note that a barrier mailbox can be reverted to the non-barrier status by resetting its capacity with *setLimit* (Sect. 9.2.3) before being bound.

9.4.2   Device Mailboxes

To bind a mailbox to a device, the first argument of *connect* should be *DEVICE* combined with at least one of *READ*, *WRITE*, and possibly with *RAW* or *COOKED*. The *COOKED* flag is not necessary: it is assumed by default if *RAW* is absent. The second argument should be a character string representing the name (file path) of the device to which the mailbox is to be bound. The third argument (*port*) is ignored, and the fourth one (*bsize*) declares the buffer size. This number should be at least equal to the maximum size of a message (in bytes) that will be written to or retrieved from the mailbox with a single *read/write* operation (see Sect. 9.4.7). The specified buffer size can be zero (the default), in which case the current capacity of the mailbox (Sect. 9.2.1) will be used. The final length of the buffer, however arrived at, must be strictly greater than zero.

The last three arguments of *connect* are only relevant when the device is attached via a serial port and the first argument includes the *RAW* flag. They specify the port speed, the parity, and the number of stop bits, respectively. The acceptable values for speed (serial rate) are: 0, 50, 75, 110, 134, 150, 200, 300, 600, 1200, 1800, 2400, 4800, 9600, 19,200, 38,400, 57,600, 115,200, 230,400. Value 0 indicates that the current speed setting for the port, whatever it is, should not be changed.

The specified parity can be 0 (no parity), 1 (indicating odd parity), or 2 (selecting even parity). If parity is nonzero, the character size is set to seven bits. The last argument can be 0 (automatic selection of the stop bits), 1, or 2.

For illustration, with the following call:

m3->connect (DEVICE+RAW+READ+WRITE, "/dev/ttyS1", 0, 256);

the mailbox pointed to by *m3* is bound to device */dev/ttyS1*. The interface is raw (no tty-specific preprocessing will be carried out by the system), and the mailbox will be used both for input and output. The buffer size is 256 bytes.

Below is the code of a (not very efficient) process that copies one file to another.

```
mailbox File (int);

. . .

process Reader {
        File *Fi, *Fo;
        int outchar;
        setup (const char *finame, const char *foname) {
                Fi = create File;
                Fo = create File;
                if (Fi->connect (DEVICE+READ, finame, 0, 1))
                        excptn ("cannot connect input mailbox");
                if (Fo->connect (DEVICE+WRITE, foname, 0, 1))
                        excptn ("cannot connect output mailbox");
        };
        states { GetChar, PutChar };
        perform {
                state GetChar:
                        if (Fi->empty ( )) {
                                if (Fi->isActive ( )) {
                                        Fi->wait (UPDATE, GetChar);
                                        sleep;
                                } else
                                        Kernel->terminate ( );
                        }
                        outchar = Fi->get ( );
                transient PutChar:
                        if (Fo->put (outchar) != ACCEPTED) {
                                if (Fo->isActive ( )) {
                                        Fo->wait (OUTPUT, PutChar);
                                        sleep;
                                } else
                                        excptn ("write error on output file");
                        }
                        proceed GetChar;
        };
};
```

See Sects. 9.2.3 and 9.4.5 for the semantics of methods *empty* and *isActive*.

As we see, bound mailboxes can be used to access files, but because of the non-blocking character of I/O, one must be careful there. For example, it usually does not make sense to bind the same mailbox to a file with both *READ* and *WRITE* flags, because SMURPH will try to read ahead a portion of the file into the buffer, even if no explicit *get* or *read* operation has been issued by the program. This may confuse the file position for a subsequent *write* operation.

If a mailbox is bound to a file with *WRITE* (but not *READ*), the file is automatically erased, i.e., truncated to zero bytes.

The *connect* operation will fail, if the specified device/file cannot be opened for the required type of access. Such a failure is non-fatal: the method returns *ERROR*.

9.4.3   Client Mailboxes

By a socket mailbox we understand a mailbox bound to a TCP/IP port, possibly across the Internet, or to a UNIX-domain socket restricted to the current host. A client connection of this kind assumes that the other party (the server) has made its end of the connection already available. For a TCP/IP session, that end is visible as a port number on a specific host, whereas a local server is identified via the file path name of a local socket.

For a client-type connection, the first argument of *connect* must be either *INTER-NET+CLIENT* or *LOCAL+CLIENT*. The explicit *CLIENT* specification is not required as it is assumed by default when the *SERVER* flag is absent.

If the connection is local, the second argument of *connect* should specify the path name of the server's socket in the UNIX domain. For an Internet connection, that argument will be interpreted as the Internet name of the host on which the server is running. The host name can be specified as a DNS name, e.g., *sheerness.cs.ualberta.ca*, or as an IP address (in a string form), e.g., 129.128.4.33. The third argument (*port*) is ignored for a local connection. For an Internet connection, it specifies the server's port number on the remote host.

The fourth argument (*bsize*) indicates the mailbox buffer size. The rules are the same as for a device mailbox (Sect. 9.4.2). In particular, the buffer size for a client mailbox must be strictly greater than zero. The remaining arguments do not apply to socket mailboxes and are ignored.

For illustration, the following call:

m1->connect (CLIENT+INTERNET, "sheerness.cs.ualberta.ca", 2002, 256);

binds the mailbox pointed to by *m1* to port 2002 on the host *sheerness.cs.ualberta.ca*.

A local client-type connection is illustrated by the following call:

m2->connect (LOCAL, "/home/pawel/mysocket", 0, 1024);

Note that the *CLIENT* specification is not necessary, because it is assumed by default.

A *connect* operation will fail, if the specified port is not available on the remote host, the remote host cannot be located, or the path name does not identify a local UNIX-domain socket (for a local connection). In such a case, *connect* returns *ERROR*.

9.4.4 Server Mailboxes

With a server mailbox, a SMURPH program can accept incoming connections from Internet or UNIX-domain sockets. To set up a server mailbox, the first argument of *connect* should be *INTERNET+SERVER* (for an Internet mailbox) or *LOCAL+SERVER* (for a UNIX-domain connection). Additionally, at most one of two flags *WAIT* and *MASTER* can be thrown into the first argument.

For a TCP/IP mailbox, the second argument of *connect* (h) is ignored and the third argument (*port*) specifies the port number for accepting incoming connections. For a UNIX-domain setup, the second argument specifies the path name of the local socket and the third argument is ignored. The following shorthand variant of *connect* sets up a server mailbox bound to the Internet:

```
int connect (int tp, int port, int bsize = 0) {
        return connect (tp, NULL, port, bsize);
};
```

If all that is expected from the server mailbox is to accept one connection, its operation is simple and closely resembles the operation of its client counterpart. This type of a bound mailbox will be called an immediate server mailbox. It is selected by default if the first argument of *connect* does not include the *MASTER* flag.

For an immediate server mailbox, the *bsize* argument of *connect* is relevant and indicates the buffer size at the server end. The same rules apply as for a client/device mailbox (Sect. 9.4.1). The remaining arguments are ignored.

As soon as it has been successfully bound, an immediate server mailbox becomes usable in exactly the same way as a client mailbox. Until the other party connects to it and sends some data, the mailbox will appear empty to reading operations (Sect. 9.4.7). Also, any data deposited in the mailbox by writing operations will not be expedited until the other end of the connection becomes alive.

While binding an immediate server mailbox, the program can request the *connect* operation to block until the other party connects. This is accomplished by adding *WAIT* to the list of flags contributing to the first argument of *connect*.

For illustration, the following two operations:

m1->connect (INTERNET+SERVER, 3345, 4096);
m2->connect (LOCAL+SERVER+WAIT, "/home/pawel/mysocket", 1024);

bind two immediate server mailboxes. The first mailbox is bound to a TCP/IP port (number 3345) with the buffer size of 4096 bytes. In the second case, the mailbox is local (a UNIX-domain socket) and the operation blocks until the client connects. The buffer size is 1024 bytes.

An immediate server mailbox can be recycled for multiple connections, but those connections must be serviced sequentially, one at a time. A connection can be closed without unbinding the mailbox (see operation *disconnect* in Sect. 9.4.5), which will make the mailbox ready to accommodate another client.

To handle multiple client parties requesting service at the same time, the program must set up a master mailbox. This is accomplished by adding *MASTER* to the flags specified in the first argument of *connect*.

The sole purpose of a master mailbox is to accept connections. Once it has been accepted, the connection will be represented by a separate mailbox which has to be associated with the master mailbox. As a master mailbox is only used for accepting connections (rather than sustaining them), it needs no buffer storage for incoming or outgoing data. Thus, the buffer size specification for a master mailbox (argument *bsize*) is ignored. Moreover, the *WAIT* flag cannot be specified for a master mailbox.

The following method of *Mailbox* tells whether a master mailbox is connection pending (meaning that an incoming connection is awaiting acceptance):

int isPending ();

The method returns *YES*, if there is a pending connection on the mailbox, and *NO* otherwise. It also returns *NO*, if the mailbox is not bound, or it is not a master mailbox.

When a master mailbox becomes connection pending, it triggers a *NEWITEM* (*UPDATE*) event (Sect. 9.2.2). The same event triggered on a regular bound mailbox indicates the availability of input data.

The following variant of *connect* is used to accept a connection on a master mailbox:

int connect (Mailbox *m, int bsize = 0);

The first argument specifies the master mailbox on which the connection is pending. The second argument (*bsize*) specifies the buffer size for the connection. The method associates its mailbox with the connection and makes it capable to receive or send data from/to the connecting party. The Internet/local mode of this association is inherited from the master mailbox.

For illustration, we show below the code of a process accepting and distributing connections arriving on a master mailbox.

```
process Server {
        Mailbox *Master;
        void setup (int Port) {
                Master = create Mailbox;
                if (Master->connect (INTERNET+SERVER+MASTER, Port) != OK)
                        excptn ("Server: cannot setup MASTER mailbox");
        };
        states { WaitConnection };
                perform {
                        state WaitConnection:
                                if (Master->isPending ( )) {
                                        create Service (Master);
                                        proceed WaitConnection;
                                }
                                Master->wait (UPDATE, WaitConnection);
                };
};
```

Note that the *Server* process creates a separate process (an instance of *Service*) for every new connection. The recommended structure of such a process is this:

```
process Service {
        Mailbox *Local;
        void setup (Mailbox *master) {
                Local = create Mailbox;
                if (Local->connect (master, BUFSIZE) != OK) {
                        delete Local;
                        terminate ( );
                }
        };
        states {...};
        perform;
        ...
};
```

The *connect* operation that associates a mailbox with a master mailbox returns *OK* upon success. It will fail, returning *REJECTED*, if the mailbox specified as the first argument is not bound, is not a master mailbox, or is not connection pending.

9.4.5   Unbinding and Determining the Bound Status

A bound mailbox can be unbound with the following *Mailbox* method:

int disconnect (int how = SERVER);

Depending on the type of the bound mailbox, the method behaves in the following way:

Device mailbox	The device is closed, the buffers are deallocated, the mailbox is reverted to its unbound status
Client mailbox	The client end of the socket is closed, the buffers are deallocated, the mailbox is reverted to its unbound status
Immediate server mailbox	The current connection (the socket) is closed. If the method has been called with *SERVER* as the argument (the default), the buffers are deallocated and the mailbox is reverted to its unbound status. If the argument is *CLIENT*, only the current connection is closed. The mailbox remains bound and ready to accommodate another connection
Master mailbox	If the method has been called with *SERVER* (the default), the master socket is closed and the mailbox is reverted to its unbound status. If the argument is *CLIENT*, the operation is ignored

Upon a successful completion, the method returns *OK*. This also happens when *disconnect* has been called for an unbound mailbox, in which case it does nothing. The method returns *ERROR* (and fails to perform its task) in the following circumstances:

1. there are some processes waiting on the mailbox
2. the mailbox is nonempty, i.e., the input buffer contains pending data that hasn't been retrieved
3. the output buffer contains some data that hasn't been expedited from the mailbox.

It is possible to force unbinding even if one or more of the above conditions hold. For that, *disconnect* should be called with *CLEAR* specified as the argument. In such a case, the method carries out the following actions:

1. every process waiting on the mailbox for one of the following events: *NEWITEM* (*UPDATE*), *SENTINEL*, *OUTPUT*, or a count event (Sects. 9.4.6 and 9.4.7) receives the awaited event
2. both buffers are erased and reset to the empty state (this operation is void for a master mailbox)
3. the operation continues as if *disconnect* were invoked with *CLIENT*.

Note that *disconnect* (*CLEAR*) does not change the status of a master mailbox. For an immediate sever mailbox, it only terminates the current connection. This kind of a formal disconnection is forced automatically by SMURPH when it detects that the connection has been dropped by the other party or broken because of a networking problem. The program can monitor such events and respond to them, e.g., with the assistance of the following method:

int isConnected ();

which returns *YES*, if the mailbox is currently bound, and *NO* otherwise. Note that *isConnected* will return *YES* for a bound immediate server mailbox, even though there may be no connection on the mailbox at the time. The following method:

int isActive ();

takes care of this problem: it returns *YES*, only if the mailbox is in fact connected to some party. In particular, it always returns *NO* for a master mailbox.

To find out that there is a problem with the connection, the program should invoke *isActive* in response to a *NEWITEM*, *SENTINEL*, or a count event, and also whenever a *read* or *write* operation (Sect. 9.4.7) fails. If *isActive* returns *NO*, it means that the connection has been closed by the other party or broken. Note that the mailbox is automatically unbound in such a case: there is no need to perform *disconnect* on it, but it will do no harm.

A drastic but effective way to completely close a connection and forget about it is to simply delete the mailbox.[6] This has the effect of closing any connection currently active on the mailbox, deallocating all buffers, and waking up all processes waiting on it. An awakened process can find out that the mailbox does not exist anymore by examining *TheMailbox*, which will be *NULL* in that case.

9.4.6 Wait Requests

Practically all events that can be awaited on a fifo mailbox (Sect. 9.2.2) also make sense for a bound mailbox, with the interpretation of its input buffer as the item queue, where the items are individual bytes awaiting extraction. For example, for *NONEMPTY*, the mailbox is assumed to be nonempty, if its input buffer contains at least one byte.

A *NEWITEM* (aka *UPDATE* or *PUT*) event is triggered when anything new shows up in the mailbox's input buffer. Also, for a master mailbox, this event indicates a new connection awaiting acceptance.

For a bound mailbox, the *RECEIVE* event occurs when the input buffer contains at least one byte. The first byte is removed from the buffer and stored in *TheItem* before the process responding to *RECEIVE* is run.

The count event has a slightly different semantics for a bound mailbox. Namely, the event is triggered when the number of bytes in the input buffer reaches or exceeds the specified count. Its primary purpose is to be used together with *read* (Sect. 9.4.7) for extracting messages from the mailbox in a non-blocking manner.

Two events, *OUTPUT* and *SENTINEL*, can only be awaited on a bound mailbox. The first one is triggered whenever some portion (at least one byte) of the output buffer associated with the mailbox is emptied and sent to the other party. The event is

---

[6]Such a mailbox must be owned by a process rather than by a station (Sect. 9.2.1).

used to detect situations when it makes sense to retry a *write* operation (Sect. 9.4.7) that failed on a previous occasion.

One possible way to retrieve a block of data from a bound mailbox is to use *readToSentinel*, which operation reads a sequence of bytes up to a declared marker (sentinel) byte (Sect. 9.4.7). A process that wants to know when it is worthwhile to try this operation may issue a wait request for the *SENTINEL* event. This event will be triggered as soon as the input buffer either contains at least one sentinel byte or becomes completely filled. If this condition holds at the time when the wait request is issued, the event occurs immediately.

The reason why the *SENTINEL* event is (also) triggered by a completely filled buffer (even if it does not contain the sentinel byte) is that in such a situation there is no way for the buffer to receive the sentinel byte until some portion of the buffered data is removed. Sentinels are used to separate blocks of data with some logical structure (e.g., lines of text read from the keyboard), and if an entire block does not fit into the input buffer at once, its portions must be read in chunks. Only the last of those chunks will be terminated by the sentinel byte.

When a *SENTINEL* event is received, the environment variable *TheCount* of type int (an alias for *Info01*) contains the number of bytes in the input buffer preceding and including the first occurrence of the sentinel. If the sentinel does not occur in the buffer (the event has been triggered because the buffer is full), *TheCount* contains zero.

When a bound mailbox becomes forcibly disconnected by *disconnect* (*CLEAR*) (Sect. 9.4.5), which will also happen when the connection is closed or broken, or the mailbox is deleted, all the awaited *NEWITEM*, *SENTINEL*, *OUTPUT*, and count events are triggered. This way, a process waiting to receive or expedite some data through the mailbox will be awakened and it will be able to respond to the new status of the mailbox. Consequently, to make its operation foolproof, the process cannot assume that any of the above events always indicates the condition that it was meant for. The recommended way of acquiring/expediting data via a bound mailbox is discussed in Sect. 9.4.7.

9.4.7 Operations on Bound Mailboxes

In principle, the standard operations *put* and *get* described in Sect. 9.2.3 are sufficient to implement network communication via a bound mailbox. For a bound mailbox, *put* behaves as follows. If the mailbox is typed (*put* takes an argument), the least significant byte from the argument is extracted. If the output buffer of the mailbox (Sect. 9.4.1) is not completely filled, the byte is deposited at the end of the outgoing portion of the buffer and *put* returns *ACCEPTED*. If the buffer is full, nothing happens and *put* returns *REJECTED*. If the bound mailbox is a counting (untyped) mailbox (*put* takes no argument), a zero byte is expedited in exactly the same way. Generally, this makes little sense, so counting mailboxes are not good candidates for binding.

An extraction by *get* retrieves the first byte from the input buffer. If the mailbox is a counting (untyped) one, the byte is removed and ignored, so the only information returned by *get* in such a case is whether a byte was there at all (the method returns *YES* or *NO*, Sect. 9.2.3). For a typed mailbox, the extracted byte is returned in the least significant byte of the method's value. The operation of *first* mimics the behavior of *get*, except that it only peeks at the byte without removing it from the buffer.

If a typed bound mailbox defines *inItem* and/or *outItem*, the methods will be called as described in Sects. 9.2.1 and 9.2.3, with the deposited/extracted bytes interpreted as the items. This only happens when the communication involves *put* and *get*. Also, *erase*, which can be used to completely erase the input buffer of a bound mailbox will call *outItem* for every single byte removed from the buffer. The methods *empty*, *nonempty*, *full*, *notfull* tell the status of the input buffer, and their meaning is obvious. Also, *getCount* returns the number of bytes present in the input buffer.

While a mailbox remains bound, the operations *setLimit* and *putP* are illegal. The priority put operation makes no sense (because the event is not received by the SMURPH program). The size of the mailbox buffer can be enlarged dynamically, at any time, with this *Mailbox* method:

int resize (int newsize);

where the argument specifies the new size. If the new size is less than or equal to the present buffer size (*limit*), the method does nothing. Otherwise, it reallocates the buffers according to the new increased size and preserves their contents. In both cases, the method returns the old size. Note that buffer size of a bound mailbox can only be increased.

Sending and receiving one byte at a time may be inefficient for serious data exchange. Therefore, the following additional methods have been made available for a bound mailbox:

int read (char *buf, int nc);
int readAvailable (char *buf, int nc);
int readToSentinel (char *buf, int nc);
int write (const char *buf, int nc);

The first operation attempts to extract exactly *nc* bytes from the input buffer of the mailbox and store them in *buf*. If the input buffer does not hold that many bytes at the time, *read* returns *REJECTED* and does nothing. Otherwise, it returns *ACCEPTED* and removes the acquired bytes from the input buffer. It is an error to specify *nc* larger than the mailbox buffer size (Sect. 9.4.1) because the operation would have no chance of ever succeeding.

If the operation fails (i.e., returns *REJECTED*), the issuing process may decide to wait for a count event equal to the requested number of bytes (Sect. 9.4.6). For a bound mailbox, this event will be triggered as soon as the input buffer becomes filled with at least the specified number of bytes.

The second operation (*readAvailable*) extracts at most *nc* bytes from the mailbox. The method does not fail (as *read* does), if less than *nc* bytes are available. Its value returns the actual number of extracted bytes. That number will be zero if the mailbox is empty when the method is invoked.

The third read operation (*readToSentinel*) attempts to extract from the input buffer a block of bytes terminated with the sentinel byte (the sentinel will be included in the block). Prior to using this operation, the program should declare a sentinel byte for the mailbox by calling the following method:

void setSentinel (char sentinel);

The default sentinel, assumed when *setSentinel* has never been called, is the null byte. The second argument of *readToSentinel* specifies the limit on the number of bytes that can be extracted from the mailbox (the capacity of the buffer pointed to by the first argument). The operation succeeds, and acquires data from the mailbox, if the mailbox buffer either contains at least one occurrence of the sentinel or it is completely filled (Sect. 9.4.6). In such a case, the function returns the number of extracted bytes. If the second argument of the method is less than the number of bytes preceding and including the sentinel, only an initial portion of the block will be read. If the sentinel never occurs in the mailbox buffer and that buffer is not completely filled, the function fails and returns zero. The issuing process can wait for a *SENTINEL* event (Sect. 9.4.6) and then retry the operation.

Note that *readToSentinel* succeeds when the mailbox buffer is completely filled, even if no sentinel is present in the buffer. The reason for this behavior was given in Sect. 9.4.6. It makes no sense to wait for the sentinel in such a case, as some room must be reclaimed in the buffer before the sentinel can possibly materialize. The calling process can easily determine what has happened (without searching the acquired data for the sentinel), by examining the *Boolean* environment variable *TheEnd* (an alias for *Info01*). Its value is *YES* if the sentinel is present in the acquired data chunk, and *NO* otherwise.

The following *Mailbox* method:

int sentinelFound ();

checks whether the mailbox buffer contains at least one sentinel byte. It returns the number of bytes preceding and including the first sentinel, or zero if no sentinel is present.

The *write* operation acts in the direction opposite to *read*. It succeeds (returns *ACCEPTED*) if the specified number of bytes could be copied from buf to the output buffer of the mailbox. Otherwise, if there is not enough room in the buffer, the method returns *REJECTED* and leaves the output buffer intact.

Note that, generally, it is more natural to fail for *read* than for *write*. In the first case, the expected message may be some sensor data arriving from a remote component of the control system, and such data only arrives when there is something relevant to report, possibly at very sparse intervals. Therefore, it is natural for a process monitoring a bound mailbox to mostly wait, for a count, *NEWITEM*, or *SENTINEL*, until some data arrives. On the other hand, the inability to write, e.g., expedite a status change message to a remote component (e.g., an actuator) is something much less natural and usually abnormal. If the output buffer of a bound mailbox is filled and unable to accommodate a new outgoing message, it typically means that there is something wrong with the network: it cannot respond as fast as the control program would like it to respond. Because of this asymmetry, there is no count event that would indicate the ability of the output buffer to accept a specific number of bytes. Instead, there is a single event, *OUTPUT* (Sect. 9.4.6), which is triggered whenever any portion of the output buffer is flushed to the network. A process that cannot send its message immediately (because *put* or *write* has failed) may use this event to determine when the operation can be sensibly retried. In most cases, the *OUTPUT* event indicates that the entire buffer (or at least a substantial part of it) has been freed and is now available.

It is also possible to check whether the output buffer contains any pending data that has not been sent yet to the other party. For example, before deleting a bound mailbox (which operation immediately discards all buffered data), one may want to wait until all output has been flushed. This *Mailbox* method:

Long outputPending ();

returns the outstanding number of bytes in the output buffer that are still waiting to be sent out.

For illustration, assume that we would like to set up a process mapping sensor messages arriving from some remote physical equipment into simple events occurring on an internal mailbox. The following mailbox type:

mailbox NetMailbox (int);

will be used to represent the remote sensor, and the following process will take care of the conversion:

```
process PhysicalToVirtual {
        SensorMailbox *Sm;
        NetMailbox *Nm;
        void setup (SensorMailbox, const char*, int, int);
        states { WaitStatusChange };
};
```

We assume that *SensorMailbox* is a simple internal mailbox on which the status messages arriving from the physical sensor will be perceived in a friendly and uniform way. The *setup* method of our process is listed below.

```
void PhysicalToVirtual::setup (SensorMailbox *s, const char *host, int port, int bufsize) {
        Sm = s;
        Nm = create NetMailbox;
        if (Nm->connect (INTERNET+CLIENT, host, port, bufsize) != OK)
                excptn ("Cannot bind client mailbox");
        Nm->put (START_UPDATES);
};
```

Its argument list consists of the pointer to the internal mailbox, the host name to connect to, the port number on the remote host, and the buffer size for the network mailbox. The method copies the mailbox pointer to its internal attribute, creates the network mailbox, and connects the mailbox to the host (in the client mode, Sect. 9.4.3). Then it sends a single-byte message to the remote host. We assume that this message will turn on the remote sensor and force it to send (periodically) the updates of its status.

Here is the code method of our process:

```
PhysicalToVirtual::perform {
          char msg [STATMESSIZE];
          state WaitStatusChange:
                    if (Nm->read (msg, STATMESSIZE) == ACCEPTED) {
                              Sm->put (msg [STATUSVALUE0] * 256 + msg [STATUSVALUE1])
                              proceed WaitStatusChange;
                    } else
                              Nm->wait (STATMESSIZE, WaitStatusChange);
};
```

In its single state, the process expects a message from the remote sensor. If *read* succeeds, the process extracts the new value sent by the sensor and passes it to the internal mailbox. If the message is not ready (*read* fails), the process suspends itself until *STATMESSIZE* bytes arrive from the sensor.

The above code assumes that nothing ever goes wrong with the connection. To make it foolproof, we should check the reason for every failure of the *read* operation, e.g.,

```
...
state WaitStatusChange:
          if (Nm->read (msg, STATMESSIZE) == ACCEPTED) {
                    Sm->put (msg [STATUSVALUE0] * 256 + msg [STATUSVALUE1]);
                    proceed WaitStatusChange;
          } else if (Nm->isActive ( )) {
                    Nm->wait (STATMESSIZE, WaitStatusChange);
          } else {
                    // The connection has been dropped
                    ...
                    terminate;
          }
...
```

It is formally legal to issue a *read* or *write* operation (including the other two variants of *read*) to an unbound mailbox, but then the operation will fail returning *ERROR*.

Automatic unbinding (*disconnect (CLEAR)*) is forced by SMURPH when the connection is closed by the other party or broken (Sect. 9.4.5). Also, a process waiting for a count event on a mailbox that has been forcibly disconnected will receive that event (Sect. 9.4.6). One way to find out that there has been a problem with the connection is to execute *isActive* (Sect. 9.4.5).

9.5 Journaling

A mailbox bound to a TCP/IP port or a device can be journaled. This operation consists in saving the information about all transactions performed on the mailbox in a file. This file, called a *journal file*, can be used later (in another run) to feed the same or a different mailbox whose input would be otherwise acquired from a TCP/IP port or a device. The journaling capability is only available if the program has been compiled with the *–J* option of *mks* (Sect. B.5), and only if one of the options *–R* or *–W* has been selected as well.

Applications of journal files are two-fold. First, journals can be used to collect event traces from real physical equipment, e.g., to drive realistic models of the emulated (or simulated) physical system. Another use for journal files is to recover from a partial crash of a distributed control program.

Imagine two SMURPH programs, P_1 and P_2, such that P_1 sends some information to P_2. The two programs communicate via a pair of mailboxes, one mailbox operating at each party. Assume that the mailbox used by P_1 is journaled. This means that all the data expedited by the program via that mailbox are saved in a local file. If for some reason P_2 fails at some point, and all the information it has received from P_1 is destroyed, then P_2 can be rerun with its mailbox driven by the journal file. The net effect of this operation will be the same as if the mailbox of P_2 were dynamically filled, preserving the original timing, with the data originally sent by P_1. Of course, P_2 is free to journal its mailbox as well, saving all data received from P_1 in its local file. If that file is available after the crash, it can be used to the same end.

9.5.1 Declaring Mailboxes to Be Journaled

The operations of journaling and driving a mailbox from a journal file are transparent to the control program, i.e., the program does not have to be modified before its mailboxes can be journaled. When the program is called, the mailboxes to be journaled can be specified through a sequence of call arguments with the following syntax (see Sect. B.5):

–J mailboxident

where "mailboxident" is either the standard name or the nickname of the mailbox (Sect. 3.3.2). Nicknames are generally more useful for this purpose, as the standard

names (Sect. 9.2.1) may be long, contrived, and inconvenient. Therefore, it is recommended to assign unique nicknames (Sect. 3.3.8) to the mailboxes that will be
journaled.

Note that "mailboxident" should appear as a single token on the call argument
list; therefore, if the mailbox name includes blanks or other problem characters, then
either each of those characters must be escaped, or the entire name must be quoted.

For each mailbox marked in the above way, SMURPH opens a journal file in the
directory in which the program has been called. The name of this file is obtained
from "mailboxident" by replacing all non-alphanumeric characters with underscores
('_') and adding the extension *.jnj* to the result of this operation. As all journal files
must have distinct names, the user should be careful in selecting the nicknames of
journaled mailboxes. The journal file must not exist; otherwise, the program will be
aborted. SMURPH is obsessive about not overwriting existing journal files.

Formally, the idea behind journaling is very simple. Whenever the mailbox is
connected (i.e., the *connect* operation, Sect. 9.4.1, is performed on the mailbox), a
new *connect block* is started in its journal file. Whenever something is expedited
from the mailbox to the other party (which may be a TCP/IP port or a device) that
data is written to the journal file along with the time of this event. Whenever some
data arrives from the other party and is put into the input buffer of the mailbox, that
data is also written to the journal file. Of course, sent and received data are tagged
differently. Thus, the contents of the journal file describe the complete history of
all external activities on the mailbox. Note, however, that the details of the specific
internal operations performed by the program (*get*, *put*, *read*, *write*, etc.) are not
stored.

9.5.2   Driving Mailboxes from Journal Files

To drive a mailbox from a journal file, the following argument must be included
in the call argument list of the program (one per each mailbox to be driven from a
journal):

–I mailboxident

(we mean "minus capital aye") or

–O mailboxident

(we mean "minus capital oh, not zero"). In the first case, the mailbox input will be
matched with the input section of the journal file. This option makes sense if the file
was created by journaling the same mailbox. In the second case, the mailbox input
will be fed by the output data extracted from the journal file. This makes sense if the
journal comes from a mailbox that was used by some other program to send data to
the specified mailbox.

The name of the journal file to drive the indicated mailbox is obtained in a way
similar to that described in Sect. 9.5.1. The file extension is *.jnx*. Note that it is

different from the suffix of a created journal file; thus, the original file must be renamed or copied before it can be used to feed a mailbox.

A mailbox declared as being driven from a journal file is never connected to a TCP/IP port or a device. The only effect of a *connect* operation on such a mailbox is to advance the feeding journal file to the next connect block (Sect. 9.5.1). The input buffer of the mailbox is fed with the data extracted from the file, at the time determined by the time stamps associated with those data. Whatever is written to the mailbox by the program is simply discarded and ignored. Although this may seem strange at first sight, it is the only sensible thing to do.

By default, the time stamps in a journal file refer to the past, compared to the execution time of the program whose mailbox is driven by that file. Note that this past may be quite distant. This will have the effect of all the data arriving at the mailbox being deemed late and thus made available immediately, i.e., as soon as some data are extracted from the mailbox, new data will be fed in from the journal file. This need not be harmful. Typically, however, during a recovery, the program whose mailboxes are driven from journal files is run in shifted real time. This can be accomplished by using the *–D* call argument (Sect. B.6). To make sure that the time is shifted properly, the program can be called with *–D J*, which has the effect of setting the real time to the earliest creation time of all journal files driving any mailboxes in the program.

A mailbox being driven by a journal file can be journaled at the same time (note that the suffixes of the two files are different, although their prefixes are the same). This is less absurd than it seems at first sight if we recall that the output of such a mailbox is otherwise completely ignored.

9.5.3 Journaling Client and Server Mailboxes

Journaling is primarily intended for mailboxes that do not change their bound status during the execution of the control program. However, it may also make sense for mailboxes that are bound and unbound dynamically, possibly several times. For a client-type mailbox (Sect. 9.4.3), the semantics of journaling is quite simple. If the mailbox is being journaled (i.e., its transactions are backed up), each new connection starts a new connect block in the journal file. When the same (or another) mailbox is later driven from the journal file, every new connection advances the file to the next connect block. Having exhausted the data in the current connect block, the mailbox will behave as if the other party has closed the connection, i.e., the mailbox will be disconnected (Sect. 9.4.5). A subsequent *connect* operation on the mailbox will position the journal file at the beginning of the next connect block. If there are no more connect blocks in the file, the connect operation will fail returning *REJECTED*, as for a failed connection to a nonexistent server.

Server mailboxes are a bit trickier. If an immediate server mailbox (Sect. 9.4.4) is being backed up to a journal file, the data written there are not separated into multiple connection blocks, even if the session consists of several client connections.

Only a *SERVER*-type disconnection and reconnection creates a new connect block. Note that for an immediate mailbox, the program may not be able to tell apart two separate client connections directly following one another. Although the active status of the mailbox (determined by *isActive*, Sect. 9.4.5) changes when the connection is dropped, if another connection comes up before the program inquires about that status, the transition will pass unnoticed. Therefore, it is assumed that all these connections appear as a single connection and the journal file does not try to separate them. When an immediate server mailbox is driven from a journal file, it receives a continuous stream of data that appears to have arrived from a single connection.

When a journaled immediate server mailbox is unbound (see Sect. 9.4.5), by *disconnect* (*SERVER*), the current connect block is terminated. When the mailbox is bound again (by *connect*) a new connect block is started. Therefore, from the viewpoint of journaling, an immediate server mailbox behaves like a client mailbox, with every explicit *connect* operation marking a new session.

When a master server mailbox is journaled, the only information written to the journal file is the timing of the incoming connections. The only sensible journaling setup involving a master mailbox is one in which all mailboxes subsequently bound to the master mailbox are also journaled. The right way to feed them from their respective journal files is to create them in the proper order, based on the timing of connections fed from the journal file to the master mailbox. Their names (preferably nicknames, Sects. 3.3.2 and 9.5.1) should be distinct and generated in the right order, to match the corresponding journal files, according to their creation sequence.

Chapter 10
Measuring Performance

One objective of modeling physical systems in SMURPH is to investigate their quantifiable performance. Some performance measures are calculated automatically by the SMURPH kernel. The user can easily collect additional statistical data which augment or replace the standard measurements. The calculation of both the standard and user-defined performance measures is based on the concept of *random variables* represented by a special data type.

10.1 Type RVariable

An *RVariable* is a data structure for incremental calculation of the empirical distribution parameters of a random variable whose values are discretely sampled. The following measures are automatically taken care of by the built in functionality:

- the number of samples
- the minimum value encountered so far
- the maximum value encountered so far
- the mean value, i.e., the sampled average
- the sampled variance and standard deviation
- higher-order central moments.

Type *RVariable* is not extensible by the user and its attributes are hidden. For efficiency reasons, they don't represent the above-listed parameters directly and explicit operations are required to turn them into a presentable form. Type *RVariable* declares a few publicly visible methods for performing typical operations on random variables and for presenting their parameters in a legible form. The latter methods are based on the concept of exposing, which is common to all *Objects* (see Sect. 12.1).

From now on, objects of type *RVariable* will be called random variables.

© Springer Nature Switzerland AG 2019
P. Gburzyński, *Modeling Communication Networks and Protocols*,
Lecture Notes in Networks and Systems 61,
https://doi.org/10.1007/978-3-030-15391-5_10

10.1.1   Creating and Destroying Random Variables

Random variables are *Objects* (Sect. 3.3.1): they must be *created* before they are used, and they may be deallocated when they are no longer needed. A random variable is created in the standard way (Sect. 3.3.8), e.g.:

rv = create RVariable (ct, nm);

where *rv* is an *RVariable* pointer and both setup arguments are integer numbers. The first argument (denoted by *ct*) specifies the type of the sample counter. The value of this argument can be either *TYPE_long*, in which case the counter will be stored as a *LONG* number, or *TYPE_BIG*, in which case the counter will be an object of type *BIG (TIME)*.

The second argument (*nm*) gives the number of central moments to be calculated for the random variable. No moments are calculated if this number is 0. Note that the mean value and variance are moments number 1 and 2, respectively. The maximum number of moments is 32.

The *setup* method of *RVariable* is declared in the following way

void setup (int ct = TYPE_long, int nm = 2);

Thus, if no *setup* arguments are provided at *create*, the random variable has a *LONG* sample counter and keeps track of two central moments, i.e., the mean value and variation.

Being an *Object*, a random variable is assigned an identifier (Sect. 3.3.2), which is the variable's serial number in the order of creation. The *Id* attributes of random variables are not useful for identification. Unlike other *Objects*, there is no operation that would convert an *RVariable Id* into the object pointer. A random variable can be assigned a nickname Sect. 3.3.2) upon creation. The nickname version of *create* (Sect. 3.3.8) can be used for this purpose.

A random variable that is no longer needed can (and should) be erased by the standard C++ operation *delete*.

10.1.2   Operations on Random Variables

When a new random variable is created, the value of its sample counter is initialized to zero. Whenever a new sample, or a number of samples, is to be added to the history of a random variable, the following *RVariable* method should be called:

void update (double val, CTYPE cnt = 1);

where "CTYPE" is either *LONG* or *BIG*, depending on the type of the random variable's sample counter. The function increments the sample counter by *cnt* and updates the parameters of the random variable according to *cnt* occurrences of value *val*. For illustration, consider the following sequence of operations:

r = create RVariable (TYPE_long, 3);
r-> update (2.0, 1);
r-> update (5.0, 2);

The first call to *update* adds to the variable's history one sample with value 2, the second call adds two more samples with the same value of 5. Thus, after the execution of the above three statements, *s* represents a random variable with three values: 2, 5, 5 and these distribution parameters:

Minimum value	2.0
Maximum value	5.0
Mean value	4.0
Variance	2.0
Skewness (3rd moment)	−2.0

Owing to the fact that the attributes of an *RVariable* do not represent directly the distribution parameters, a special operation is required to pull them out. That operation is implemented by the following method of *RVariable*:

*void calculate (double &min, double &max, double *m, CTYPE &c);*

All arguments of *calculate* are return arguments. The type of *c* (denoted by "CTYPE") must be the same as the counter type of the random variable, i.e., either *LONG* or *BIG*. The arguments are filled (in this order) with the minimum value, the maximum value, the moments, and the sample count. Argument *m* should point to a *double* array with no less elements than the number of moments declared when the random variable was created. The first element of the array (element number 0) will contain the mean value, the second, the variance, the third, the third central moment, and so on.

Two random variables can be combined into one in such a way that the combined parameters describe a single sampling experiment in which all samples belonging to both source variables have been accounted for. The function:

*void combineRV (RVariable *a, RVariable *b, RVariable *c);*

combines the random variables *a* and *b* into a new random variable *c*. The target random variable must exist, i.e., it must have been created before.

If the two random variables being combined have different numbers of moments, the resulting random variable has the lesser of the two numbers of moments. If the counter types of the source random variable are different, the resulting counter type is *BIG*. The target random variable must have been created with the right counter type and the right number of moments. Note that *combineRV* is a global function, not a method of *RVariable*.

It is possible to erase the contents of a random variable, i.e., initialize its sample counter to zero and reset all its parameters. This is done by the following method of *RVariable*:

void erase ();

Each random variable is initially erased upon creation.

10.2   Client Performance Measures

Several performance measures are automatically collected for each traffic pattern, unless the program decides to switch them off by selecting *SPF_off* when the traffic pattern is created (Sect. 6.6.2). These measures consist of several random variables (type *RVariable*, Sect. 10.1.1) and counters, which are updated automatically under certain circumstances.

10.2.1   Random Variables

Type *Traffic* declares six pointers to random variables. Those random variables are created together with the traffic pattern, provided that the standard performance measures are not switched off, and are used to keep track of the following measures (in parentheses we give the variable name of the corresponding *RVariable* pointer):

Absolute message delay (*RVAMD*)

> The absolute delay of a message *m* is the amount of time in *ETUs* (Sect. 3.1.2) elapsing from the moment *m* was queued at the sender (that moment is indicated by the contents of the *Message* attribute *QTime*, see Sect. 6.2) until the moment when the last packet of *m* has been received at its destination. A packet is assumed to have been received when *receive* (Sects. 7.1.6 and 8.2.3) is executed for the packet. From the viewpoint of this measure, each reception of a complete message (*receive* for its last packet) belonging to the given traffic pattern generates one sample.

Absolute packet delay (*RVAPD*)

> The absolute delay of a packet *p* is calculated as the amount of time in *ETUs* elapsing from the moment the packet became ready for transmission (the message queuing time is excluded) until *p* is received at its destination. The time when a packet becomes ready for transmission is determined as the maximum (later) of the following two moments:

> - the time when the buffer into which the packet is acquired was last *released* (Sect. 6.3),
> - the time when the message from which the packet is acquired was queued at the station (*Message* attribute *QTime*).

> Note that it is illegal to acquire a packet into a full (non-released) buffer (Sect. 6.7.1), and the above prescription for determining the ready time of a packet is sound. Intuitively, as soon as a packet buffer is emptied (by *release*) the buffer becomes ready to accommodate the next packet.[1] In this sense, the next packet becomes automatically ready for transmission, provided that it is already pending, i.e., a message is queued at the station. At first sight it might seem natural to assume that the packet delay should be measured from the moment the packet is actually fetched into the buffer. However, the operation of acquiring the packet into one of the station's buffers can be postponed by the protocol until the very moment of commencing the packet's transmission (there is nothing fundamentally wrong about keeping a buffer free, even though there is a packet in the queue that could be stored

---

[1] The *TTime* attribute of an empty packet buffer (see Sect. 6.2) is used to store the time when the buffer was last released.

in it right away). Thus, the numerical value of the packet delay would be dependent on the programming style of the protocol implementer.

For this measure, every reception (by *receive*) of a complete packet belonging to the given traffic pattern generates one sample.

Weighted message delay (RVWMD)

The weighted message delay (also called the message bit delay) is the average delay in *ETUs* of a single information bit carried by the message measured from the time the message was queued at the sender, until the packet containing that bit is completely received at the destination. Whenever a packet *p* belonging to the given traffic pattern is received, *p-> ILength* samples are added to *RVWMD*, all with the same value equal to the difference between the current time (*Time*) and the time when the message containing the packet was queued at the sender. This difference, as all other delays, is expressed in *ETUs*.

Message access time (RVMAT)

The access time of a message *m* is the amount of time in *ETUs* elapsing from the moment when the message was queued at the sender, to the moment when the last packet of the message is *released* by the sender. Every operation of releasing the last packet of a message generates a data sample for this measure.

Packet access time (RVPAT)

The access time of a packet *p* is the amount of time in *ETUs* elapsing from the moment the packet becomes ready for transmission (see *absolute packet delay*) to the moment when the packet is released by the sender. Every operation of releasing a packet generates a data sample for this measure.

Message length statistics (RVMLS)

Whenever a message belonging to the given traffic pattern is queued at a sender, one data sample containing the message length is generated for this measure. Thus, the random variable pointed to by *RVMLS* collects statistics related to the length of messages generated according to the given traffic pattern. For a standard arrival process (Sect. 6.6.2), these statistics can be inferred from the traffic parameters and are thus redundant. The random variable can be used to verify the distribution of message length for a non-standard message generator (Sect. 6.6.3).

If the standard performance measures have been switched off upon the traffic pattern's creation, the random variables listed above are not created and their pointers are set to *NULL*.

10.2.2   Counters

Besides the random variables, type *Traffic* defines the following user-accessible counters:

Long NQMessages, NTMessages, NRMessages, NTPackets, NRPackets;

BITCOUNT *NQBits, NTBits, NRBits*;

All these counters are initialized to zero when the traffic pattern is created. If the standard performance measures are effective for the traffic pattern, the counters are updated in the following way:

- Whenever a message is generated and queued at the sender (by the standard method *genMSG*, Sect. 6.6.3), *NQMessages* is incremented by 1 and *NQBits* is incremented by the message length in bits. See Sect. 3.1.6 for the definition of type *BITCOUNT*.
- Whenever a packet is released, *NTPackets* is incremented by 1. At the same time *NTBits* is incremented by the length of the packet's payload (*Packet* attribute *ILength*) and *NQBits* is decremented by the same number. If the packet is the last packet of a message, *NTMessages* is incremented by 1 and *NQMessages* is decremented by 1.
- Whenever a packet is received, *NRPackets* is incremented by 1 and *NRBits* is incremented by the length of the packet's payload (attribute *ILength*). If the packet is the last packet of its message, *NRMessages* is incremented by 1.

It is possible to print (or display) the standard performance measures associated with a given traffic pattern or the combination of performance measures for all traffic patterns viewed together as a single message arrival process. These tools are described in Sect. 12.5.4.

10.2.3 Virtual Methods

To facilitate collecting non-standard statistics, type *Traffic* defines a few *virtual* methods whose default (prototype) bodies are empty. Those virtual methods can be redefined in a user extension of type *Traffic*. They are invoked automatically at certain events related to the processing of the traffic pattern. In the list below, we assume that "mtype" and "ptype" denote the message type and the packet type associated with the traffic pattern (Sect. 6.6.1).

void pfmMQU (mtype **m);*	The method is called whenever a message is generated (by the standard method *genMSG*, Sect. 6.6.3) and queued at the sender. The argument points to the newly generated message
void pfmPTR (ptype **p);*	The method is called whenever a packet is released (Sect. 6.3). The argument points to the packet
void pfmPRC (ptype **p);*	The method is called whenever a packet is received (Sects. 7.1.6 and 8.2.3). The argument points to the received packet
void pfmMTR (ptype **p);*	The method is called whenever the last packet of a message is released. The argument points to that packet
void pfmMRC (ptype **p);*	The method is called whenever the last packet of a message is received (interpreted as the reception of the entire message). The argument points to the received packet

The virtual methods listed above are called even if the standard performance measures for the given traffic pattern are switched off. If the standard performance measures are effective, the virtual methods supplement the standard functions for collecting performance data.

Two more methods, formally belonging to the same category (of *virtual* methods redefinable in *Traffic* subtypes) but not used for measuring performance, are mentioned in Sect. 6.6.4.

10.2.4 Resetting Performance Measures

The performance statistics of a traffic pattern can be reset at any time. The operation corresponds to starting the collection of these statistics from scratch and is implemented by the following *Traffic* method:

void resetSPF ();

The method does nothing for a traffic pattern created with *SPF_off* (Sect. 6.6.2), i.e., if no automatic performance measures are collected for the traffic pattern. Otherwise, it erases (Sect. 10.1.2) the contents of the traffic pattern's random variables (Sect. 10.2.1) and resets its counters (Sect. 10.2.2) in the following way:

NTMessages = NTMessages − NRMessages;
NRMessages = 0;
NTPackets = NTPackets − NRPackets;
NRPackets = 0;
NTBits = NTBits − NRBits;
NRBits = 0;

The new values of the counters (note that *NQMessages* and *NQBits* are not changed) are obtained from the previous values by assuming that the number of received messages (also packets and bits) is now zero. To maintain consistency, the counters reflecting the number of transmitted items are not zeroed, but decremented by the corresponding counts of received items. This way, after they have been reset, these counters reflect the number of items "in transit." By the same token, the number of queued items is left intact.

Whenever *resetSPF* is executed for a traffic pattern, its attribute *SMTime* of type *TIME* is set to the time when the operation was performed. Initially, *SMTime* is set to 0. This (user-accessible) attribute can be used, e.g., to correctly calculate the throughput for the traffic pattern. For example, the following formula:

thp = (double) NRBits / (Time − tp -> SMTime);

gives the observed throughput[2] (in bits per *ITU*) for the traffic pattern *tp* measured from the last time the performance statistics for *tp* were reset. This method of *Traffic*:

double throughput ();

returns the same value normalized to the *ETU*, i.e., multiplied by *Etu* (the number of *ITUs* in one *ETU*, Sect. 3.1.2). A similar method of the *Client* returns the global throughput over all traffic patterns expressed in bits per *ETU*. If the measurement time turns out to be zero (the method is called immediately after reset, and the division doesn't work) the method returns zero.

It is possible to reset the standard performance statistics globally for the entire *Client*, by executing the *resetSPF* method of the *Client*. This version of the method calls *resetSPF* for every traffic pattern and also resets the global counters of the *Client* keeping track of the number of items (messages, packets, bits) that have been generated, queued, transmitted, and received so far. These counters are reset in a similar way as the local, user-accessible counters for individual traffic patterns.

The global effective throughput calculated by the *Client* (Sects. 10.2.2 and 12.5.4) is produced by relating the total number of received bits (counted globally for all traffic patterns combined) to the time during which these bits have been received. This time is calculated by subtracting from *Time* the contents of an internal variable (of type *TIME*) that gives the time of the last *resetSPF* operation performed globally for the *Client*. If (some) traffic patterns are reset individually, but the *Client* is never reset, the global throughput measure may be useless.

If the simulation termination condition is based on the total number of messages received (Sect. 10.5.1), that condition will be affected by the *Client* variant of *resetSPF*. Namely, each call to *resetSPF* zeroes the global counter of received messages, so that it will be accumulating towards the limit from scratch. Thus, if *Client's reset-SPF* is executed many times, e.g., in a loop, the termination condition may never be met, even though the actual number of messages received during the simulation run may be huge. The semantics of the −c program call option (Sects. 10.5.1 and B.6) are also affected by the global variant of *resetSPF*. The total number of generated messages is reduced by the number of messages that have been counted as received when the method is invoked. Note that the global counters of the *Client* are not affected by the *Traffic* variant of *resetSPF*.

Every *resetSPF* operation executed on a traffic pattern triggers an event that can be perceived by a user-defined process. This way the user may define a customized action to be carried out as part of the operation, e.g., resetting non-standard random variables and/or private counters. The event, labeled *RESET*, occurs on the traffic pattern, also if the traffic pattern was created with *SPF_off*. A global *resetSPF* operation performed on the *Client* forces the *RESET* event for all traffic patterns, in addition to triggering it for the *Client*.

---

[2]Assuming the denominator is nonzero.

10.3 Link Performance Measures

Similar to traffic patterns, certain standard performance statistics are automatically collected for links. Unlike the *Traffic* measures, the link statistics are not random variables, but merely counters (similar to those associated with traffic patterns, Sect. 10.2.2) keeping track of how many bits, packets, and messages have passed through the link. The program may opt out from collecting these measures at the time when the link is created (Sect. 4.2.2).

The following publicly available *Link* attributes are related to measuring link performance:

char FlgSPF;

This attribute can take two values: *ON* (1), if the standard performance measures are to be calculated for the link, or *OFF* (0), if the standard measurements are switched off.

BITCOUNT NTBits, NRBits;
Long NTJams, NTAttempts;
Long NTPackets, NRPackets, NTMessages, NRMessages;
Long NDPackets, NHDPackets;
BITCOUNT NDBits;

All the above counters are initialized to zero when the link is created. Then, if the standard performance measures are effective for the link, the counters are updated in the following way:

- Whenever a jamming signal is inserted into a port connected to the link, *NTJams* is incremented by 1.
- Whenever a packet transmission is started on a port connected to the link, *NTAttempts* is incremented by 1.
- Whenever a packet transmission on a port connected to the link is terminated by *stop* (Sect. 7.1.2), *NTPackets* is incremented by 1 and *NTBits* is incremented by the packet's payload length (attribute *ILength*). If the packet is the last packet of its message, *NTMessages* is incremented by 1.
- Whenever a packet is received from the link (Sect. 7.1.6), *NRPackets* is incremented by 1 and *NRBits* is incremented by the packet's payload length. If the packet is the last packet of its message, *NRMessages* is incremented by 1. Note that it is possible to invoke receive without the second argument (identifying the link). In such a case, the counters are not incremented.
- Whenever a damaged packet (Sect. 7.3) is inserted into the link,[3] *NDPackets* is incremented by 1 and the packet's payload length (attribute *ILength*) is added to *NDBits*. If the packet happens to be header-damaged, *NHDPackets* is also incremented by 1. Note that a header-damaged packet is also (plain) damaged; thus, *NHDPackets* can never be bigger than *NDPackets*.

---

[3]The counters and methods related to damaged packets are only available if the program has been created with the −z option of *mks* (Sect. B.5).

Type *Link* declares a few *virtual* methods whose prototype bodies are empty and can be redefined in a user extension of a standard link type. These methods can be used to collect nonstandard performance statistics related to the link. Their signatures are identical: no returned value and a single argument of type *Packet**, so we just list their names.

pfmPTR	The method is called when a packet transmission on a port belonging to the link is terminated by *stop*. The argument points to the terminated packet
pfmPAB	The method is called when a packet transmission on a port belonging to the link is terminated by *abort*. The argument points to the aborted packet
pfmPRC	The method is called when a packet is received from the link (see above). The argument points to the received packet
pfmMTR	The method is called when a packet transmission on a port belonging to the link is terminated by *stop*, and the packet turns out to be the last packet of its message. The argument points to the terminated packet
pfmMRC	The method is called when the last packet of its message is received from the link. The argument points to the received packet
pfmPDM	The method is called when a damaged packet is inserted into the link. The method can examine the packet flags (Sects. 6.2 and 7.3) to determine the nature of the damage

It is possible to print out or display the standard performance measures related to a link. This part is covered in Sect. 12.5.6.

10.4   RFChannel Performance Measures

The set of performance-related counters defined by a radio channel closely resembles that for *Link* (Sect. 10.3), and can be viewed as its subset. Like for a link, those counters are updated automatically, unless the program decided to switch them off when the radio channel was created (Sect. 4.4.2).

The following publicly available attributes of *RFChannel* are related to measuring performance:

char FlgSPF;

This attribute can take two values: *ON* (1), if the standard performance measures are to be calculated for the radio channel, or *OFF* (0), if the standard measurements are switched off.

BITCOUNT NTBits, NRBits
Long NTAttempts, NTPackets, NRPackets, NTMessages, NRMessages;

All the above counters are initialized to zero when the radio channel is created. Then, if the standard performance measures are effective for the channel, the counters are updated in the following way:

- Whenever a packet transmission is started on a transceiver interfaced to the radio channel, *NTAttempts* is incremented by 1.
- Whenever a packet transmission on a transceiver interfaced to the channel is terminated by *stop* (Sect. 8.1.8), *NTPackets* is incremented by 1, and *NTBits* is incremented by the packet's payload length (attribute *ILength*). If the packet is the last packet of its message, *NTMessages* is incremented by 1.
- Whenever a packet is received from the radio channel (Sect. 8.2.3), *NRPackets* is incremented by 1, and *NRBits* is incremented by the packet's payload length. If the packet is the last packet of its message, *NRMessages* is incremented by 1.[4]

Similar to *Link*, type *RFChannel* declares a few *virtual* methods whose prototype bodies are empty and can be redefined in a user extension of the class. These methods can be used to collect nonstandard performance statistics related to the channel. Their signatures are identical: no returned value, single argument of type *Packet**, so we just list their names.

pfmPTR	The method is called when a packet transmission on a transceiver interfaced to the channel is terminated by *stop*. The argument points to the terminated packet
pfmPAB	The method is called when a packet transmission on a transceiver interfaced to the channel is terminated by *abort*. The argument points to the aborted packet
pfmPRC	The method is called when a packet is received from the channel (see above). The argument points to the received packet
pfmMTR	The method is called when a packet transmission on a transceiver interfaced to the channel is terminated by *stop* and the packet turns out to be the last packet of its message. The argument points to the terminated packet
pfmMRC	The method is called when the last packet of its message is received from the channel. The argument points to the received packet

It is possible to print out or display the standard performance measures related to a radio channel. This part is discussed in Sect. 12.5.8.

10.5 Terminating Execution

There are two broad categories of SMURPH programs: ones whose objective is to collect measurements (performance data) from a closed model of some system, and others whose purpose is to interact with something (or someone). Control programs driving real-life equipment or visualization models emulating real-life networks fall into the second category, but that category also covers simulation models intended for visualization (Sects. 1.3.1 and 5.3.4).

The entire content of the present section refers to the first category of models. While some of the performance measures, and tools for their collection, discussed

---

[4]If *receive* is called without the second argument (identifying the radio channel, Sect. 8.2.3), the counters are not incremented.

in the preceding sections may make sense for control programs and visualization models, the main reasons for running such models is not to obtain numbers, but to carry out some action. Another difference between the two types of programs is that a program of the first kind is expected to terminate after some definite amount of time, when enough performance data have been collected. Typically, such a program does not interact with the user (executing in a batch mode),[5] so it must decide on its own when to stop. The decision as to whether a simulator has been running long enough (having processed sufficiently many events) to claim that the performance data it has collected are representative and meaningful belongs to the art of simulation [1]. It depends on the characteristics of the modeled system and the nature of the answers expected from the model.

The execution of a SMURPH program can be terminated explicitly upon request from the program (Sects. 2.1.3 and 5.1.8). The model can also hit an error condition, be aborted by the user, or run out of events.[6] The program is also terminated automatically when any of three simple exit conditions specified by the user (called *limits*) has been met. These limits are declared by calling the following (global) functions:

setLimit (Long MaxNM, TIME MaxTime, double MaxCPUTime);
setLimit (Long MaxNM, TIME MaxTime);
setLimit (Long MaxNM);

If *setLimit* is never used, the run will continue indefinitely, until it is terminated explicitly by the protocol program, aborted by the user, or until the simulator runs out of events (if that is possible at all). The primary purpose of the function is to define hard (and easy) exit conditions for simulation experiments. Control programs driving real physical systems usually have no exit conditions and they don't call *setLimit*.

The first two variants of *setLimit* are also available with the second argument type being *double*, i.e.:

setLimit (Long MaxNM, double MaxTimeETU, double MaxCPUTime);
setLimit (Long MaxNM, double MaxTimeETU);

and accept the time limit specified in *ETUs* rather than in *ITUs*.

Typically, *setLimit* is called from the *Root* process before the simulation is started (Sects. 2.1.4, 2.2.5 and 2.4.6), but it can be called at any moment during the protocol execution. If the parameters of such a call specify limits that have been already reached or exceeded, the experiment is terminated immediately.

---

[5]The user can still look (peek) into the execution of such a program, which may make sense when its run time is long, and monitor its progress (see Sect. C.1).

[6]The last situation occurs when all processes (including system processes) get into a state where no possible event can wake them up. Note that such a situation can be intercepted by a process (or multiple processes) waiting for the *STALL* event on the *Kernel* (Sect. 5.1.8). In such a case, running out of events does not trigger termination. A control program running in the real mode (Sects. 1.3.1 and 5.1.8) is not terminated when it runs out of events. It just waits for them to arrive from whatever external sources the program is connected to.

10.5.1 *Maximum Number of Received Messages*

The first argument of *setLimit* specifies the number of messages to be entirely received (with *receive* executed for their last packets, Sects. 7.1.6 and 8.2.3) at their destinations. The simulation experiment will be terminated as soon as the number is reached.

When the simulation program is run with the –*c* option (Sect. B.6), the message number limit is interpreted as the maximum number of messages to be generated by the *Client* (Sect. 6.2). In that case, when the limit is reached, the *Client* stops and the execution continues until all the messages that remain queued at the sending stations have been received at their destinations.

If this global function:

void setFlush ();

is called before the protocol execution is started (i.e., while the *Root* process is in its initial state, Sect. 5.1.8), it has the same effect as if the program were invoked with the –*c* option.

10.5.2 *Virtual Time Limit*

The second argument of *setLimit* declares the maximum interval of virtual time in *ITUs* or *ETUs*.[7] The execution will stop as soon as the time (the contents of variable *Time*) has reached the limit.

If the virtual time limit has been reset to the end time of the tracing interval with –*t* (see Sect. B.6), *setLimit* can only reduce that limit. An attempt to extend it will be quietly ignored.

10.5.3 *CPU Time Limit*

The last argument of *setLimit* declares the limit on the CPU time used by the program. As SMURPH checks against a violation of this limit every 5000 events, the CPU time limit may be slightly exceeded before the experiment is eventually stopped.

If the specified value of any of the above limits is 0, it stands for "no limit," i.e., any previous setting of this limit is canceled. If no value is given for a limit at all (the last two versions of *setLimit*), the previous setting of the limit is retained.

All three limits are viewed as an alternative of exit conditions, i.e., the execution stops as soon as any of the limits is reached.

---

[7] As determined by the argument type: type *double* selects the *ETU* interpretation.

10.5.4   The Exit Code

The protocol program, specifically the *Root* process (Sect. 5.1.8), can learn why it has been terminated. When the *DEATH* event for the *Kernel* process is triggered, the integer environment variable *TheExitCode* (an alias for *Info01*, see Sect. 5.1.6) contains a value that tells what has happened. It can be one of the following:

EXIT_msglimit	the message number limit has been reached
EXIT_stimelimit	the simulated time has reached the declared limit
EXIT_rtimelimit	the CPU time limit has been reached
EXIT_noevents	there are no more events to process
EXIT_user	the *Kernel* process has been explicitly killed
EXIT_abort	the program has been aborted due to a fatal error condition

The actual numerical values represented by the symbolic constants are 0–5, in the order in which the constants are listed.

Reference

1. A.M. Law, W. D. Kelton, *Simulation Modeling & Analysis* (McGraw-Hill, 2007)

Chapter 11
Tools for Testing and Debugging

SMURPH comes equipped with a few tools that can assist the user in the process of debugging and validating a protocol program, especially in those numerous cases when formal correctness proofs are infeasible. The incorrectness of a protocol or a control program can manifest itself in one of the following two ways:

- under certain circumstances, the protocol crashes and ceases to operate
- the protocol seems to work, but it does not exhibit certain properties expected by the designer and/or the implementer.

Generally, problems of the first type are rather easy to detect in SMURPH by extensive testing, unless the "certain circumstances" occur too seldom to be caught. My experience indicates that the likelihood of finding a crashing error decreases much more than linearly with the running time of the protocol prototype. Notably, the finite-state-machine paradigm of programming in SMURPH facilitates self-documenting code, which, in turn, makes simple bugs relatively difficult to introduce and also easy to spot, if they manage to sneak in after all. This is not an empty marketing statement. When the same approach was adopted to programming real-life reactive systems [1], we saw that most of those programs worked practically "out of the box," with little debugging following the first case of successful compilation. On the other hand, the specific nature of protocol programs, especially low-level (medium access control) protocols, makes them susceptible for errors of the second type, which may be more difficult to detect and diagnose.

11.1 User-Level Tracing

One simple way to detect run time inconsistencies is to assert *Boolean* properties involving variables in the program. SMURPH offers some tools to this end (see Sect. 3.2.7). Once an error has been found, the detection of its origin may still pose

© Springer Nature Switzerland AG 2019

P. Gburzyński, *Modeling Communication Networks and Protocols*,
Lecture Notes in Networks and Systems 61,
https://doi.org/10.1007/978-3-030-15391-5_11

a tricky problem. Often, one must trace the protocol behavior for some time prior to the occurrence of the error to identify the circumstances leading to the trouble.

The following function writes to the output file a line describing the contents of the indicated variables at the time of its call:

void trace (const char*, ...);

It accepts a format string as the first argument, which is used to interpret the remaining arguments (in the same way as in *printf* or *form*). The printout is automatically terminated by a new line, so the format string shouldn't end with one. A line produced by *trace* is preceded by a header, whose default form consists of the virtual time in *ITU*. For example, this command:

trace ("Here is the value: %1d", n);

may produce a line looking like this:

Time: 1184989002 Here is the value: 93

The most useful feature of *trace*, which gives it an advantage over standard output tools, is that it can be easily disabled or restricted to a narrow interval of modeled time by a program call argument. This is important because tracing problems that manifest themselves after lengthy execution may generate tones of intimidating output. The –*t* program call argument (described in Sect. B.6) can disable or restrict the output of *trace*. It is also possible to confine that output to the context of a few stations. By default, i.e., without –*t*, every *trace* command is effective.

The program can also affect the behavior of *trace* dynamically, from within, by calling the following global functions:

void settraceFlags (FLAGS opt);
void settraceTime (TIME begin, TIME end = TIME_inf);
void settraceTime (double begin, double end = Time_inf);
void settraceStation (Long s);

The first argument of *settraceFlags* is a collection of additive flags that describe the optional components of the header preceding a line written by *trace*. If *opt* is zero, *trace* is unconditionally disabled, i.e., its subsequent invocations in the program will produce no output until a next call to *settraceFlags* selects some actual header components. Here are the options (represented by symbolic constants):

TRACE_OPTION_TIME	This option selects the standard header, i.e., current simulation time in *ITUs*. It is the only option being in effect by default
TRACE_OPTION_ETIME	Includes the simulated time presented in *ETUs* as a floating point number
TRACE_OPTION_STATID	Includes the numerical *Id* of the current station (Sect. 4.1.2)
TRACE_OPTION_PROCESS	Includes the standard name of the current process (Sect. 5.1.3)
TRACE_OPTION_STATE	Includes the symbolic name of the state in which the current process has been awakened. This option is independent of the previous option

For example, having executed:

settraceFlags (TRACE_OPTION_ETIME + TRACE_OPTION_PROCESS);

followed by:

trace ("Here is the value: %1d, n);

we may see something like this:

Time: 13.554726 /Transmitter/Here is the value: 93234

Note that the default component of the trace header, represented by *TRACE_OPTION_TIME*, will be removed if the argument of *settraceFlags* does not include this value. However, there is no way to produce a *trace* line without a single header component because *settraceFlags (0)* completely disables *trace* output.

With the two variants of *settraceTime*, we can confine the output of *trace* to a specific time interval. For the first variant, that interval is described as a pair of *TIME* values in *ITUs*, while the second variant accepts *double ETU* values. If the second value is absent, the end of the tracing period is undetermined, i.e., *trace* will remain active until the end of run (see Sect. 3.1.3 for the meaning of constants *TIME_inf* and *Time_inf*). The specified time interval is interpreted in such a way that *begin* describes the first moment of time at which *trace* is to be activated, while *end* (if not infinite) points to the first instant of time at which *trace* becomes inactive. This means that *trace* invoked exactly at time *end* will produce no output. Every call to *settraceTime* (any variant) overrides all previous calls. The only way to define multiple tracing intervals is to issue a new *settraceTime* call for the next interval at the end of the previous one.

Tracing can be restricted in space as well as in time. The argument of *settraceStation* specifies the station to which the output of *trace* should be confined. This means that the environment variable *TheStation* (Sect. 4.1.3) must point to the indicated station at the time when *trace* is invoked. Multiple calls to the function add stations to the tracing pool. This way it is possible to trace an arbitrary subset of the network. By calling *settraceStation* with *ANY* as the argument, we revert to the default case of tracing all stations. A subsequent call to *settraceStation* identifying a station will initialize the set of stations restricting the range of *trace* to the single specified station.

Notably, the *–t* program call argument (Sect. B.6) is interpreted after the network has been built (Sect. 5.1.8), and thus overrides any *settraceTime* and *settraceStation* calls possibly made from the first state of *Root*. This delay in the processing of *–t* is required because the interpretation of an *ETU* value (e.g., used as a time bound) can only be meaningful after the *ETU* has been defined by the program (Sect. 3.1.2). The *–t* argument, if it appears on the program call line, effectively cancels the effect of all *settraceTime* and *settraceStation* calls issued in the first state of *Root*. However, any calls to those functions issued later (after the network has been built)[1] will override the effects of *–t*.

---

[1] The network build can be forced with *rootInitDone* (Sect. 5.1.8) which also forces the processing of *–t*, and any trace options set after the call to *rootInitDone* will be applied after the *–t* argument has been processed.

By the same token, a sensible *settraceTime* call with *ETU* arguments can only be made after the *ETU* has been set. Before then, the function will transform the specified values into *TIME* using the default setting of 1 *ITU* = 1 *ETU*.

The user can easily take advantage of the restriction mechanism for *trace* in any debugging code inserted into the protocol program. The global macro *Tracing*,[2] with the properties of a *Boolean* variable, produces *YES*, if *trace* would be active in the present context, and *NO* otherwise.

11.2   Simulator-Level Tracing

In addition to explicit tracing described in Sect. 11.1, SMURPH has a built-in mechanism for dumping the sequence of process states traversed by the simulator during its execution to the output file. This feature is only available when the program has been compiled with –g or –G (see Sect. B.5).

One way of switching on the state dump is to use the –T program call argument (Sect. B.6). Its role is similar to –t (which controls the user-level tracing, Sect. 11.1), except that the simulator-level tracing is disabled by default. The options available with –T can also be selected dynamically from the program, by invoking functions from this list:

void setTraceFlags (Boolean full);
void setTraceTime (TIME begin, TIME end = TIME_inf);
void setTraceTime (double begin, double end = Time_inf);
void setTraceStation (Long s);

They look similar to the user-level tracing functions from Sect. 11.1 (their names differ on the case of the "t" in "trace"). For the first function, if *full* is *YES*, it selects the "full" version of state dumps, as described below. The remaining functions restrict the tracing context in the same way as for the corresponding user-level functions discussed in Sect. 11.1.

Any episode of a process being awakened that falls into the range of the tracing parameters (in terms of time and station context) results in one line written to the output file. That line is produced before the awakened process gains control and includes the following items:

- the current simulated time
- the type name and the *Id* of the *AI* generating the event
- the event identifier (in the *AI*-specific format, see Sect. 12.5)
- the station *Id*
- the output name of the process (Sects. 3.3.2 and 5.1.3)
- the identifier of the state to be assumed by the process.

---

[2]The macro expands into a call to an argument-less, global function.

With full tracing, that line is followed by a dump of all links and radio channels accessible via ports and transceivers from the current station. This dump is obtained by requesting mode 2 exposures of the *Link* (Sect. 12.5.6) and *RFChannel* objects (Sect. 12.5.8).

The global macro *DebugTracing* has the appearance of a *Boolean* variable whose value is *YES*, if (1) the program has been compiled with *–g* or *–G*, i.e., debugging is enabled, and (2) state dumping is active in the current context. Otherwise, *Debug-Tracing* evaluates to *NO*.

Normally, the waking episodes of SMURPH internal processes are not dumped under simulator-level tracing (they are seldom interesting to the user). To include them alongside the protocol processes, use the *–s* program call option (Sect. B.6).

11.3 Observers

Observers [2, 3] are tools for expressing global assertions possibly involving combined actions of many processes. An observer is a dynamic object that resembles a regular protocol process. In contrast to processes, observers never respond directly to events perceived by regular processes, nor do they generate events that may be perceived by regular processes. Instead, they react to the so-called *meta-events*. By a meta-event we understand the operation of awakening a regular process. An observer may declare that it wants to be awakened whenever a regular process is run at some state. This way, the observer can monitor the behavior of the protocol viewed as a collection of interacting finite-state machines.

An observer is an object belonging to an *Observer* type. Such a type is defined in the following way:

```
observer otype : itypename {

    ...

    local attributes and methods

    ...

    states { list of states };

    ...

    perform {
            the observer code
    };
};
```

where "otype" is the name of the declared observer type and "itypename" identifies an already known observer type (or types), according to the general rules of SMURPH type declaration described in Sect. 3.3.3.

The above layout closely resembles a process type declaration (Sect. 5.1.2). One difference is that the structure of observers is flat, i.e., an observer does not belong to any particular user station, has no formal parents and no children. Observer types are extensible and they can be derived from other user-defined observer types.

A *setup* method can be declared in the standard way (see Sect. 3.3.8) to initialize the local attributes when an observer instance is created. The default observer *setup* method is empty and takes no arguments. Observers are created by *create*, in the standard way.

The observer's code method has the same layout as a process code method, i.e.:

perform {

   state OS_0:

       ...

   state OS_1:

       ...

   state OS_{p-1}:

       ...

};

where $OS_0, \ldots, OS_{p-1}$ are state identifiers listed with the *states* statement within the observer type declaration.

An awakened observer, like a process, performs some operations, specifies its future waking conditions, and goes to sleep. The waking conditions for an observer are described by executing the following operations:

inspect (s, p, n, ps, os);
timeout (t, os);

Operation *inspect* identifies a class of scenarios when a regular process is run. The arguments of *inspect* have the following meaning:

s	identifies the station to which the process belongs; it can be either a station object pointer or a station *Id* (an integer number)
p	is the process type identifier
n	is the character string interpreted as the process's nickname (Sect. 3.3.2)
ps	is the process's state identifier
os	identifies the observer's state where the observer wants to be awakened

Let us assume that an observer has issued exactly one *inspect* request and put itself to sleep. The *inspect* request is interpreted as a declaration that the observer wants to remain dormant until SMURPH wakes up a process matching the parameters of the *inspect* request. Then, immediately after the process completes its action, the observer will be run and the global variable *TheObserverState* will contain the value that was passed to *inspect* through *os*. This value will be used by SMURPH to select the state within the observer's *perform* method.

If any of the first four arguments of *inspect* is *ANY*, it acts as a wildcard indicating that the actual value of that attribute in the process awakening scenario described by the statement is irrelevant. For example, with the request:

inspect (ANY, transmitter, ANY, ANY, WakeMeUp);

the observer wants to be restarted after any process of type *transmitter* is awakened in any state.

If the observer wants to monitor the behavior of a single process, and the type of that process has multiple instances, the process can be assigned a unique nickname (Sect. 3.3.2). That nickname (if it is unique among all processes of the given type) will identify exactly one process. It is also possible to identify a subclass of processes within a given type (or even across different types) by assigning the same nickname to several processes.

The process state identifier can only be specified, if the process type identifier has been provided as well (i.e., it is not *ANY*). A state is always defined within the scope of a specific process type, and specifying a state without a process type makes no sense.

There exist abbreviated versions of *inspect* accepting 1, 2, 3 and 4 arguments. In all these versions, the last argument specifies the observer's state (corresponding to *os* in the full-fledged, five-argument version), and all the missing arguments are assumed to be *ANY*. Here is the correspondence between the abbreviated and complete variants:

inspect (os);	*inspect (ANY, ANY, ANY, ANY, os);*
inspect (s, os);	*inspect (s, ANY, ANY, ANY, os);*
inspect (s, p, os);	*inspect (s, p, ANY, ANY, os);*
inspect (s, p, ps, os);	*inspect (s, p, ANY, ps, os);*

Thus, the single-argument version of *inspect* says that the observer wants to be restarted after *any* event awaking *any* protocol process at *any* station. In the two-argument version, the first argument identifies the station owning the awakened process. The three-argument version accepts the station identifier as the first argument and the process type identifier (*p*) as the second argument. For the four-argument version, the first two arguments are the same as in the two-argument case, and the third argument identifies the process state (*ps*).

When an observer is created (by *create*), it is initially started in the first state from its *states* list, in a manner similar to a process. The analogy between observers and processes extends onto *inspect* requests which are somewhat similar to *wait* requests. For example, an observer can issue multiple *inspect* requests before it puts itself to sleep.

Like a process, an observer puts itself to sleep by exhausting the list of commands at its current state or by executing *sleep*. Unlike the multiple *wait* requests issued by the same process, the order in which multiple *inspect* requests are issued by an observer *is* important. The pending *inspect* requests are examined in the order in which they have been issued, and the first one that matches the current process wakeup scenario is applied. Then the observer is awakened in the state determined by the value of the *os* argument specified with that *inspect*.

For illustration, assume that an observer has issued the following sequence of inspect requests:

inspect (ThePacket->Receiver, receiver, ANY, Rcv, State1);
inspect (TheStation, ANY, ANY, ANY, State2);
inspect (ANY, ANY, ANY, ANY, State3);

The observer will wake up at *State1*, if the next process restarted by SMURPH:

- is of type *receiver*
- belongs to the station determined by the *Receiver* attribute of *ThePacket*
- wakes up in state *Rcv*.

The three conditions must hold all at the same time. Otherwise, if the restarted process belongs to the current station (as of the instant when the *inspect* request was issued), the observer will be awakened at *State2*. Finally, if the process wakeup scenario does not match the arguments of the first two *inspects*, the observer will be awakened in State3 (which covers absolutely all possible process wakeup scenarios not covered by the first two *inspect* requests).

The three requests above can be rewritten using the abbreviated versions of *inspect*, in the following way:

inspect (ThePacket->Receiver, receiver, Rcv, State1);
inspect (TheStation, State2);
inspect (State3);

A restarted observer has access to the environment variables of the process that has been awakened. Those variables reflect the configuration of the process environment after the process completes its action in the last state.

Being in one state, an observer can branch directly to another state by executing

proceed nstate;

where "nstate" identifies the observer's new state. Despite the same look, the semantics of this operation is different from that of *proceed* for a process (Sect. 5.3.1). The observer variant of *proceed* branches to the indicated state immediately, and absolutely nothing else can happen in the meantime. For another difference, the *proceed* statement for an observer cannot be executed from a function (invoked from the observer), but must be issued directly from the observer's code method.

Whenever an observer is restarted, its entire *inspect* list is cleared, so its new waking conditions will be specified from scratch. This also applies to the *timeout* operation, which implements alarm clocks. By executing

timeout (t, ms);

the observer sets up an alarm clock that will wake it up unconditionally t *ITUs* after the current moment, if no process matching any of the outstanding *inspect* requests is awakened in the meantime. In contrast to process clocks, which may operate with a tolerance (Sect. 5.3.3), observer clocks are always accurate, i.e., the *timeout* interval is always precisely equal to the specified number of *ITUs*.

An arbitrary number of observers can be defined and active at any moment. Different observers are completely independent, in the sense that the behavior of a given observer is not affected by the presence or absence of other observers. An observer can terminate itself in the same way as a process, i.e., by executing *terminate* (Sect. 5.1.3), or by going to sleep without issuing a single *inspect* or *timeout* request.

When the protocol program is built with the $-v$ option of *mks* (see Sect. B.5) all observers are deactivated; all *create* operations for them are then void. This is an easy way of disabling all observers, whose presence may otherwise adversely affect the execution time of the model.

References

1. P. Gburzyński, W. Olesiński, On a practical approach to low-cost ad hoc wireless networking. J. Telecommun. Inf. Technol. **2008**(1), 29–42 (2008)
2. M. Berard, P. Gburzyński, P. Rudnicki, Developing MAC protocols with global observers, in *Proceedings of Computer Networks '91* (1991)
3. J.M. Ayache, P. Azéma, M. Diaz, Observer: a concept for on-line detection of control errors in concurrent systems, in *Proceedings of the 9th Symposium on Fault-Tolerant Computing*, Madison, 1979

Chapter 12
Exposing Objects

By exposing an object, we mean either printing[1] out some information related to the object or displaying that information on the terminal screen. In the former case, the exposing is done exclusively by SMURPH; in the latter, the action is carried out by a separate display program (possibly running on a different host) communicating with the SMURPH program via IPC tools.

12.1 General Concepts

Each *Object* type (Sect. 3.3.1) carrying some dynamic information that may be of interest to the user can be made exposable. An exposable type defines a special method that describes how the information related to the objects of this type should be printed out and/or how it should be displayed on the terminal screen. By "printing out" we mean including that information in the output (results) file (Sect. 3.2.2), By "displaying on the terminal screen" we understand sending the information to DSD, a display program (Sect. C.1) that organizes it into a collection of windows presented on the screen. In the latter case, the information is displayed dynamically, in the sense that it is updated periodically and, at any moment, reflects a snapshot in the middle of execution of the protocol program. Printing out, in the sense of the above definition, will also be called exposing "on paper," whereas displaying will be called exposing "on screen." The property of an exposure that says whether the information is to be "printed" or "displayed" is called the *exposure form*.

The way an object is exposed is described by the *exposure* declaration associated with the object type. Such a declaration specifies the code to be executed when the object is exposed. An object can be exposed in several ways, regardless of the form (i.e., whether the exposure is "on paper" or "on screen"). Each of those ways is

---

[1]In a figurative sense, as explained below.

© Springer Nature Switzerland AG 2019
P. Gburzyński, *Modeling Communication Networks and Protocols*,
Lecture Notes in Networks and Systems 61,
https://doi.org/10.1007/978-3-030-15391-5_12

called an *exposure mode* and is identified by a (typically small) nonnegative integer number. Different exposure modes can be viewed as different fragments (or types) of information associated with the object. Moreover, some exposure modes may be (optionally) station-relative. In such a case, besides the mode, the user may specify the station to which the information printed (or displayed) is to be related. If no such station is specified, the global variant of the exposure mode is used.

For most of the standard *Object* types, the display modes coincide with the printing modes, so that, for a given mode, the same information is sent to the "paper" and to the "screen." However, each mode for any of the two exposure forms is defined separately. For example, the standard exposure of the *Client* defines the following four "paper" modes:

0	Information about all processes that have pending *wait* requests to the *Client*. If a station-relative variant of this mode is selected, the information is restricted to the processes belonging to the given station
1	Global performance measures taken over all traffic patterns combined. This mode cannot be made station-relative
2	Message queues at all stations, or at one specified station (for the station-relative variant of this mode)
3	Traffic pattern definitions. No station-relative variant of this mode exists

The first three modes are also applicable for exposing the *Client* "on screen" (and they present the same information as their "paper" counterparts), but the last mode is not available for the "screen" exposure. This is because the information shown in that exposure is static, so presenting it as a snapshot makes little sense.

In general, the format of the "paper" exposure need not be similar to the format of the corresponding "screen" exposure. The number of modes defined for the "paper" exposure need not be equal to the number of the "screen" exposure modes.

12.2 Making Objects Exposable

In principle, objects of any *Object* type (Sect. 3.3.1) can be exposed, but they must be made exposable first. All standard *Object* types have been made exposable internally. The user may declare a non-standard subtype of *Object* as exposable and describe how objects of this type are to be exposed. It is also possible to declare a non-standard exposure for an extension of an exposable standard type. This non-standard exposure can either replace or supplement the standard one.

The remainder of this section, and the entire next section, may be uninteresting to the user who is not interested in creating non-standard exposable types. However, besides instructing on how to expose objects of non-standard types, this section also explains the mechanism of exposing objects of the standard types. This information may make it easier to understand the operation of the display program presented in Appendix C.

All user-created subtypes of *Object* must belong to type *EObject* (Sect. 3.3.1). To make such a subtype exposable, one must declare an exposure for the subtype. The exposure declaration should appear within the type (class) definition. It resembles the declaration of a regular method and has the following general format:

```
exposure {
        onpaper {
                exmode mp₀:
                        ...
                exmode mp₁:
                        ...
                        ...
                exmode mpₖ₋₁
                        ...
        };
        onscreen {
                exmode ms₀:
                        ...
                exmode ms₁:
                        ...
                        ...
                exmode msₙ₋₁:
                        ...
        };
};
```

As for a regular method, it is possible to merely announce the exposure within the definition of an *Object* type to be exposed and specify it later. An *Object* type is announced as exposable by putting the keyword *exposure;* into the list of its publicly visible attributes. For example, the following declaration defines a non-standard, exposable *Object* type:

```
eobject MyStat {
        Long NSamples;
        RVariable *v1, *v2;
        void setup ( ) {
                v1 = create RVariable;
                v2 = create RVariable;
        };
        exposure;
};
```

The exposure definition, similar to the specification of a method announced in a class declaration, must appear below the type definition in one of the program's files, e.g.:

```
MyStat::exposure {
        onpaper {
                exmode 0: v1->printCnt ( );
                exmode 1: v2->printCnt ( );
        }
        onscreen {
                exmode 0: v1->displayOut (0);
                exmode 1: v2->displayOut (1);
        }
}
```

An exposure specification consists of two parts: the "paper" part and the "screen" part. Any of the two parts can be omitted; in such a case the corresponding exposure form is undefined and the type cannot be exposed that way. Each fragment starting with *exmode m:* and ending at the next *exmode*, or at the closing brace of its form part, contains code to be executed when the exposure with mode *m* is requested for the given form.

The exposure code has immediate access to the attributes of the exposed object: it is effectively a method declared within the exposed type. Additionally, the following two variables are accessible from that code (they can be viewed as standard arguments passed to every exposure method):

Long SId;
char *Hdr;

If *SId* is not *NONE* (-1), it stores the *Id* of the station to which the exposed information is to be related. It is up to the exposure code to interpret that value, or ignore it, e.g., if it makes no sense to relate the exposed information to a specific station.

Variable *Hdr* is only relevant for a "paper" exposure. It points to a character string representing the header to be printed along with the exposed information. If *Hdr* contains *NULL*, it means that no specific header is requested. Again, the exposure code must perform an explicit action to print out the header; it may also ignore the contents of *Hdr*, print a default header, if *Hdr* contains *NULL*, etc. The contents of *Hdr* for a "screen" exposure are irrelevant and should be ignored.

Any types, variables, and objects needed locally by the exposure code can be declared immediately after the opening *exposure* statement.

If an exposure is defined for a subtype of an already exposable type, it overrides the supertype's exposure. The user may wish to augment the standard exposure by the new definition. In such a case, the new definition may include a reference to the supertype's exposure in the form:

supertypename::*expose;*

Such a reference is equivalent to invoking the supertype's exposure in the same context as the subtype's exposure. The most natural place where the supertype exposure can be referenced is at the very beginning of the subtype exposure declaration, e.g.:

```
exposure {
        int MyAttr;
        SuperType::expose;
        onpaper {
            ...
        };
        onscreen {
            ...
        };
};
```

If the supertype's exposure does not contain the mode for which the subtype's exposure has been called, the reference to the supertype's exposure has no effect. Thus, the subtype exposure can just add new modes, ones that are not serviced by the supertype exposure.

12.3 Programming Exposures

The "paper" part of the exposure body should contain statements that output the requested information to the results file (Sect. 3.2.2). The use of the *print* function is recommended.

The issues related to programming the "screen" part are somewhat more involved, although, in most cases that part is shorter and simpler than the "paper" part. The main difference between the two exposure forms is that while the "paper" exposure can be interpreted as a regular function that is called explicitly by the user program to write some information to the output file exactly at moment of call, the "screen" exposure may be called many times to refresh information displayed on the terminal screen, in an asynchronous way from the viewpoint of the SMURPH program. One simplification with respect to the "paper" case is that the exposure code need not be concerned with formatting the output: it just sends out "raw" data items which will be interpreted and organized by the display program.

Below we list the basic system functions that can be accessed by "screen" exposures:

void display (LONG ii);
void display (double dd);
void display (BIG bb);
void display (const char *tt);
void display (char c);

Each of the above functions sends one data item to the display program. For the first three functions, the data item is a numerical value (integer, *double*, or *BIG/TIME*, Sect. 3.1.3). The data item sent by the fourth method is a character string terminated by a null byte. The last method sends a simple character string consisting of a single character, e.g.,

display ('a');

is equivalent to:

display ("a");

A data item passed to the display program by one of the above functions is converted by that program into a character string and displayed within a window (Sect. C.4). It is possible to send to the display program graphic information in the form of a collection of curves to be shown within a rectangular area of a window. Such an area is called a *region*. The following functions:

void startRegion (double xo, double xs, double yo, double ys);
void startRegion ();

can be used to start a region, i.e., to indicate that the forthcoming *display* operations (see below) will build the region contents. The first function starts the so-called scaled region. The arguments have the following meaning:

xo	the starting value for the x co-ordinate, i.e., the x co-ordinate of the left side of the region's rectangle
xs	the terminating value for the x co-ordinate, i.e., the x co-ordinate of the right side of the region's rectangle
yo	the starting value for the y co-ordinate, i.e., the y co-ordinate of the bottom side of the region's rectangle
ys	the terminating value for the y co-ordinate, i.e., the y co-ordinate of the top side of the region's rectangle

Calling:

startRegion ();

is equivalent to calling:

startRegion (0.0, 1.0, 0.0, 1.0);

i.e., by default, the region comprises a unit square at the beginning of the right upper quadrant of the coordinate system. A region is terminated by calling:

void endRegion ();

which informs the display program that nothing more will be added to the region.

Display statements executed between *startRegion* and *endRegion* send to the display program information describing one or more segments. A segment can be viewed as a single curve delineated by a collection of points. The following function starts a segment:

void startSegment (Long att = NONE);

where the argument, if specified, is a bit pattern defining segment attributes (described in Sect. C.6.3). If the attribute argument is not specified (or if its value is *NONE*), default attributes are used.

A segment is terminated by invoking:

void endSegment ();

The contents of a segment are described by calling the following function:

void displayPoint (double x, double y);

which provides the co-ordinates of one point of the segment. If (based on the segment attribute pattern) the segment points are to be connected, a line will be drawn between each pair of points consecutively displayed by *displayPoint*. The first and the last points are not connected, unless they are the only points of the segment.

It is possible to display an entire segment with a single function call, by having all the point co-ordinates prepared in advance in two arrays. One of the following two functions can be used to this end:

void displaySegment (Long att, int np, double *x, double *y);
void displaySegment (int np, double *x, double *y);

The first argument of the first function specifies the segment attributes; default attributes are applied with the second function. The two *double* arrays contain the *x* and *y* co-ordinates of the segment points. The expected size of these arrays, i.e., the number of points in the segment, is equal to *np*.

It is also possible to display an entire region with a single statement, using the following variants of the *display* function introduced above:

void display (double xo, double xs, double yo, double ys, Long att, int np,
    double *x, double *y);
void display (Long att, int np, double *x, double *y);
void display (int np, double *x, double *y);

Such a region must consist of exactly one segment (which is not uncommon for a region). The first variant specifies the scaling parameters (the first four arguments), the attribute pattern, the number of points, and the point arrays. The second variant uses default scaling, and the last one assumes default scaling and default segment attributes.

For illustration, consider the following declaration of a non-standard exposable type:

```
eobject MyEType {
        TIME WhenCreated;
        int NItems;
        void setup ( ) { WhenCreated = Time; NItems = 0; };
        exposure;
};
```

with the following exposure definition:

```
MyEType::exposure {
        onpaper {
                exmode 0:
                        if (Hdr)
                                Ouf << Hdr << '\n';
                        else
                                Ouf << "Exposing " << getSName ( ) << '\n';
                        print (WhenCreated, "Created at");
                        print (NItems, "Items =");
        };
        onscreen {
                exmode 0:
                        display (WhenCreated);
                        display (NItems);
        };
};
```

Only one exposure mode is defined for both forms. The data items sent to the display program are the same as those printed out with the "paper" exposure.

For another example, suppose that we want to have an exposable object representing a few random variables. The only exposure mode of this object will display the mean values of those variables in a single-segment region. This is how the type of our object can be defined:

```
eobject RVarList {
        int VarCount, MaxVars;
        RVariable **RVars;
        void setup (int max) {
                VarCount = 0;
                RVars = new RVariable* [MaxVars = max];
        };
        void add (RVariable *r) {
                Assert (VarCount < MaxVars, "Too many random variables");
                RVars [VarCount++] = r;
        };
        exposure;
};
```

The setup method of *RVarList* initializes the random variable list as empty. New random variables can be added to the list with the *add* method. This is how their mean values can be exposed in a region:

```
RVarList::exposure {
        double min, max, mean, Min, Max;
        int i;
        LONG count;
        onscreen {
                exmode 0:
                        if (VarCount == 0) return;
                        Min = HUGE; Max = —HUGE;
                        for (i = 0; i < VarCount; i++) {
                                RVars [i] -> calculate (&min, &max, &mean, &count);
                                if (min < Min) Min = min;
                                if (max > Max) Max = max;
                        }
                        startRegion (0.0, (double)(VarCount—1), Min, Max);
                        startSegment (0xc0);
                        for (i = 0; i < VarCount; i++) {
                                RVars [i] -> calculate (&min, &max, &mean, &count);
                                displayPoint ((double)i, mean);
                        }
                        endSegment ( );
                        endRegion ( );
        }
};
```

The first loop calculates the scaling parameters for the vertical coordinate: the *y*-range is set to the difference between the minimum and maximum values assumed by all variables in the list. The second loop displays the mean values of all variables. We could avoid re-calculating the parameters of the random variables in the second loop by saving the mean values calculated by the first loop in a temporary array.

Sometimes, in the "screen" part of an exposure, one would like to tell when the exposure has been called for the first time. This is because "screen" exposures are often invoked repeatedly, to update the contents of a window. In the above example, we could use a temporary array for storing the mean values calculated by the first loop, so that we wouldn't have to recalculate them in the second one. It would make sense to allocate this array once, when the exposure is called for the first time, possibly deallocating it at the last invocation.

Generally, a "screen" exposure code may need some dynamically allocatable data structure. Technically, it is possible to allocate such a structure within the object to be exposed (e.g., by its *setup* method), but methodologically, it is better to do it in the right place, i.e., within the exposure code. This way, the object will not have to worry about the irrelevant from its perspective issues of exposing, and the data structure will only be allocated when needed.

The following global variables, useful for this purpose, are available from the exposure code:

int DisplayOpening, DisplayClosing;
*void *TheWFrame*;

If *DisplayOpening* is *YES*, it means that the exposure mode has been invoked for the first time. Otherwise, *DisplayOpening* is *NO*. If, upon recognizing the first invocation, the exposure code decides to create a dynamic data structure, the pointer to this structure should be stored in *TheWFrame*. In subsequent invocations of the same mode *TheWFrame* will contain the same pointer.

If *TheWFrame* is not *NULL*, then, when the window corresponding to a given exposure mode is closed, the exposure code is called for the last time with *DisplayClosing* set to *YES*. No display information should be sent in that case: the last invocation should be used as the opportunity to deallocate the structure pointed to by *TheWFrame*. Note that the closing invocation is only made when *TheWFrame* is not *NULL*.

For illustration, let us rewrite the previous example to make the code more efficient:

```
RVarList::exposure {
        double min, max, mean, *Means, Min, Max;
        int i;
        LONG count;
        onscreen {
                exmode 0:
                        if (DisplayOpening)
                                TheWFrame = (void*) (Means = new double [MaxVars]);
                        else if (DisplayClosing) {
                                delete (double*) TheWFrame;
                                return;
                        } else
                                Means = (double*) TheWFrame;
                        if (VarCount == 0) return;
                        Min = HUGE; Max = —HUGE;
                        for (i = 0; i < VarCount; i++) {
                                RVars [i] -> calculate (&min, &max, &mean, &count);
                                if (min < Min) Min = min;
                                if (max > Max) Max = max;
                                Means [i] = mean;
                        }
                        startRegion (0.0, (double)(VarCount—1), Min, Max);
                        startSegment (0xc0);
                        for (i = 0; i < VarCount; i++)
                                displayPoint ((double)i, Means [i]);
                        endSegment ( );
                        endRegion ( );
        }
};
```

Note that it is impossible for *DisplayOpening* and *DisplayClosing* to be both set at the same time. The only action performed by the exposure when *DisplayClosing* is set is the deallocation of the temporary array.

12.4 Invoking Exposures

An object is exposed by calling one of the following three methods defined within *Object*:

void printOut (int m, const char *hdr = NULL, Long sid =NONE);

void printOut (int m, Long sid);
void displayOut (int m, Long sid = NONE);

The first two methods expose the object "on paper," the last one renders a "screen" exposure. The second method behaves identically to the first one with *hdr* equal *NULL*. In each case, the first argument indicates the exposure mode. Depending on whether the exposure is "on paper" (the first two methods) or "on screen" (the last method), the appropriate code fragment from the object's *exposure* definition, determined by the form and mode (m), is selected and invoked. For the first method, the contents of the second and the third arguments are made available to the exposure code via *Hdr* and *SId*, respectively (Sect. 12.2). With the second method, *Hdr* is set to *NULL*, and the second argument is stored in *SId*.

Usually, the protocol program never requests "screen" exposures directly, i.e., it does not call *displayOut*. One exception is when the "screen" exposure code for a compound type T requests exposures of subobjects (see Sect. 12.2). In such a case, the exposure code for T may call *displayOut* for a subtype: this will have the effect of sending the subobject's display information to the display program.

The "screen" exposure of an object can be requested practically at any moment. SMURPH cooperates with the display program (DSD) using a special protocol and, generally, that communication is transparent to the user. There exist variables and functions that make it visible to the extent of providing some potentially useful tools.

If the global integer variable *DisplayActive* contains *YES*, it means that DSD is connected to the protocol program. During such a connection, DSD maintains possibly multiple windows whose contents are to be refreshed periodically. A single window corresponds to one display mode of a single object. SMURPH keeps an internal description of the set of windows currently requested by the display program. From the point of view of a SMURPH program, such a window corresponds to a pair: an object pointer and a display mode (an integer number). The SMURPH program is not concerned with the window layout: it just sends to DSD the raw data prescribed by the object's "screen" exposure. That information is sent at some intervals that can be changed upon a request from the display program. If SMURPH operates in the virtual mode (as a simulator), the default interval separating two consecutive updates is 5000 events. In the real mode, this interval is expressed in milliseconds, and its default length is 1000 (i.e., one second). In the visualization mode, the display interval coincides with the "resync grain" (Sect. 5.3.4) and cannot be changed.

The current length of the display interval (i.e., the amount of time separating two consecutive updates sent by the SMURPH program to DSD, expressed in events or in milliseconds) is kept in the global integer variable *DisplayInterval*.

A protocol program may request connection to the display program explicitly, e.g., upon detecting a situation that may be of interest to the user. This is done by calling the following function:

int requestDisplay (const char *msg = NULL);

which halts the simulation until a connection to the display program is established (see Sect. C.1). The function returns *OK* when the connection has been established and *ERROR*, if the connection was already established when the function was called. Note that *requestDisplay* blocks the protocol program and waits indefinitely for DSD to come in. Therefore, the function is not recommended for control programs, at least it should be used with diligence.

The optional argument is a character string representing the textual message to be sent to the display program immediately after establishing connection. The protocol program may also send a textual message to the display program at any moment during the connection by calling the function:

void displayNote (const char *msg);

where *msg* is a pointer to the message. The function does nothing, if DSD is not connected to the protocol program at the time of its invocation.

While the protocol program is connected to the display program, it may request that the screen contents be updated before the end of the display interval is reached. By calling the function:

void refreshDisplay ();

the protocol program immediately sends the exposure information for the currently defined configuration of windows to the display program.

The protocol for sending window updates to DSD is based on credits. The SMURPH program is typically allowed to send several successive updates before receiving anything back from the display program. Each update takes one credit, and if the program runs out of credits, it must receive some from DSD to continue sending window updates. This way, if DSD cannot cope with an avalanche of updates, it may reduce the frequency of those updates below the nominal rate determined by the current value of *DisplayInterval*.

12.5 Standard Exposures

In this section, we list the standard exposures defined for the built-in *Object* types of SMURPH. There exist a few conventions that are obeyed (with minor exceptions) by all standard exposures. To save space, and cut on the monotony of repeating the same description twice, unless explicitly stated, we will assume (by default) that the number of "screen" exposure modes is the same as the number of "paper" exposure modes, and that the corresponding modes of both forms send out identical items of information. Thus, we will not discuss "screen" exposures, except for the few special cases when their information contents differ from the information contents of the corresponding "paper" exposures, or when there is some feature related to the screen exposure that warrants a separate discussion.

The default header (Sect. 12.2) of a standard "paper" exposure contains the output name of the exposed object (Sect. 3.3.2). In most cases, the simulated time of the exposure is also included in the header.

All standard exposable types define mnemonic abbreviations for requesting "paper" exposures. For example, calling:

prt- > printRqs ();

where *prt* points to a *Port*, is equivalent to calling:

prt- > printOut (0);

i.e., to requesting the port's "paper" exposure with mode 0 (Sect. 12.4). The mode 0 exposure for a port prints out information about the pending wait requests to the port.

In general, the header specification of a method that provides an abbreviated way of requesting "paper" exposures may have one of the following forms:

*print*xxx *(const char *hdr = NULL, Long sid = NONE);*
*print*xxx *(const char *hdr = NULL);*

where "xxx" stands for three or four letters related to the character/nature of the information contents of the exposure. With the first version, by specifying a station *Id* as the second argument, it is possible to make the exposure station-relative (Sects. 12.1 and 12.2). No station-relative exposure is available for the second version. For both versions, the optional first argument may specify a non-standard header.

While describing the information content of an exposure, we will ignore headers and other "fixed" text fragments. The user will have no problems identifying these parts in the output file (they are naturally descriptive and impossible to confuse with data). Fixed text fragments are never sent to the display program as part of a "screen" exposure. We will discuss the individual data items in the order in which they are output, which, unless stated otherwise, will be the same for the two exposure forms.

Quite often the output size of a list-like exposure may vary, depending on the number of elements in the list. All such exposures organize the output information into a sequence of rows obeying the same layout rules. In such cases, we will describe the layout of a single row, understanding that the number of rows is variable.

Many standard exposures produce information about (pending) *wait* requests and various events. By default, only the wait requests issued by protocol processes and events perceptible by those processes are visible in such exposures. There are a few internal processes of SMURPH that issue internal wait requests and respond to some internal events, which are rather exotic from the user's point of view and generally completely uninteresting.[2] To include information about these internal requests and events in the exposed data, the protocol program should be called with the *−s* option. This also concerns the event information generated by the simulator-level tracing (see Sect. 11.2).

---

[2]For example, the action of purging from links or rfchannels the activities that have reached the end of their stay (Sects. 7.1.1 and 8.1.1), is carried out by internal processes responding to *Timer* events.

Information produced by standard exposures often contains event identifiers. Event identifiers are *AI*-specific: they provide textual representations for events triggered by the *AIs*. The format of those event identifiers is discussed below, by the *AI* type.

Timer	A *Timer* event (Sect. 5.3.1) is triggered when the delay specified in a *Timer wait* request elapses. If the setting of *BIG_precision* (Sects. 3.1.3 and B.5) makes *BIG* (*TIME*) values representable directly as simple (possibly *long long*) numbers, the event identifier is the numerical value of the delay in *ITUs*. Otherwise, the character string *wakeup* represents all events
Mailbox	*Mailbox* events (Sect. 9.2.2) are generated when items are stored in mailboxes or removed from mailboxes. The event identifier can be one of the character strings: *EMPTY*, *GET*, *NEWITEM*, *NONEMPTY*, *OUTPUT*, *RECEIVE*, or a number identifying a count event
Port	The identifier of a port event is the symbolic name of the event, according to Sect. 7.1.3, i.e., *SILENCE*, *ACTIVITY*, *COLLISION*, etc.
Transceiver	The identifier of a transceiver event is the symbolic name of the event, according to Sect. 8.2.1, i.e., *SILENCE*, *ACTIVITY*, *BOP*, etc.
Client	The event identifier for the *Client* is one of the following character strings: *ARRIVAL*, *INTERCEPT* (see Sect. 6.8), *SUSPEND*, *RESUME* (Sect. 6.6.3), *RESET* (Sect. 10.2.4), or *arr Trf* xxx. The sequence "xxx" in the last string stands for the *Id* of the traffic pattern whose message arrival is awaited. This case corresponds to a *Client* wait request with the first argument identifying a traffic pattern (Sect. 6.8)
Traffic	The event identifier for a *Traffic AI* is one of the following four strings: *ARRIVAL*, *INTERCEPT* (see Sect. 6.8), *SUSPEND*, *RESUME* (Sect. 6.6.3), or *RESET* (Sect. 10.2.4)
Process	The event identifier for a *Process AI* (Sect. 5.1.7) is one of the following character strings: *START*, *DEATH*, *SIGNAL*, *CLEAR*, or a state name. The first string represents the virtual event used to start the process; a process that appears to be waiting for it is a new process that has just been created, within the current *ITU*. The *DEATH* event is triggered by a process termination. Similar to *START*, it occurs only once in the lifetime of a process. Events *SIGNAL* and *CLEAR* are related to the process's signal repository. A state name represents the event that will be generated by the *Process AI* when the process gets into the named state

Neither links nor rfchannels generate any events that can be directly awaited by protocol processes. There exists a system process called *LinkService*, which takes care of removing obsolete activities from links and their archives. Two system events, *LNK_PURGE* and *ARC_PURGE*, are generated by links, and perceived by the *LinkService* process, to handle the internal shuffling of activities between the link repositories (Sect. 7.1.1). Another system process, called *RosterService*, takes care of the various administrative duties related to rfchannels. The following system events perceived by that process formally originate at transceivers: *DO_ROSTER*, responsible for advancing activity stages at perceiving transceivers (Sect. 8.1.1), *BOT_TRIGGER*, formally initiating proper packet transmission at the sending transceiver following the end of preamble, *RACT_PURGE*, responsible for cleaning out obsolete activities from the channels.

The following two internal events are generated by the *Client AI* and sensed by the system process called *ClientService* which is responsible for traffic generation: *ARR_MSG*, to indicate that a new message is to be generated and queued at a station, and *ARR_BST*, to indicate that a new burst should be started.

One more system process responding to an internal event is *ObserverService*. That process is responsible for restarting an observer after its *timeout* delay has elapsed (Sect. 11.3). The event is formally triggered by a dummy *AI* called *Observer*, and its identifier is *TIMEOUT*.

12.5.1 Timer Exposure

Mode 0: full request list

<u>Calling</u>: *printRqs (const char *hdr = NULL, Long sid = NONE);*

The full list of processes waiting for the *Timer* is produced. The list has the form of a table with each row consisting of the following entries (in this order):

1. The virtual time in *ITUs* when the *Timer* event will occur.
2. A single-character flag that describes the status of the *Timer* request: * means that according to the current state of the simulation, the *Timer* event will restart the process, blank means that another waking event will occur earlier than the *Timer* event, and ? says that the *Timer* event may restart the process, but another event has been scheduled at the same *ITU* and has the same order as the *Timer* event.
3. The *Id* of the station owning the process or *Sys*, for a system process. This data item is not included, if the station-relative variant of the exposure is selected.
4. The process type name.
5. The process *Id*.
6. The identifier of the process's state associated with the *Timer* request, i.e., the state where the process will be restarted by the *Timer* event, should this event in fact restart the process.
7. The type name of the activity interpreter that, according to the current state of the simulation, will restart the process.
8. The *Id* of the activity interpreter or blanks, if the *AI* has no *Id* (*Client, Timer, Monitor*).
9. The identifier of the event that will wake the process up.
10. The identifier of the state where, according to the current configuration of activities and events, the process will be restarted.

The station-relative version of the exposure restricts the list to the processes belonging to the indicated station. The same items are sent to the display program with the "screen" version of the exposure.

For all *AIs*, the mode 0 exposure produces similar information, according to the above layout. When the time of the event occurrence (item 1) is not known at the time of exposure (this cannot happen for a *Timer wait* request), the string *undefined* appears in its place.

Mode 1: abbreviated request list

Calling: *printARqs (const char *hdr = NULL, Long sid = NONE);*

The processes waiting for *Timer* events are listed in the abbreviated format. The list has the form of a table with each row consisting of the following entries (in this order):

1. The virtual time in *ITUs* when the *Timer* event will occur.
2. A character flag describing the status of the *Timer* request (see above).
3. The *Id* of the station owning the process. This data item is not included, if the station-relative variant of the exposure is selected.
4. The process type name.
5. The process *Id*.
6. The identifier of the process's state associated with the *Timer* request.

Only user processes are included in the abbreviated list, even if the *–s* call option is used. The station-relative version of the exposure restricts the list to the processes belonging to the indicated station. The same items are sent to the display program with the "screen" version of the exposure.

12.5.2   *Mailbox Exposure*

Mode 0: full request list

Calling: *printRqs (const char *hdr = NULL, Long sid = NONE);*

The full list of processes awaiting events on the *Mailbox*. The description is the same as for the *Timer* exposure with mode 0, except that the word *Timer* should be replaced by *Mailbox*.

Mode 1: abbreviated request list

Calling: *printARqs (const char *hdr = NULL, Long sid = NONE);*

The processes awaiting events on the *Mailbox* are listed in the abbreviated format. The description is the same as for the *Timer* exposure with mode 1, except that the word *Timer* should be replaced by *Mailbox*.

Mode 2: mailbox contents

Calling: *printCnt (const char *hdr = NULL);*

The exposure produces information about the mailbox contents. The following items are printed in one line:

1. the header argument or the mailbox output name (Sect. 3.3.2), if the header argument is not specified
2. the number of elements currently stored in the mailbox (for a bound mailbox, Sect. 9.4, this is the number of bytes in the input buffer)
3. the mailbox capacity (for a bound mailbox, this is the size of the input buffer)
4. the number of pending wait requests issued to the mailbox.

The line with the above values is preceded by a header line.

The "screen" version of the exposure sends to the display program the last three items, i.e., the mailbox name is not sent. Of course, the header line is also absent.

Mode 3: short mailbox contents

Calling: *printSCnt (const char *hdr = NULL);*

The "paper" version of the exposure produces the same output as the mode 2 exposure (see above), except that no header line is printed. Thus, the output can be embedded into a longer output, e.g., a table listing the contents of several mailboxes. No "screen" version of this exposure is provided. Note that the "screen" version of the mode 2 exposure is already devoid of the header line.

12.5.3   RVariable Exposure

Mode 0: full contents

Calling: *printCnt (const char *hdr = NULL);*

The exposure outputs the contents of the random variable (Sect. 10.1). The following data items are produced (in this order):

1. the number of samples
2. the minimum value
3. the maximum value
4. the mean value (i.e., the first central moment)
5. the variance (i.e., the second central moment)
6. the standard deviation (i.e., the square root of the second central moment)
7. the relative confidence margin for probability $P_a = 0.95$; this value is equal to half the length of the confidence interval at $P_a = 0.95$ divided by the absolute value of the calculated mean
8. the relative confidence margin for probability $P_a = 0.99$.

If the random variable was created with more than two moments (Sect. 10.1.1), the above list is continued with their values. If the number of moments is 1, items 5–8 do not appear; if the number of moments is 0, item 4 is skipped as well.

The "screen" version of the exposure contains the same numerical items as the "paper" exposure without item 8. This list is preceded by a region (Sect. 12.3) that displays graphically the history of the 24 last exposed mean values of the random variable. The region consists of a single segment including up to 24 points. It is scaled by $(0, 23)$ in the x axis and (min, max) in the y axis, where min and max are the current minimum and maximum values of the random variable.

Mode 1: abbreviated contents

Calling: *printACnt (const char *hdr = NULL);*

The exposure outputs the abbreviated contents of the random variable. The following data items are produced (in this order):

1. the number of samples
2. the minimum value
3. the maximum value
4. the mean value, i.e., the first central moment (not included if the random variable does not have at least one moment).

The values are preceded by the header line.

The "screen" version of the exposure contains the same numerical items as the mode 0 "paper" exposure, i.e., the mode 0 "screen" exposure without the region.

Mode 2: short contents

Calling: *printSCnt (const char *hdr = NULL);*

The exposure outputs the short contents of the random variable. The items produced are identical to those for mode 1. The difference is that no header line is printed above the items, which allows the program to expose multiple random variables in a table. Most naturally, this can be done by exposing the first random variable with mode 1 and the remaining ones with mode 2.

The "screen" version of the exposure produces the same numerical items as the "paper" exposure followed additionally by the standard deviation. If the random variable has less than two standard moments, the missing values are replaced by dashes (---).[3]

12.5.4   Client and Traffic Exposure

Mode 0: request list

Calling: *printRqs (const char *hdr = NULL, Long sid = NONE);*

---

[3]The typing of simple (non-region) items sent to the display program is flexible, so there is no problem when a textual item is sent instead of a number (see Sect. C.4.2).

This exposure produces the list of processes waiting for *Client* and *Traffic* events. The description is the same as for the *Timer* exposure with mode 0, except that the word *Timer* should be replaced by *Client* or *Traffic*.

Mode 1: performance measures

Calling: *printPfm (const char *hdr = NULL);*

This exposure lists the standard performance measures for all standard traffic patterns viewed globally, as if they constituted a single traffic pattern. The exposure is produced by combining (Sect. 10.1.2) the standard random variables belonging to the individual traffic patterns (Sect. 10.2.1) into a collection of global random variables reflecting the performance measures of the whole standard *Client*, and then exposing them with mode 0 (*printCnt*, Sect. 12.5.3) in the following order:

1. *RVAMD*: the absolute message delay
2. *RVAPD*: the absolute packet delay
3. *RVWMD*: the weighted message delay
4. *RVMAT*: the message access time
5. *RVPAT*: the packet access time
6. *RVMLS*: the message length statistics.

The data produced by the above list of exposures are followed by the *Client* statistics containing the following items:

1. the number of all messages ever generated and queued at their senders
2. the number of messages currently queued awaiting transmission
3. the number of all messages completely received (Sects. 7.1.6 and 8.2.3)
4. the number of all transmitted packets (terminated by *stop*, Sects. 7.1.2 and 8.1.8)
5. the number of all received packets
6. the number of all message bits queued at stations awaiting transmission; this is the combined length of all messages currently queued (item 2)
7. the number of all message bits successfully transmitted so far; this is the combined length of all packets from item 4
8. the number of all message bits successfully received so far; this is the combined length of all packets from item 5
9. the global throughput of the network calculated as the ratio of the number of received bits (item 8) to the simulated time in *ETU* (see Sect. 10.2.4).

The "screen" version of the exposure does not include the last 9 items. The six random variables representing the global performance measures are exposed with mode 1 (Sect. 12.5.3).

Mode 2: message queues

Calling: *printCnt (const char *hdr = NULL, Long sid = NONE);*

This exposure produces information about message queues at all stations, or at one indicated station, for the station-relative variant. The global variant prints, for each station, one line consisting of the following three items:

1. the station *Id*
2. the number of messages queued at the station
3. the number of message bits queued at the station, i.e., the combined length of all queued messages.

The station-relative variant of the exposure produces one row for each message queued at the indicated station. That row contains the following data:

1. the time in *ITUs* when the message was queued
2. the *Id* of the traffic pattern to which the message belongs
3. the message length in bits
4. the *Id* of the station to which the message is addressed, or *bcast* for a broadcast message.

The global variant of the "screen" version produces a region that displays graphically the length of message queues at all stations. The region consists of a single segment containing *NStations* points (Sect. 4.1.2) scaled by $(0, N Stations-1)$ on the x axis and $(0, L_{max})$ on the y axis, where L_{max} is the maximum length of a message queue at a station. The y co-ordinate of a point is equal to the number of messages queued at the station whose *Id* is represented by the x co-ordinate.

Mode 3: traffic definition/client statistics

Calling: *printDef (const char *hdr = NULL);*

The "paper" variant of this exposure prints out information about the definition of all traffic patterns. This information is self-explanatory. The "screen" variant of mode 3 displays the *Client* statistics containing the same items as the statistics printed by the mode 2 exposure. The order of these items is also the same, except that the throughput is moved from the last position to the first.

For *Traffic* objects, the exposure modes, their "paper" abbreviations and contents are identical to those of the *Client* exposures, with the following exceptions:

1. with mode 0, the list of processes is limited to the ones awaiting events from the given traffic pattern
2. the performance measures produced by mode 1 refer to the given traffic pattern
3. the *Client* statistics printed by mode 1 and displayed by mode 3 refer to the given traffic pattern, and they do not contain the throughput item
4. the output produced by the station-specific variant of mode 2 does not contain the traffic pattern *Id*; this output, similar to the output of the station-relative variant, is restricted to the given traffic pattern
5. the information printed by the mode 3 exposure relates to the definition of the given traffic pattern.

12.5.5 Port Exposure

Mode 0: request list

Calling: *printRqs (const char *hdr = NULL);*

The exposure produces the list of processes waiting for events on the port. The description is the same as for the *Timer* exposure with mode 0, except that the word *Timer* should be replaced by *Port*. Note that no station-relative exposures are defined for ports: a port always belongs to a specific station, and all its exposures are implicitly station-relative.

Mode 1: list of activities

Calling: *printAct (const char *hdr = NULL);*

This exposure produces the list of all activities currently present in the link to which the port is connected, including the link's archive. The list is sorted in the non-decreasing order of the time when the beginning of the activity was, is, or will be heard on the port. Each activity takes one row consisting of the following items:

1. a single-letter activity type designator: *T*—a transmission (possibly an aborted transmission attempt), *J*—a jamming signal, *C*—a collision (see below)
2. the starting time of the activity in *ITUs*, as perceived by the port
3. the end time of the activity in *ITUs*, as perceived by the port; if this time is not known at present (i.e., the activity is still being inserted into the link by the originating port), the string *undefined* is written
4. the *Id* of the station that started the activity
5. the station-relative port number (Sect. 4.3.1)
6. the *Id* of the receiver station or *bcst* for a broadcast packet
7. the traffic pattern *Id*
8. the packet signature (see Sect. 6.2).

Items 6–8 are only meaningful for packets. If the activity is not a packet transmission, then the string --- appears in place of each of the three items. In the "paper" version of the exposure, item 8 is only printed if the program has been compiled with the –*g* (or –*G*) option of *mks* (Sect. B.5), so packet signatures are available. In the "screen" version, item 8 is always displayed, but unless the program has been compiled with –*g* (or –*G*), it is filled with dashes.[4]

If, according to the port's perception, the current configuration of activities results in a future or present collision, a dummy (artificial) activity representing the collision is included in the activity list. Only the first two attributes of this activity (i.e., the activity type designator and the starting time) are meaningful; the remaining items are printed as ---.

The same items are displayed with the "screen" form of the exposure.

---

[4]The display program is not aware of the differences between SMURPH compilation options, and its display templates (Sect. C.4) include placeholders for the maximum number of items that the simulator may want to send within the exposure.

Mode 2: event timing

<u>Calling</u>: *printEvs (const char *hdr = NULL);*

The exposure describes the timing of future or present events on the port (Sect. 7.1.3). The description of each event takes one row consisting of the following items:

1. the event identifier (e.g., *ACTIVITY*, *BOT*, *COLLISION*)
2. the time when the event will occur, or ---, if no such event is predicted at this moment
3. the type of activity triggering the event (*T* for a packet transmission, *J* for a jamming signal)
4. the *Id* of the station that started the triggering activity
5. the station-relative number of the port on which the activity was started
6. the *Id* of the receiving station, if the triggering activity is a packet transmission (--- is printed otherwise)
7. the traffic pattern *Id* of the triggering activity, if the activity is a packet transmission (--- is printed otherwise)
8. the total length of the packet in bits, if the triggering activity is a packet transmission (--- is printed otherwise)
9. the packet signature (see the comments regarding item 8 for mode 1 *Port* exposure).

The list produced by this exposure contains exactly ten rows, one row for each of the ten event types discussed in Sect. 7.1.3. If, based on the current configuration of link activities, no event of the given type is predicted, all entries in the corresponding row, except for the first one, contain ---.

The information displayed by the "screen" form of the exposure contains the same data as the "paper" form.

12.5.6 Link Exposure

Mode 0: request list

<u>Calling</u>: *printRqs (const char *hdr = NULL, Long sid = NONE);*

The exposure produces the list of processes waiting for events on ports connected to the link. The description is the same as for the *Timer* exposure with mode 0, except that the word *Timer* should be replaced by "a port connected to the link." If the −*s* run-time option is selected (Sect. B.6), the system processes waiting for internal events generated directly by the link *AI* (Sect. 12.5) are included in the list.

The station-relative version of the exposure restricts the output to the processes belonging to the indicated station.

Mode 1: performance measures

<u>Calling:</u> *printPfm (const char *hdr = NULL);*

The exposure outputs the values of the link counters discussed in Sect. 10.3. The following items are produced (in this order):

1. the total number of jamming signals ever inserted into the link (*NTJams*)
2. the total number of packet transmissions ever started in the link (*NTAttempts*)
3. the total number of packet transmissions terminated by *stop* (*NTPackets*)
4. the total number of payload bits transmitted via the link (*NTBits*)
5. the total number of packets received on the link (*NRPackets*)
6. the total number of payload bits received on the link (*NRBits*)
7. the total number of messages transmitted via the link (*NTMessages*)
8. the total number of messages received on the link (*NRMessages*)
9. the number of damaged packets inserted into the link (*NDPackets*)
10. the number of header-damaged packets inserted into the link (*NHDPackets*)
11. the number of damaged payload bits inserted into the link (*NDBits*)
12. the received throughput of the link determined as the ratio of the total number of bits received on the link (item 6) to the simulated time expressed in *ETUs*
13. the transmitted throughput of the link determined as the ratio of the total number of bits transmitted on the link (item 4) to the simulated time expressed in *ETUs*.

The "screen" version of the exposure displays the same items. The two throughput measures are displayed first and followed by the remaining items, according to the above order.

Items 9, 10, and 11 do not appear in the link exposure if the program was not compiled with the *−z* option of *mks* (Sect. B.5).

Mode 2: activities

<u>Calling:</u> *printAct (const char *hdr = NULL, Long sid = NONE);*

This exposure produces the list of activities currently present in the link and the link's archive. The description of each activity takes one row that looks like a row produced by the *Port* exposure with mode 1, except for the following differences:

1. type *C* virtual activities (collisions) are omitted, as collisions occur on ports, not in links
2. the starting and ending times reflect the starting and ending times of the activity at the port responsible for its insertion
3. in the station-relative version of the exposure, the *Id* of the station inserting the activity is not printed; only the activities inserted by the indicated station are printed in this mode.

The "screen" form of the exposure displays the same items as the "paper" form.

12.5.7 *Transceiver Exposure*

Mode 0: request list

<u>Calling</u>: *printRqs (const char *hdr = NULL);*

   The exposure produces the list of processes waiting for events on the transceiver. The description is the same as for the *Timer* exposure with mode 0, except that the word *Timer* should be replaced by *Transceiver*. Note that no station-relative exposures are defined for transceivers: a transceiver always belongs to a specific station, and this way all its exposures are implicitly station-relative.

Mode 1: list of activities

<u>Calling</u>: *printRAct (const char *hdr = NULL);*

   This exposure produces the list of all activities currently perceived by the transceiver. The title line specifies the total number of those activities and the total signal level (as returned by *sigLevel*, Sect. 8.2.5). If the transmitter is active, the title also includes the word *XMITTING*. Each activity is described by a single line with the following items:

1. the station *Id* of the sender
2. the station *Id* of the intended recipient, *none* (if the recipient is unspecified), or *bcst* for a broadcast packet
3. the packet type, i.e., the *TP* attribute of the packet (Sect. 6.2)
4. the total length of the packet (attribute *TLength*)
5. the status consisting of two characters: if the activity is *dead*, i.e., has been assessed as non-receivable (Sects. 8.1.4 and 8.2.5), the first character is an asterisk, otherwise it is blank; the second character can be: *D* for "done" (the packet is past its last bit and it does not count any more), *P*—the packet is before stage *P* (Sect. 8.1.1), *T*—before stage *T*, *A*—before stage *T*, but that stage will not occur because the preamble has been aborted, *E*—past stage *T*, *O*—own packet, i.e., transmitted by this transceiver
6. the time when the transmission of the proper packet (the first bit following the preamble) commenced at the sender; note that this time may be *undefined*
7. perceived signal level of the activity; for an "own" activity, this is the current transmission power (*XPower*) of the transceiver
8. the current interference level experienced by the activity
9. the maximum interference level experienced by the activity, according to its present stage (Sect. 8.1)
10. the average interference level experienced by the activity
11. the packet's signature (Sect. 6.2) included if the simulator has been compiled with –*g* or –*G*.

   Items 8–10 do not show up for an "own" activity and are replaced with dashes.

The same items are displayed with the "screen" form of the exposure, except that item 11 is always included. Unless the program has been created with –g or –G, item 11 is filled with dashes.

Mode 2: the neighborhood

Calling: *printNei (const char *hdr = NULL);*

The exposure lists the current population of neighbors (as determined by *RFC_cut*, Sect. 4.4.3) of the transceiver. The title line shows the transceiver's location (two or three coordinates in *DUs*) followed by one line per neighbor, including:

1. the station *Id*
2. the station-relative transceiver *Id*
3. the distance in *DUs* from the current transceiver
4. the coordinates (in *DUs*) describing the neighbor's location; these can be two or three numbers depending on the dimensionality of the node deployment space (Sect. 4.5.2).

The information displayed by the "screen" form of the exposure contains the same data as the "paper" form. Owing to the rigid structure of the window template (Sect. C.4.2), the list of node coordinates always includes three items (with the third item being cropped out by the standard geometry of the window). In the 2d case, the third coordinate is blank.

12.5.8   RFChannel Exposure

Mode 0: request list

Calling: *printRqs (const char *hdr = NULL, Long sid = NONE);*

The exposure produces the list of processes waiting for events on transceivers connected to the rfchannel. The description is the same as for the *Timer* exposure with mode 0, except that the word *Timer* should be replaced by "a transceiver interfaced to the radio channel." If the –s run-time option is selected (Sect. B.6), system processes waiting for internal events related to the rfchannel (Sect. 12.5) are included in the list.

The station-relative version of the exposure restricts the output to the processes belonging to the indicated station.

Mode 1: performance measures

Calling: *printPfm (const char *hdr = NULL);*

The exposure outputs the values of the counters associated with the rfchannel and discussed in Sect. 10.4. The following items are produced (in this order):

1. the total number of packet transmissions ever attempted in the channel (*NTAttempts*)
2. the total number of packet transmissions terminated by *stop* (*NTPackets*)
3. the total number of information bits transmitted over the channel (*NTBits*)
4. the total number of packets received on the channel (*NRPackets*)
5. the total number of information bits received on the channel (*NRBits*)
6. the total number of messages transmitted via the channel (*NTMessages*)
7. the total number of messages received on the channel (*NRMessages*)
8. the received throughput of the channel determined as the ratio of the total number of bits received on the channel (item 5) to the simulated time expressed in *ETUs*
9. the transmitted throughput of the link determined as the ratio of the total number of bits transmitted on the link (item 3) to the simulated time expressed in *ETUs*.

The "screen" version of the exposure displays the same items. The two throughput measures are displayed first and followed by the remaining items, according to the above order.

Mode 2: activities

<u>Calling</u>: *printRAct (const char *hdr = NULL, Long sid = NONE);*

This exposure produces the list of activities currently present in the rfchannel. Each activity is represented by a single line presenting the following items:

1. the station *Id* of the sender; this item is not included with a station-relative variant of the exposure
2. the station-relative transceiver *Id* of the sender
3. the starting time of the proper packet transmission (this time can be *undefined*)[5]
4. the ending time of the packet transmission (this time can be *undefined*)
5. the station *Id* of the intended recipient, *none* (if the recipient is unspecified), or *bcst*, for a broadcast packet
6. the packet (traffic) type, i.e., the *TP* attribute (Sect. 6.2)
7. the total length of the packet (attribute *TLength*)
8. the packet's signature (Sect. 6.2), included if the simulator has been compiled with *–g* or *–G* (Sect. B.5).

An asterisk following the ending time of transmission indicates an aborted packet.
    The same items are displayed with the "screen" form of the exposure, except that item 8 is always included. Unless the program has been created with *–g* or *–G*, item 8 is filled with dashes.

Mode 3: topology

<u>Calling</u>: *printTop (const char *hdr = NULL);*

---

[5]This contrasts with ports and links, because *RFChannel* activities include preambles.

This exposure presents the configuration of transceivers interfaced to the channel. The title line shows the number of transceivers, the coordinates of the minimum rectangle encompassing them all, and the length of the diagonal of that rectangle. This is followed by the list of transceivers (one line per transceiver), followed in turn by the list of transceiver neighborhoods. One entry from the transceiver list contains the following items:

1. the *Id* of the station to which the transceiver belongs
2. the station-relative *Id* of the transceiver
3. the transceiver's transmission rate, i.e., attribute *XRate* (Sect. 4.5.2)
4. the current setting of the transmission power (attribute *XPower*)
5. the current setting of the receiver gain (attribute *RPower*)
6. the number of neighbors, i.e., transceivers in the neighborhood, as determined by *RFC_cut* (Sect. 4.4.3)
7. the x and y coordinates of the transceiver in *DUs*; one more coordinate (z) is included for the 3d variant of the node deployment space.

A neighborhood description begins with the identifier of a transceiver, which is followed by the list of identifiers of all transceivers in its neighborhood along with their distances from the current transceiver in *DUs*. Each transceiver identifier is a pair: station *Id*, station-relative transceiver *Id*. A list like this is produced for every transceiver configured into the radio channel.

The "screen" variant of this exposure presents a region with different transceivers marked as points (black circles). There exists a station-relative variant of this exposure, which marks the transceivers belonging to the indicated station red and their neighbors green.

For the 3d variant of the node deployment space, the z coordinate is ignored, i.e., the nodes are projected onto the 2d plane with $z = 0$.

12.5.9   Process Exposure

Mode 0: request list

Calling: *printRqs (const char *hdr = NULL, Long sid = NONE);*

The exposure produces the list of processes waiting for events to be triggered by the *Process AI* (see Sect. 5.1.7). The description is the same as for the *Timer* exposure with mode 0, except that the word *Timer* should be replaced with *Process*.

The station-relative version of the exposure restricts the list to the processes belonging to the indicated station.[6]

---

[6]In most cases, a process waiting for an event from a *Process AI* belongs to the same station as the *Process AI*; thus, there seems to be little need for the station-relative variant of the exposure.

Mode 1: wait requests of this process

<u>Calling</u>: *printWait (const char *hdr = NULL);*

The exposure outputs the list of pending wait requests issued by the process. Each wait request is described by one row containing the following items (in this order):

1. the type name of the activity interpreter to which the request has been issued
2. the *Id* of the activity interpreter or blanks, if the *AI* has no *Id* (e.g., *Timer*)
3. the identifier of the awaited event (see Sect. 12.5)
4. the identifier of the process state to be assumed when the event occurs
5. the time in *ITUs* when the event will be triggered or *undefined*, if the time is unknown at present.

The "screen" version of the exposure displays the same items in the same order.

12.5.10   Kernel Exposure

The *Kernel* process (Sect. 5.1.8) defines a separate collection of exposure modes. These modes present information of a general nature.

Mode 0: full global request list

<u>Calling</u>: *printRqs (const char *hdr = NULL, Long sid = NONE);*

This exposure prints out the full list of processes waiting for events at the time. This is the SMURPH view of the global event queue. Each *wait* request is represented by one row containing the following data (in this order):

1. the *Id* of the station owning the waiting process or *Sys*, for a system process; this item is not printed by the station-relative version of the exposure
2. the process type name (Sect. 3.3.2)
3. the process *Id*
4. the simulated time in *ITUs* when the event will occur or *undefined*, if this time is unknown at present
5. a single-character flag that describes the status of the wait request: * means that, according to the present state of the simulation, the awaited event will restart the process, blank means that another waking event will occur earlier than the awaited event, and ? indicates that the awaited event may restart the process, but another event awaited by the process is scheduled at the same *ITU* with the same order
6. the type name of the activity interpreter expected to trigger the event
7. the *Id* of the activity interpreter or blanks, if the *AI* has no *Id* (*Client, Timer, Monitor*)
8. the event identifier (Sect. 12.5)

9. the state where the process will be restarted by the awaited event, should this
    event in fact restart the process.

The station-relative version of this exposure lists only the wait requests issued by the
processes belonging to the indicated station.
For multiple wait requests issued by the same process, the first three items are printed
only once, at the first request of the process, and blanks are inserted instead of them
in the subsequent rows describing other requests of the same process.
   The "screen" version of the exposure displays the same items in the same order.

Mode 1: abbreviated global request list

Calling: *printARqs (const char *hdr = NULL, Long sid = NONE);*

   This is an abbreviated variant of the mode 0 exposure, which produces a single
row of data per one process awaiting events. That row has the same layout as for the
mode 0 exposure. It describes the wait request that, according to the current state of
the simulation, will restart the process.
   The same items are produced by the "screen" version of the exposure.

Mode 2: simulation status

Calling: *printSts (const char *hdr = NULL);*

   The exposure prints out global information about the status of the simulation run.
The following data items are written (in this order):

1. the (UNIX/Windows) process Id of the SMURPH program
2. the CPU execution time in seconds
3. the current virtual time in *ITUs*
4. the total number of events processed by the program from the beginning of run
5. the event queue size (the number of queued events)
6. the total number of messages ever generated and queued at the senders by the
    standard *Client*
7. the total number of messages entirely received at destinations (Sects. 7.1.6 and
    8.2.3)
8. the total number of queued bits, i.e., the combined length of all messages queued
    at the time
9. the global throughput determined as the total number of bits received so far
    divided by the simulation time in *ETUs* (see Sect. 1.2.40)
10. the output name (Sect. 3.3.2) of the last active station, i.e., the station whose
    process was last awakened
11. the output name of the last awakened process
12. the output name of the *AI* that awakened the process
13. the waking event identifier (see Sect. 12.5)
14. the state at which the process was restarted.

   Only items 1 through 10 (inclusively) are displayed by the "screen" version of
the exposure, in the above order.

Mode 3: last event

No "paper" form.

This mode exists in the "screen" form only. It displays the information about the last awakened process, i.e., the last five items from the "paper" form of mode 3, in the same order.

12.5.11 Station Exposure

Mode 0: process list

Calling: *printPrc (const char *hdr = NULL);*

This exposure prints out the list of processes belonging to the station and waiting for events. Note that an alive process always waits for some event(s), unless it is the process requesting the exposure, and it hasn't (yet) issued any wait request in its present state. Each wait request is represented by one row containing the following data (in this order):

1. the process output name (Sect. 3.3.2)
2. the type name of the activity interpreter expected to trigger the awaited event
3. the *Id* of the activity interpreter or blanks, if the *AI* has no *Id* (*Client, Timer, Monitor*)
4. the event identifier (Sect. 12.5)
5. the state where the process will be restarted by the awaited event, should this event in fact restart the process
6. the simulated time in *ITUs* when the event will occur or *undefined*, if this time is unknown at present
7. a single-character flag that describes the status of the wait request: * means that, according to the present state of the simulation, the awaited event will restart the process, blank means that another waking event will occur earlier than the awaited event, and ? indicates that the awaited event may restart the process, but another event awaited by the process has been scheduled at the same *ITU* with the same order.

For multiple wait requests issued by the same process, the first item is printed only once, at the first request of the process, and replaced with blanks in the subsequent rows representing more requests of the same process.

The "screen" version of the exposure displays the same items in the same order.

Mode 1: buffer contents

Calling: *printBuf (const char *hdr = NULL);*

This exposure prints out the contents of the packet buffers at the station. One row of information is printed per each buffer, in the order in which the buffers have been declared (Sect. 6.3). The following data items are included:

1. the buffer number (from 0 to $n-1$), where n is the total number of packet buffers owned by the station
2. the time when the message from which the packet in the buffer has been acquired was queued at the station
3. the time when the packet became ready for transmission (attribute *TTime*, see Sects. 6.3 and 10.2.1)
4. the *Id* of the packet's receiver or *bcst*, for a broadcast packet
5. the *Id* of the traffic pattern to which the packet belongs
6. the payload length of the packet (attribute *ILength*, see Sect. 6.2)
7. the total length of the packet (attribute *TLength*)
8. two standard flags of the packet: *PF_broadcast* and *PF_last* (Sect. 7.1.6); the third standard flag (*PF_full*, Sect. 6.3), associated with the packet buffer, is not shown (when this flag is 0, the buffer is empty and no packet information is printed at all, see below); this item is a piece of text consisting of up to two possibly combined letters: *B* (for a broadcast packet) and *L* (for the last packet of a message); if neither of the two packet flags is set, the text is empty
9. the packet signature (see the comments regarding item 8 for mode 1 *Port* exposure).

Items 2 through 9 are only shown if the packet buffer is nonempty; otherwise, the string *empty* is written in place of item 2, and --- replaces all the remaining items.

The "screen" version of the exposure produces the same contents.

Mode 2: mailbox contents

Calling: *printMail (const char *hdr = NULL);*

The exposure prints information about the contents of all mailboxes owned by the station. One row of data is produced for each mailbox with the following items (in this order):

1. the station-relative serial number of the mailbox reflecting its creation order, or the nickname (Sect. 3.3.2), if one is defined for the mailbox
2. the number of elements currently stored in the mailbox; for a bound mailbox (Sect. 9.4), this is the number of bytes in the input buffer
3. the mailbox capacity (or the input buffer size, if the mailbox is bound)
4. the number of pending wait requests issued to the mailbox.

The "screen" version of the exposure produces the same items in the same order.

Mode 3: link activities

Calling: *printAct (const char *hdr = NULL);*

The exposure prints out information about port (link) activities started by the station. All links are examined and all activities that were originated by the station and are still present in the link or the link archive are exposed. One row of data per activity is produced with the following contents:

1. the station-relative number of the port (Sect. 4.3.1) on which the activity was started
2. the *Id* of the link to which the port is connected
3. the starting time of the activity
4. the end time of the activity or *undefined*, if the activity has not been finished yet
5. the activity type: *T* for a packet transmission, *J* for a jam
6. the *Id* of the receiver (or *bcst*), if the activity is a packet transmission
7. the *Id* of the traffic pattern to which the packet belongs
8. the total length (attribute *TLength*, Sect. 6.2) of the packet in bits
9. the packet signature (see the comments regarding item 8 for mode 1 *Port* exposure).

Items 6 through 9 are only printed, if the activity represents a packet transmission; otherwise, those items are replaced with ---.

The same data are displayed by the "screen" version of the exposure.

Mode 4: port status

Calling: *printPort (const char *hdr = NULL);*

The exposure produces information about the status of the station's ports. One row of ten items is printed for each port. The first item is the port identifier. It can be either the station-relative serial number reflecting the port creation order, or the port's nickname, if one is defined for the port (Sect. 3.3.2). Each of the remaining items corresponds to one event that can be present or absent on the port: *** means present, i.e., the port is currently perceiving the event, ... means absent. The nine events are (in this order): *ACTIVITY*, *BOT*, *EOT*, *BMP*, *EMP*, *BOJ*, *EOJ*, *COLLISION*, *ANYEVENT* (see Sect. 7.1.3).

The "screen" version of the exposure displays the same information, except that *** is reduced to * (a single asterisk) and ... to. (a single dot), respectively.

Mode 5: radio activities

Calling: *printRAct (const char *hdr = NULL);*

The exposure prints out information about radio activities started by the station. It examines all radio channels (*RFChannels*) and produces one line for each activity found there that was transmitted on one of the station's transceivers. The line contains (in this order):

1. the numerical *Id* of the radio channel
2. the station-relative *Id* of the originating transceiver
3. the starting time of the proper packet transmission (this time can be *undefined*)
4. the ending time of the packet transmission (this time can be *undefined*)
5. the station *Id* of the intended recipient, *none* (if the recipient is unspecified), or *bcst* for a broadcast packet
6. the packet (traffic) type, i.e., the *TP* attribute (Sect. 6.2)
7. the total length of the packet (attribute *TLength*)

8. the packet's signature included if the simulator has been compiled with –g or –G.

An asterisk following the ending time of a transmission indicates an aborted packet.

The same items are displayed with the "screen" form of the exposure, except that item 8 is always included. Unless the program has been compiled with –g or –G, item 8 is filled with dashes.

Mode 6: transceiver status

Calling: *printTransceiver (const char *hdr = NULL);*

The exposure produces information about the status of activities on the station's transceivers. One row of thirteen items is printed for each transceiver. The first item is the transceiver identifier. It can be either the station-relative serial number reflecting the transceiver's creation order or the nickname, if one is defined for the transceiver (Sect. 3.3.2). The remaining items have the following meaning:

1. the receiver status: + (on) or − (off); note that if the receiver is off, the remaining indications (except for the possible "own activity") appear like no activity is perceived by the transceiver
2. the busy status of the receiver, as returned by *RFC_act* (Sect. 4.4.3): *Y* (busy), *N* (idle)
3. the own activity indicator (Sect. 8.1.1): *P*—preamble being transmitted (packet before stage *T*), *T*—packet past stage *T*, blank—no own activity (no packet being transmitted)
4. the total number of activities perceived by the transceiver at the time
5. the total number of simultaneous *BOP* events (Sect. 8.2.1)
6. the total number of simultaneous *BOT* events
7. the total number of simultaneous *EOT* events
8. the total number of simultaneous *BMP* events
9. the total number of simultaneous *EMP* events
10. the total number of "any" events (see *ANYEVENT*, Sect. 8.2.1)
11. the total number of preambles being perceived simultaneously
12. the total number of packets (past stage *T*) being perceived simultaneously (including non-receivable ones)
13. the total number of non-receivable (killed) packets based on the (current) assessment by *RFC_eot* (Sect. 8.1.2).

The "screen" version of the exposure displays the same information in the same order.

12.5.12   System Exposure

The *System* station has its own two exposure modes to print out information about the network layout. These modes have no "screen" versions.

Mode 0: full network description

Calling: *printTop (const char *hdr = NULL);*

Full information about the network configuration is printed out in a self-explanatory form.

Mode 1: abbreviated network description

Calling: *printATop (const char *hdr = NULL);*

Abbreviated information about the network configuration is printed. The output is self-explanatory.

Note that the contents of the mailboxes owned by the *System* station (Sect. 4.1.2) can be printed/displayed using the mode 2 *Station* exposure (for the *System* station).

12.5.13 Observer Exposure

Mode 0: information about all observers

Calling: *printAll (const char *hdr = NULL);*

Information about all observers is printed. This information includes the *inspect* lists and pending *timeouts*. A single row contains the following items:

1. the output name of the observer (Sect. 3.3.2)
2. the *Id* of the station specified in the inspect request or *ANY* (see Sect. 11.3)
3. the process type name or *ANY*
4. the process nickname or *ANY*
5. the process state or *ANY*
6. the observer state to be assumed when the inspect fires.

A pending *timeout* is printed in the form: *Timeout at xxxxxx*, where the text *Timeout at* is printed formally as item 2 and the following time value (in *ITUs*) as item 3.[7] The remaining items are blank.

For multiple entries corresponding to one observer, the observer's name (item 1) is printed only once (at the first entry) and it is blank in the subsequent entries referring to the same observer.

The same items are displayed by the "screen" version of the exposure.

---

[7]This is important for the "screen" exposure because the display program expects several rows with the same layout. Thus, the *timeout* row is handled in the same way as an *inspect* row and must fit the same slots for items.

Mode 1: inspect list

<u>Calling</u>: *printIns (const char *hdr = NULL);*

The *inspect* list of the observer is printed (possibly including the *timeout* request, see above). The layout of the list is exactly as for mode 0, except that the first item (the observer's output name) is absent.

The same information is produced by the "screen" version of the exposure.

Appendix A
Library Models of Wireless Channels

The standard include library of SMURPH (Sects. B.2 and B.5) offers, among other components,[1] three built-in channel models. Those models are structured around a common, generic, base model providing a collection of tools that may be of interest to a wide class of more specific models derived from the base. In terms of SMURPH components, the base model is an *RFChannel* extension (Sect. 4.4) accompanied by a few auxiliary type declarations. The three specific channel models are derived from the same base subtype. Extending existing models (Sect. 3.3.3) is a very natural way to build new wireless channel models, because the main aspects of functionality of those models are captured by the virtual assessment methods (Sect. 4.4.3). Thus, by redefining some virtual methods in a derived type of an *RFChannel* subtype, one can easily transform one channel model into another one, quite often with just a few simple tweaks.

A.1 The Base Model

The base model is described by type *RadioChannel*, which is a direct extension of *RFChannel*. A program willing to use the type should include the file *wchan.h* in the header section. Then, one of the *.cc* files in the program's directory should also include the file *wchan.cc* containing the code (the full definitions of the methods) of the base model, as explained in Sect. B.5.

Type *RadioChannel* does not define a directly usable RF channel model; it is still open ended, like the built-in type *RFChannel*. It brings in a few tools providing a framework for a class of "typical" wireless channel models, where the bit error rate, dependent on the signal-to-interference ratio (and possibly other factors), is the primary criterion affecting packet reception. The tools that come with the base model facilitate:

---

[1] Many of those other components were described in [1].

© Springer Nature Switzerland AG 2019
P. Gburzyński, *Modeling Communication Networks and Protocols*,
Lecture Notes in Networks and Systems 61,
https://doi.org/10.1007/978-3-030-15391-5

- Defining continuous functions via tables that specify the function values as a set of discrete points. A function defined this way accepts any argument between some minimum and maximum and, for the arguments that do not exactly match the entries in the table, interpolates the function value from the discrete points in the table. One application of this mechanism is the representation of the SIR to BER function (see Sect. 2.4.5 for an example) which, given a signal-to-interference ratio, returns the corresponding bit error rate. The most natural way to specify such a function in the input data to the channel model is by providing a representative set of sample points, e.g., as shown in Fig. 2.7.
- A standard interpretation of *Tags* associated with wireless signals (see Sect. 4.4.1) where a *Tag* encodes two attributes: the channel number and the bit rate. This way, it is possible to naturally accommodate wireless bands[2] offering multiple channels with definite crosstalk characteristics. Also, the model can account for different bit/symbol encoding rates and make the assessment methods (Sect. 4.4.3), aware of those differences in an easily comprehensible and parameterizable way.

The class of channels that can be naturally built on top of *RadioChannel* covers most of the narrow band solutions, e.g., non-spread-spectrum ISM [2], where (1) a baseband signal is modulated (keyed) into bits/symbols, (2) a number of selectable channels separated by relatively narrow margins can coexist within the same "channel" model, (3) multiple bit rates are possible which may result in different BER under the same SIR. Additionally, the model supports a few simple, boilerplate implementation features, e.g., transmit power setting based on multiple, indexed, discrete levels, and received signal strength indications (RSSI) calculated as an arbitrary, parameterizable function of the received signal strength (and presented as a possibly small range of discrete values). The role of those features is to make the "application-level" look and feel of the virtual implementation close to its real-life counterpart.[3]

A.1.1   The Mapper Classes

Table-based functions are implemented through two classes, named *IVMapper* and *DVMapper*, which are both produced from the same class template named *XVMapper* and parameterized by the type of the function argument (domain). The argument can be *unsigned short* (for *IVMapper*) and *double* (for *DVMapper*). The function value is *double* in both cases. The idea behind *IVMapper* is to describe functions over simple integer domains representing selections of discrete options.

---

[2]See Footnote 17 on p. 459.
[3]The models were originally built for VUEE [3], to facilitate virtual execution of ISM-band wireless sensor networks.

For example, an RF module may accept eight discrete transmit power settings represented by ordinals from 0 to 7. Those settings can be described by an *IVMapper* transforming an integer number between 0 and 7 into a *double* strength of the transmitted signal, e.g., in dBm. At first sight, this may sound unnecessarily complicated and make one wonder whether devoting a special, nontrivial type to describing something as simple as an array (to be sold as a "function over an integer domain") is truly such a terrific feature. The advantage of *IVMapper* over a simple array is primarily in the following two *actual* features:

- The integer domain need not cover a range starting from zero, and the table need not cover the entire range. Function values for the missing entries are automatically interpolated (if allowed to be).
- *IVMapper* (as well as *DVMapper*) provides for an efficient reverse mapping, i.e., range to domain.

For the latter, the function must be invertible, i.e., strictly monotonic, which is the case for most of the interesting applications of the mappers in the models of wireless channels. If the function is not monotonic, the mapper can still produce its interpolated values (from domain to range), but it won't be able to do the reverse interpolation.

The two mapper types are defined as regular C++ classes, so their instances are initialized with standard constructors. Here is the generic header from the template class:

XVMapper (unsigned short n, RType *x, double *y, Boolean lg = NO);

where *RType* is either *unsigned short* or *double* and refers to the domain type of the function. Thus, the two effective headers of the types *IVMapper* and *DVMapper* become:

IVMapper (unsigned short n, unsigned short *x, double *y, Boolean lg = NO);
DVMapper (unsigned short n, double *x, double *y, Boolean lg = NO);

The first argument, n, gives the sizes of the two arrays following it. Those arrays provide the pairs of discrete points $(x_i, y_i), 0 \leq i < n$, defining the function. They must be listed in such a way that the values of x_i are increasing. If the reverse mapping is expected to work, then the values of y_i have to be monotonic as well, i.e., they should either strictly increase or strictly decrease. The last argument indicates whether the function values should be interpreted as logarithmic, i.e., expressed in dB or dBm, or directly (we say linearly). In the former case (if the argument is *YES*), the function value is automatically converted to linear after it has been interpolated (see below). This accounts for the frequent interpretation of the function values described by the mappers as signal levels or ratios. In such cases, it is more natural to store and interpolate the function values in the logarithmic domain, and then convert them to the linear domain for their target usage in the model.[4]

---

[4]Recall that SMURPH represents signal levels internally in the linear domain, e.g., see Sect. 4.4.3.

The functionality of a mapper is described by its *public* methods listed below.

unsigned short size ();

The method returns the number of pairs in the table, equal to the value of the first argument of the constructor.

RType row (int i, double &v);

Given the index i of a pair of values, the method returns the corresponding entries from the table. The argument (domain entry) is returned by the method's value, the corresponding function value is returned via the second argument.

double setvalue (RType x, Boolean lin =YES);

The method calculates the function value for the given value of the argument x (from the function's domain). If the argument x is less than the minimum argument value, x_0, stored in the table (in the first entry), the function value associated with the minimum argument, $y = f(x_0) = y_0$, is selected. Similarly, if x is greater than the maximum argument value, x_{n-1}, stored in the table, the function value associated with the maximum argument, $y = f(x_{n-1}) = y_{n-1}$ is used. Otherwise, x falls between some argument values x_{i-1} and x_i with the corresponding function values y_{i-1} and y_i. The function value $f(x) = y$ is obtained by interpolating between the values of y_{i-1} and y_i.

The second argument of *setvalue* is only meaningful, if the mapper was created with *lg = YES*, i.e. the function values are to be interpreted on the logarithmic scale. Then, by default (*lin = YES*), the value of y obtained in the previous step is converted to linear (see Sect. 3.2.10), i.e., the value returned by *setvalue* is *dBToLin (y)*. If *lin* is *NO* (which requires an explicit mention of that argument in the invocation of *setvalue*), the logarithmic-scale value of y is returned directly, as it appears after the interpolation step. If the mapper was created with *lg = NO*, the second argument of *setvalue* is ignored.

Note that the function values are never extrapolated outside the domain explicit in the discrete set of points, as specified when the mapper was created. While the admissible arguments to *setvalue* can fall below the minimum and above the maximum argument values available in the table, the function is constant (equal to y_0) below x_0, and also constant (equal to y_{n-1}) above x_{n-1} (ignoring the optional linear conversion in the last step).

RType getvalue (double y, Boolean db = YES);

This method does the reverse mapping to that performed by *setvalue*. The mapping is carried out exactly as described above, with x replaced by y and vice versa. This will only work if the function values are strictly monotonic, so the mapping is well defined. Otherwise, the result of *getvalue* may be incorrect. If

RType is *unsigned short* (i.e., this is an *IVMapper*), the returned value (which is possibly interpolated), is rounded to an integer, which never exceeds the domain bounds in the defining set of discrete points. Regardless of the mapper type, the returned value is always between x_0 and x_{n-1}, inclusively.

The second argument of *getvalue* is only interpreted, if the mapper was created with *lg* = *YES*, i.e., the function values are assumed to be logarithmic. If this is the case, then, by default, the value of *y* passed to *getvalue* is converted to *linTodB (y)*, before the lookup and interpolation. If *db* is *NO*, that step is not performed, i.e., the specified value is assumed to be already on the logarithmic scale.

Boolean exact (RType x);

The method checks if the specified argument occurs as an exact entry in the table, i.e., it matches one of the discrete points defining the function, so there will be no need to interpolate (or truncate to the domain) when calculating the function value for the argument *x*. Its primary use is in an *IVMapper* to prevent unwanted interpolation, when the domain described by the table is hollow, but the holes should not be interpolated out. By invoking *exact* before *setvalue*, and rejecting the request when *exact* returns *NO*, we reduce the interpretation of *IVMapper* to a simple set of options exactly matching the entries in the table.

Boolean inrange (RType x);

The method checks if $x_0 \leq x \leq x_{n-1}$, i.e., the argument is within the range of the domain as specified in the table. If *inrange* returns *YES*, it means that the argument does not fall beyond the range of the defining set of discrete points.

RType lower ();
RType upper ();

The methods return the lower and upper bounds of the domain, i.e., x_0 and x_{n-1}, respectively.

A.1.2 Channel Separation

Another external type that comes as part of the generic channel model *RadioChannel*, is class *MXChannels* describing the set of channels[5] and their separation. Here is the header of its constructor:

*MXChannels (unsigned short nc, int ns, double *sep);*

where *nc* is the number of channels, *ns* (which must be strictly less than *nc*) is the number of entries in the separation table, and *sep* is the separation table itself. The

---

[5]See footnote 17 on page 111.

simple idea is that the available channels are assigned numbers from 0 up to $nc - 1$, and *sep [i]* describes the separation between all pairs of channels whose numbers differ by i. For example, the separation between channels 5 and 7 is the same as the separation between channels 19 and 17 and equal to *sep [2]*.

The separation is expressed in dB (on the logarithmic scale), such that a higher (positive) value means better separation. Note that the size of the separation table need not cover the maximum distance between channels ($nc - 1$). For channels whose numbers differ by more than ns, the separation is assumed to be perfect, i.e., equal to infinity.

Class *MXChannels* offers no *public* attributes other than its constructor plus the following methods:

double ifactor (unsigned short c1, unsigned short c2);

The method returns the inverse linear separation between the channels numbered $c1$ and $c2$. If s is the separation value specified in *sep* when the class was constructed, then the inverse linear separation corresponding to s is equal to *dBToLin (−s)*. In simple words, the value returned by *ifactor* is the linear interference factor, i.e., the multiplier to be applied to the signal from an interfering, separated channel to produce its contribution to the combined interference at the perceiving channel.

unsigned short max ();

This method returns the maximum channel number, i.e., $nc - 1$, where nc is the first argument of the constructor.

void print ();

The method writes to the output file (Sect. 3.2.2) the user contents of the class, i.e., the number of channels and the separation table, as specified in the constructor.

A.1.3   The Functions of the Base Model

While the base model, as such, does not presume any units of power (signal levels are relative and the interpretation of units is up to the user), it will make things easier to assume some power units, if only for the sake of the discussion. This way, for example, we will be able to make a clear distinction between the logarithmic and linear scales. The most natural (logarithmic) unit of RF signal power, especially for the class of applications targeted by the model, is dBm, i.e., the decibel factor relative to 1 mW (milliwatt). We shall assume this unit in all cases where we talk about definite signal levels, as opposed to ratios.

Type *RadioChannel* is intended as a base type for defining other *RFChannel* types. The *setup* method of *RadioChannel* (meant to be invoked from the *setup* methods of its subtypes) has the following header:

*void setup (Long nt, double bn, const sir_to_ber_t *st, int nst, int bpb,*
  *int frm, IVMapper **ivcc, MXChannels *mxc);*

The first argument is the number of transceivers to be interfaced to the channel. Note that this number has to be passed to the *setup* method of *RFChannel* (which must be invoked from the *setup* method of *RadioChannel*, see Sect. 4.4.2). The remaining arguments have the following meaning:

bn	The background noise level. According to what we said in the first paragraph of this section, we assume that *bn* is expressed in dBm
st	This is an array of entries of type *sir_to_ber_t* representing the SIR to BER function as an interpolated table of discrete points (in the fashion of *DVMapper*, see Sect. A.1.1). The role of the *sir_to_ber_t* structure is to simplify the specification of the function to a single array of structures (instead of two separate arrays of numbers). The structure has two *double* attributes: *sir* and *ber*, which should be set, respectively, to the logarithmic-scale signal-to-interference ratio in dB and to the corresponding bit-error rate expressed as the probability of a single-bit error (as in Sect. 2.4.5)
nst	This is the length of the *st* array, i.e., the number of entries in the SIR to BER table
bpb	Bits per byte, i.e., the number of physical bits per byte of transmitted data. The argument will determine how the logical packet length, expressed in information bits, is transformed into the physical duration of the packet. Note that transmission rates are expressed in physical bits; also, bit-error rates apply to physical bits (Sect. 8.1.1)
frm	The number of extra physical bits needed to encapsulate a packet
ivcc	This is an array of pointers to the standard set of mappers (Sect. A.1.1) used by the model
mxc	This is a pointer to the *MXChannels* object (Sect. A.1.2) describing the set of channels available in the model and their separation

The generic model implements very little, and most of the interpretation of its construction arguments is left to the final models. Thus, for example, the interpretation of the background noise level is up to the assessment methods calculating the amount of interference experienced by packets and transforming that interference into bit errors. The *nb* argument of the constructor is just stored in the *public* attribute *BNoise* of *RadioChannel* from where it can be accessed by the assessment methods. Similarly, the arguments *bpb* and *frm* are copied by the constructor to attributes *BitsPerByte* and *PacketFrameLength*, respectively.

The only assessment method defined by the generic model (note that it can still be overridden by the final model) is *RFC_xmt* (Sect. 4.4.3) calculating the amount of time needed to transmit a packet. That time amounts to:

*rate * nb * BitsPerByte + PacketFrameLength*

where *rate* is the effective transmission rate, and *nb* is the number of bytes in the packet (equal to *TLength/8*, see Sect. 6.2). Note that the packet transmission time does not account for the preamble (Sect. 8.1.1) whose transmission in logically separated from the packet transmission.

RadioChannel defines this method:

double ber (double sir);

for calculating the bit error rate, given a signal-to-interference ratio. The method interpolates the table (passed in the arguments *st* and *nst* of the constructor), assuming that its argument (the signal-to-interference ratio) is linear. It is to be invoked from the assessment methods of the final model, where the ratios and signal levels are typically linear.

The generic model assumes that the array of mappers (argument *ivcc*) provides four standard mappers named *Rates, RBoost, RSSI,* and *PS*. These are the names of the public attributes of *RadioChannel* (of type *IVMapper**) into which the mapper pointers are copied. This list also specifies the order in which the mapper pointers should be stored in the *ivcc* array, i.e., *Rates* is the entry number 0, *RBoost* is 1, *RSSI* is 2, and *PS* is 3. Note that all those mappers are *IVMappers*, i.e., they map from *unsigned short* to *double*. They are all optional: when a mapper is absent, its entry in *ivcc* should be *NULL*. The model does not attempt to interpret the mappers leaving this up to the final models, or to the protocol program, i.e., the processes implementing packet transmission and reception. In this context, calling those mappers "standard" relates more to their intentions than to the actual roles. The base model was arrived at as an intersection of a class of useful, final channel models, then it was isolated from those models to simplify their organization. This way, the intentions inherent in the standard (albeit unimplemented) features of the base model come from the final models that conceptually preceded the base model.

The reason why all mappers are passed in a single array is to simplify the list of arguments of the constructor. Note that a final type may require additional mappers; in such a case, the *ivcc* array can be longer, and the final type can use the entries starting at index 4, with the entries numbered 0-3 reserved for the mappers expected by the base type.

The following two methods of *RadioChannel*:

unsigned short tagToCh (IPointer tag);
unsigned short tagToRI (IPointer tag);

convert a *Tag* value to the channel number (*tagToCh*) and the so-called rate index (*tagToRI*). Recall that the role of *Tags* (Sect. 8.1.2) is to encode generic *Transceiver* attributes, other than the transmission power and reception gain, that affect (or may affect) packet reception and interference. The base model assumes that a *Tag* encodes two such attributes: the channel number and the bit rate, the latter as a rate index (a small integer number) selecting one of the rate options defined in the *Rates* mapper (see below) in the final model. The two entities are encoded in such a way that the channel number occupies the lowest 16 bits of the integer *Tag* value (see Sect. 3.1.1 for the definition of type *IPointer*), and the rate index is stored on the next (more significant) sequence of 16 bits. The above methods extract the respective fragments from the argument *Tag*.

The last argument of the constructor, *mxc*, passes to the model an instance of the *MXChannels* class describing the channels, i.e., their number and separation (Sect. A.1.2). The argument is copied to the *public* attribute *Channels* of *RadioChannel* (of type *MXChannels**). To obtain the (linear) separation factor for a given pair of channel numbers *c1* and *c2* the assessment method can use this function:

Channels->ifactor (c1, c2)

The value returned by *ifactor* is the multiplier for a signal on the interfering (separated) channel.

The role of the *Rates* mapper is to convert between the rate index and the actual rate. The model assumes no interpretation of the value returned by the mapper (and implements no methods for setting the transceiver rate according to the rate index). The final model, or the protocol program, should provide the pertinent function which will also act as the sole interpreter of the mapper values. For example, the rate values can be expressed in bits per *ETU* (second). In such a case, assuming that *Ether* points to the channel model, the function setting the transmission rate for a transceiver may look like this:

```
void setxrate (Transceiver xcv, unsigned short rate_index) {
    xcv->setXRate((RATE) round (Etu / Ether->Rates->setvalue (rate_index)));
}
```

Recall that *Etu* stores the number of *ITUs* in one *ETU* (Sect. 3.1.2), and the argument of *setXRate* is equal to the number of *ITUs* required to insert a single bit into the medium (Sect. 4.5.2).

The *RBoost* mapper describes a function that maps rate indexes into so-called rate boosts, i.e., rate-dependent factors used to multiply signal levels at the destination. In real life, different rates mean different encoding schemes, where, generally, a lower encoding rate gives better reception opportunities (a lower bit error rate at the same SIR). In a model, this can be achieved by applying different SIR to BER functions to signals at different rates, but the simpler (easier to parameterize) solution assumed by the base model is to give them different "boosts," i.e., multipliers that make them more or less pronounced against the interference (including the background noise). Again, the base model defines no assessment functions, except for *RFC_xmt*, so it does not incorporate the rate boosts directly; however, by including *RBoost* as a standard mapper, it suggests a way to handle different encoding rates by the final models.

The *RSSI* mapper incorporates another standard feature of many radio devices which make a measure of the received signal strength available to the application. That measure is typically some (more or less crude) function of the actual received signal strength (e.g., in dBm) whose values are in some restricted integer range (e.g., between 1 and 127).[6] Thus the mapper describes the function from the discrete RSSI indications to the corresponding signal values on the logarithmic scale, e.g., in dBm. As the function is required to be strictly monotonic, it doesn't really matter which is the domain and which is the range (the mapper works equally easily both ways, Sect. A.1.1).

The last standard mapper, *PS*, defines the transmit power settings for the model. Quite often, a real-life radio device offers a restricted number of discrete transmit power levels indexed by small numbers, like 0, 1, and so on. The mapper describes a function from those settings into the actual power levels expressed in the logarithmic domain.

As the base channel model is not directly usable, the *setup* method of *RadioChannel* is never invoked directly from the protocol program, but from the *setup* method of a final model. The method writes to the output file (Sect. 3.2.2) all the elements contributing to the definition of the base part of the model, including the mapper functions and the channel separation table.

A.2 The Shadowing Model

The shadowing model consists of two files: *wchansh.h* and *wchansh.cc*. These files must be included by the program that wants to use the model, in the way described in Sect. A.1. The model defines an rfchannel type named *RFShadow* derived from *RadioChannel* (Sect. A.1.3).

The model [5] is based on a blanket signal propagation formula where the sole environmental parameter affecting signal attenuation is the distance between the transmitter and the receiver. A randomized (lognormal) component is added, to model the impact of reflections (and "unpredictable" factors) in a generic, easily parameterizable way. The formula used in our version of the model is:

$$\left[\frac{P_r(d)}{P_r(d_0)}\right]_{dB} = -10\beta \log\left(\frac{d}{d_0}\right) + X(\sigma_{dB})$$

where $P_r(d)$ is the received signal strength at distance d, d_0 is a reference distance, β is the loss exponent, and X is a Gaussian random variable with zero mean and standard deviation σ_{dB}. As the value produced by the right-hand side is in decibels, the attenuation (or loss) is proportional to the β power of distance. The randomized component is added to the logarithm of the actual (linear) attenuation, so it effectively becomes a multiplier.

---

[6]They are often confined to a single byte.

A.2.1 Model Parameters

For our model, the above attenuation formula must be transformed into *RFC_att*, i.e., the assessment method to generate signal attenuation (Sects. 4.4.3 and 8.1.2). Thus, we turn it into:

$$\left[\frac{P_r(d)}{P_x}\right]_{\text{dB}} = -10\beta \log\left(\frac{d}{d_0}\right) + X(\sigma_{\text{dB}}) - L(d_0)$$

where P_x is the transmission power at the sender, and $L(d_0)$ is the calibrated loss at the reference distance d_0. The reference distance d_0 is expected to be small, because the above formula is only meaningful when $d \geq d_0$. The model assumes that a transmission at a distance shorter than d_0 incurs the same attenuation as a transmission at distance d_0.

Note that the value produced by *RFC_att* (and by the attenuation formula implemented by the method) is applied as a multiplier to a signal traveling between two transceivers in the network. Thus, strictly speaking, the value represents "amplification" (which is usually much less than 1) rather than "attenuation," although we shall stick to the intuitively clear (albeit formally incorrect) term. The setup method of *RFShadow* has the following header:

*void setup (Long nt, const sir_to_ber_t *st, int nst, double rd, double lo,*
  double be, double sg,double bn, double bu, double co,
  *Long mp, int bpb, int frm, IVMapper**ivcc,*
  *MXChannels*mxc, double (*gain) (Transceiver*, Transceiver*));*

The arguments *nt*, *st*, *nst*, *bn*, *bpb*, *frm*, *ivcc*, and *mxc* are passed directly to the *setup* method of *RadioChannel* (they have been discussed in Sect. A.1.3). The remaining arguments have the following meaning:

rd	The reference distance (in *DUs*, Sect. 3.1.2), corresponding to d_0 in the attenuation formula
lo	The loss at the reference distance, corresponding to $L(d_0)$, expressed in dB
be	The loss exponent, i.e., the value of β
sg	The value of σ_{dB}, i.e., the standard deviation of the lognormalGaussian random variable added to the attenuation in the logarithmic domain (or multiplying the attenuation in the linear domain)
bu	The "busy" threshold. This is the minimum level of total received signal strength at which a transceiver will be found "busy" (Sect. 8.2)
co	The cut-off threshold (Sect. 4.4.3), i.e., the signal level (specified in the logarithmic domain) below which the simulator can assume that there is no signal at all. The threshold is used to define neighborhoods beyond which signals are never propagated, to reduce the complexity (execution time) of the model
mp	This parameter defines the minimum received preamble length and is used by *RFC_bot* (Sect. 4.4.3) in assessing the beginning of packet (as explained in Sect. A.2.2)
gain	This is an optional function producing an extra linear (gain) factor applied dynamically to the received signal strength between a pair of transceivers. Its primary role is to

(continued)

(continued)

	implement models of directional antennas. If this argument is *NULL*, the *gain* function is not defined (which is equivalent to a function returning the constant 1.0)

The *setup* method of *RFShadow* writes to the output file (Sect. 3.2.2) the elements amounting to its part of the final model. This information follows the data written by the base model (Sect. A.1.3).

A.2.2 Event Assessment

The model defines its versions of all the assessment methods (Sect. 4.4.3) except for *RFC_xmt* defined in the base part (Sect. A.1.3). Signal attenuation, carried out by *RFC_att*, is described by the formula from Sect. A.2.1 with the following provisions:

1. If the channel number setting of the receiving transceiver is different from the channel number of the sender, the received signal level is multiplied by the channel separation factor (as returned by *ifactor*, Sect. A.1.3).
2. If the propagation distance is less that the reference distance (argument *rd* of the *setup* method, Sect. A.2.1), the propagation distance is set to the reference distance.
3. If the *gain* function is available, the signal level is multiplied by *gain (snd, rcv)*, where *snd* is the sending transceiver, and *rcv* is the receiving transceiver.

Signal addition at the receiver, carried out by *RFC_add*, is done in the most straightforward way. Note that the signals have been attenuated prior to the addition. If the transceiver happens to be transmitting at the time, its transmit power setting is added to the total received signal, with the channel separation factor (*ifactor*) accounting for the possible difference between the transmit and receive channels.

RFC_act and *RFC_cut* operate according to the simple thresholds determined by the *setup* arguments *bu* and *co*, respectively.

RFC_erb, determining the number of error bits in the run of *n* bits under steady interference operates as follows:

1. If the receiver *Tag* and the sender *Tag* are different, this always means "no reception opportunities at all" and the method returns *n* (all bits in error). While formally that number would be *n*/2, the case should not be treated as an attempt to receive *n* bits at random, but rather a misconfiguration of the receiver (tuned to the wrong channel/bit rate), which makes it formally impossible to recognize any packet (bits) at all.
2. Otherwise, the signal level at which the bits arrive is multiplied by the receiver gain (Sect. 4.5.2) and divided by the interference level (including the background noise). The result is applied as the argument to the *ber* method of *RadioChannel* (Sect. A.1.3) to yield the probability *p* of a single-bit error. Then,

the number of bits received in error is calculated by calling *lRndBinomial (p, n)*, i.e., from the binomial distribution (Sect. 3.2.1).

RFC_erd is set to calculate, given a steady level of interference,[7] the expected number of bits until the first bit error. If the receiver *Tag* and the sender *Tag* are different, the method returns 0 (no reception is possible at all). Then, the probability *p* of a single-bit error is calculated as for *RFC_erb*. The expected number of bits until the first error bit is obtained by calling *dRndPoisson (1.0 / p)*. The idea is essentially as presented in Sect. 8.2.4, with the same intended reception model.

The beginning of a potentially receivable packet, assessed by *RFC_bot*, is recognized when the packet preamble has at least *mp* consecutive error-free bits immediately preceding the packet's beginning. The end of packet assessment, done by *RFC_eot*, is trivial. The method simply returns YES for any packet for which the beginning of packet assessment has succeeded. The reception model assumes that the reception will fail on the first bit error (as determined by *RFC_erd*). This corresponds to the assumption of the error run length of 1 (see Sect. 8.2.4) and is compatible with the following idea of a reception process (we assume that *Xcv* points to a transceiver):

```
RCVProcess::perform {
        state WAIT_FOR_BOT:
                Xcv->wait (BOT, RCV_START);
        state RCV_START:
                skipto WAIT_FOR_EOT:
        state RCV_WAIT_FOR_EOT:
                Xcv->wait (EOT, RCV_GOT_IT);
                Xcv->wait (BERROR, WAIT_FOR_BOT);
        state RCV_GOT_IT:
                Client->receive (ThePacket, Xcv);
                sameas WAIT_FOR_BOT;
}
```

---

[7]Recall that while *RFC_erb* and *RFC_erd* carry out their calculations assuming a steady level of interference, the simulator internally uses those results to estimate the bit error rate under dynamic interference (Sect. 8.2.4).

Having responded to the beginning of a packet, the receiving process implicitly marks the packet as "followed" (see Sect. 8.2.1). Then, in state *RCV_WAIT_FOR_EOT*, the process awaits two events: an *EOT*, which will be triggered when the end of the followed packet arrives at the transceiver, and the first single-bit error, which will abort the reception. Thus, the *EOT* event is assessed trivially, because the true, positive end of packet assessment is the absence of a *BERROR* event during the packet's reception.

Note that the assessment methods can be easily replaced, individually and *en bloc*, by defining rfchannels derived from *RFShadow*. For example, if the *BERROR* event triggered by a single-bit error is too simplistic, *RFC_erd* can be modified, e.g., along the lines of Sect. 8.2.4.

A.3   The Sampled Model

The idea of the sampled model (files *wchansd.h* and *wchansd.cc*) is to replace the blanket attenuation formula used in the shadowing model with samples collected from the environment. The modeled network is assumed to match some real-life deployment, and the model attempts to get close to the operation of its real-life counterpart in terms of the observed signal strength. The model defines an rfchannel type named *RFSampled* derived from *RadioChannel* (Sect. A.1.3).

The attenuation function, as calculated by *RFC_att*, is described by two components:

- the attenuation table specifying a blanket attenuation function as a function of distance via a *DVMapper* (Sect. A.1.1)
- the file containing samples of RSSI readings between pairs of stations at specific locations (represented by station coordinates).

As the coverage by real-life samples can seldom be exhaustive, especially when the network model includes mobility, the blanket attenuation function is provided as a fallback to easily derive attenuation from distance if no other way is available. But in all those cases when the coordinates of the sender-receiver pair closely match some sample(s), those samples will take precedence.

A.3.1   Model Parameters

The *setup* method of *RFSampled* is declared with the following header:

void setup (Long nt, const sir_to_ber_t *st, int nst, double kn, double sg, double bn, double bu,

double co, Long mp, int bpb, int frm,IVMapper**ivcc, const char *sfname, Boolean sym,
MXChannels*mxc, double (*gain) (Transceiver*, Transceiver*));

The arguments *nt*, *st*, *nst*, *bn*, *bpb*, *frm*, *ivcc*, and *mxc* are passed directly to the *setup* method of *RadioChannel* (they have been described in Sect. A.1.3). Of the remaining arguments, *bu*, *co*, *mp*, and *gain* have the same interpretation as the corresponding arguments of the *setup* method of *RFShadow* (Sect. A.2.1). Here is a brief description of the arguments specific to *RFSampled*:

kn	This is the so-called sigma threshold affecting the randomized component for an attenuationvalue derived from *RSSI* samples (as explained below)
sg	This is the default value of the standard deviation of the lognormalGaussian random variable added to the attenuation obtained from the blanket distance-based function, akin to σ_{dB} in the shadowing model (Sect. A.2)
sfname	This argument provides the name of the file containing RSSI samples. The format of that file is discussed below
sym	The Boolean value (*YES* or *NO*) indicating whether the samples should be treated symmetrically with respect to the sender and the receiver (as explained below)

The file with RSSI samples is optional, so *sfname* can be *NULL*. If this is the case, then the blanket distance-based attenuation function is the only mechanism to calculate attenuation. It can be easily argued that the function still offers more flexibility than the attenuation formula from the shadowing channel model, because it can be crafted to match the real-life environment, at least statistically, when only the distance between the sender and the receiver is taken into account.

The model uses two additional mappers passed to it via pointers stored in *ivcc* as entries number 4 and 5 (above the four standard mappers assumed by the base model, Sect. A.2.1). The mappers are *DVMappers* and can be accessed as attributes *ATTB* and *SIGMA* (of *RFSampled*), respectively. *ATTB* defines the blanket, distance-based attenuation function with the domain representing distance in *DUs*, and the range describing the attenuation in dB. *SIGMA*, with the same domain as *ATTB*, produces the values of σ for randomizing the lognormal,Gaussian component of the attenuation. The function defined by *SIGMA* need not be strictly monotonic, because the reverse mapping, from σ to distance, is never used.

The *SIGMA* mapper is optional. If it is absent, i.e., its entry in *ivcc* is *NULL*, the model assumes that the same value of σ, determined by the value of *sg*, is to be applied to all cases of blanket attenuation, regardless of the distance.

A.3.2   The RSSI Samples

The file with RSSI samples consists of lines of texts and is thus perusable by the user. It can even be created manually, although it makes more sense to generate it by scripts, e.g., from logs automatically collected in the real-life network.

The only elements of a line that are interpreted are numbers. An empty line, or a line whose first non-blank character is #, is ignored. The expected content of the lines depends on whether the model of node deployment is 2d or 3d (see Sects. 4.5.1 and B.5). A line begins with the coordinates of two points representing the locations of the sender and the recipient. For a 2d model, this pair of points is represented by four numbers, i.e., x_0, y_0, x_1, y_1. In a 3d model, the coordinates amount to six numbers, i.e., $x_0, y_0, z_0, x_1, y_1, z_1$. The remaining portion of the line should contain either one or two numbers. Here is a sample line from a 2d model:

$$12.45 \quad 76.1 \quad 200.0 \quad 44.55 \quad 67$$

The fifth number (the first number following the coordinate pair) is an integer RSSI reading, and the optional sixth number indicates the transmission power setting at which the sample was taken. The above sample was collected for a packet transmitted from a node located at coordinates $x = 12.45$, $y = 76.1$ (in *DUs*, Sect. 3.1.2) to a node at $x = 200.0$, $y = 44.55$, and the RSSI reading of that sample at the recipient was 67. Note that the RSSI readings and power settings are interpreted through the respective mappers (*RSSI* and *PS*, Sect. A.1.3), which map them to signal level representations. If there is no indication of transmission power, the maximum power setting from the *PS* mapper is assumed by default.

The interpretation of the coordinates in the file is the only place in the channel model where the dimensionality of the network deployment comes directly into play.[8] Otherwise, the model deals with (metric) distance, which notion is agnostic to the number of coordinates needed to describe a point. From now on, for simplicity and clarity, we are only considering the 2d case. Whenever we talk about a point (mentioning its two coordinates), the 3d case can be immediately accounted for by throwing in the third (z) coordinate in the obvious way.

The numbers can be mixed with non-numerical text, which is ignored when the sample is read. For example, the above line could be written as:

$$XA = 12.45 \quad YA = 76.1 \quad XB = 200.0 \quad YB = 44.55 \quad RSSI = 67$$

or:

$$< 12.45, 76.1 > \ === \ > \ < 200.0, 44.55 > \ : 67$$

and it would be interpreted in the same way.

Every sample of the above form is an attenuation case of a signal transmitted from some point S (the first pair of coordinates) to another point R (the second pair). To transform the sample into the attenuation, one needs to consult the *RSSI* mapper to convert the RSSI reading into the signal level at the recipient and then divide that

---

[8]Note that it doesn't come into play in the shadowing model, because that model only cares about distances.

signal level by the transmission power of the sample (derived from the power index through the *PS* mapper), or subtract the two levels, if calculating in dB.

A sample read from the file is preprocessed by the channel model. The two-point coordinates are converted from *DUs* to *ITUs* (Sect. 3.1.2), which discretizes them into the grid of integer multiples of the *ITU*. Multiple samples whose sender-recipient points fall into the same grid squares (or cubes) are merged into one. Note that when collecting RSSI readings from a real network, it is quite natural to take multiple readings for the same sender-recipient pair. The combined single RSSI of such a merged sample is the simple average of all the contributing RSSI readings. The sampled standard deviation of those readings is also calculated and stored within the merged sample as σ to randomize the attenuation value produced from the sample. This only happens when the number of input samples contributing to the preprocessed samples is not less than the threshold *kn*. Otherwise, the standard deviation of the lognormalGaussian component of the attenuation is computed differently (see Sect. A.3.3).

The normalized attenuation value associated with a sample is determined by dividing the received signal strength by the transmit power. In the logarithmic domain (in which the samples are in fact stored in the model's database), this means subtracting from the signal level of the sample (in dBm) the signal level corresponding to the sample's transmission power attribute (also in dBm). This produces the normalized attenuation in dB, and makes it independent of the transmit power. Also, multiple samples referring to the same sender-recipient pair obtained under different transmit power levels can be merged into one. In any case, one pair of sender-recipient points (after discretizing the point coordinates to the ITU grid) results in a single sample, with a single, averaged RSSI reading. If the number of contributing samples is *kn* or more, the sampled standard deviation is also stored to be used for randomizing the attenuation cases generated from the sample.

If the *sym* argument of the *setup* method is *YES*, the channel is assumed to behave symmetrically, meaning that the ordering of the two end-points of a sample doesn't matter. In other words, the sample is going to fit a given sender-recipient scenario, as long as the two nodes occupy the same two points without telling which one is which. Otherwise, if *sym* is *NO*, the channel is considered non-symmetric, and the ordering of the two points matters. Note that in the former case, these two lines in the data file:

$$30.0 \quad 60.0 \quad 40.0 \quad 70.0 \quad 190$$
$$40.0 \quad 70.0 \quad 30.0 \quad 60.0 \quad 192$$

will be merged into a single sample, while in the latter case they will end up separate.

A.3.3 Computing the Attenuation

A preprocessed sample (being a possible merge of several samples) is stored as the following record:

$$<X_a, Y_a, X_b, Y_b, A, \sigma>$$

where X_a, Y_a, X_b, Y_b are the coordinates of the two points, A is the attenuation factor, and σ is the standard deviation of the lognormal component associated with the sample. If the preprocessed sample results from kn or more raw samples, then σ is the sampled standard deviation derived from the RSSI values of the contributors. Otherwise, σ is set to the sg argument of the channel's *setup* method (Sect. A.3.1).

Below we outline the procedure for producing signal attenuation for a transmission from an arbitrary point S to an arbitrary point R, as carried out by *RFC_att*. Note that we mean the basic attenuation factor. Other factors (separation factors and boosts) may be applied to the received signal level, on top of the basic attenuation factor, in the same way as for the shadowing model (see Sect. A.2.2).

If the two points S and R match some sample exactly (modulo the *ITU* grid), the answer is straightforward: the attenuation value is taken directly from that sample and randomized according to the corresponding value of σ. If there is no perfect match, the model averages the attenuation values from all samples weighing them by how closely the sample's endpoints fall to S and R.

Let N be the total number of samples in the model's database. Given $S = (X_s, Y_s)$ and $R = (X_r, Y_r)$, the distance of sample $i, 0 \leq i < N$, from the pair of points S and R is defined as:

$$d_i = \sqrt{\left(X_a^i - X_s\right)^2 + \left(Y_a^i - Y_s\right)^2} + \sqrt{\left(X_b^i - X_d\right)^2 + \left(Y_b^i - Y_d\right)^2}$$

i.e., it is the sum of the Euclidean distances between the starting and ending points (X_a^i is the x coordinate of the starting point of sample i). If the channel is symmetric (the *sym* argument of the *setup* method was *YES*), the distance is the minimum of d_i and:

$$\bar{d}_i = \sqrt{\left(X_b^i - X_s\right)^2 + \left(Y_b^i - Y_s\right)^2} + \sqrt{\left(X_a^i - X_d\right)^2 + \left(Y_a^i - Y_d\right)^2}$$

i.e., the same distance calculated for the two points of the sample switched. For every sample i, the model calculates its normalized weight:

$$W_i = \frac{1}{d_i \times \sum_{j=0}^{N-1} \frac{1}{d_j}}$$

reflecting a measure of the sample's proximity to the (S, R) pair. Note that:

$$\sum_{i=0}^{N-1} W_i = 1$$

Then, the following averages are calculated:

$$A_{avg} = \sum_i W_i \times A_i, \quad \sigma_{avg} = \sum_i W_i \times \sigma_i, \quad D_{avg} = \sum_i W_i \times D_i$$

where D_i is the straightforward, Euclidean distance between the two endpoints of sample i. Finally, the basic attenuation factor for the pair (S, R) is calculated as:

$$A_{SR} = A_{avg} + ATTB(D_{SR}) - ATTB(D_{avg}) + X(\sigma_{avg})$$

where D_{SR} is the Euclidean distance between S and R. Note that all attenuation values in the above formula are expressed in the logarithmic domain.

The last step scales the attenuation to the actual distance between S and R, factoring in the blanket attenuation for that distance in proportion to how it differs from the weighted average distance over the samples. For example, when the match to samples is poor, and, say, D_{SR} is significantly less than the weighted average of all D_i, the impact of the attenuation obtained from the samples should be reduced in proportion to how the two distances differ. Strictly speaking, it should be reduced in proportion to how the blanket attenuation values for the two distances differ, which is what the above formula attempts to capture. Note that the assessment method must also be able to calculate attenuation for non-receivable cases, where D_{SR} may be significantly longer than any D_i (because the collected RSSI samples may only refer to received packets). In such a case, according to the above formula, $ATTB(D_{SR})$ will (rightfully) tend to dominate.

The remaining assessment methods of *RFSampled* operate identically to their counterparts from the shadowing model.

A.4 The Neutrino Model

The neutrino model is contained in files *wchannt.h* and *wchannt.cc*. This is an unrealistic model intended for illustration and for tests. It assumes that packets never interfere, and the bit error rate is always zero. The limitations for packet reception are purely formal (unrelated to propagation, attenuation, etc.) and boil down to:

- The optional range (distance) limit: it is possible to define a sharp "propagation distance" of the model beyond which the reception is zero, and below which the reception is 100%.

- The channel number: there is no cross-channel interference, with channel separation being always perfect.
- The transmission rate: the rate of the receiver must agree with the packet's transmission rate, otherwise there is no reception (and no interference).

Note that transmission rates and distances (translating into nonzero transmission/reception time and nonzero propagation times) are honored by the neutrino model. Thus, the timing of actions involved in exchanging a single undisturbed and received packet across a given distance is going to look the same in the neutrino channel as in, say, the shadowing model.

A.4.1   Model Parameters

The model defines an rfchannel type named *RFNeutrino* derived from *RadioChannel* (Sect. A.1). The setup method of the channel is declared with this header:

*void setup (Long nt, double ran, int bpb, int frm, IVMapper **ivcc,*
    *MXChannels *mxc);*

The meaning of all arguments except *ran* is the same as for the shadowing model. Some arguments applicable to other (realistic) channel models (like *bn*, *bu*, *co*) do not apply to the neutrino model. The *ran* argument plays a role similar to *co* in that it determines the sharp range of the channel, expressed in *DUs*, beyond which reception is cut off. The argument is saved in the *public* attribute *Range* of *RFNeutrino*.

The model retains the concepts of channels and rates from the base model, but ignores rate boosts and power settings. Thus, the mappers like *RBoost*, *PS*, and *RSSI* are (basically) useless (and completely ignored by the model). We say "basically," because the protocol program can set the transmit power and monitor the received signal strength, but these measures have no impact on packet reception. The attenuation is always 0 dB (or 1, in the linear domain), unless the distance between the sender and the receiver exceeds the range threshold, in which case the attenuation is infinite (the linear factor is 0).

A.4.2   Event Assessment

The model inherits the *RFC_xmt* method from the base model; thus, the timing of packet transmission (and reception) is standard and the same as for the other models. The remaining assessment methods are mostly trivial, specifically:

RFC_add	Performs standard, straightforward signal addition. This affects the formal signal level, but is irrelevant from the viewpoint of packet reception
RFC_att	Returns 0 if the transmitter's *Tag* and the receiver's *Tag* (Sect. 8.1.2) differ, which means that the channel numbers and/or rates differ, or when the distance between the two parties exceeds *ran*. Otherwise, the method returns 1
RFC_act	Returns the constant *NO*. The primary role of the method is to check whether the medium is busy (Sect. 4.4.3), e.g., before a transmission, as to avoid interfering with another packet. There is no interference in the neutrino model, so the medium is never busy
RFC_cut	Returns the value of *ran* from the *setup* method (attribute *Range*). The cutoff distance is equal to the maximum transmission range
RFC_erb	Returns 0 if the transmitter's *Tag* is equal to the receiver's *Tag*. Otherwise (the *Tags* are different), the method returns the total number of bits in the sequence (no bit received correctly)
RFC_erd	Returns infinity (*MAX_Long*) if the transmitter's *Tag* is equal to the receiver's *Tag* (no bit error will ever happen). Otherwise (the *Tags* are different), the method returns 0 (a bit error occurs immediately)
RFC_bot	Returns *YES* if the transmitter's *Tag* is equal to the receiver's *Tag* (every packet is unconditionally receivable), otherwise returns *NO*
RFC_eot	The same as *RFC_bot*

Appendix B
SMURPH Under Linux and Windows

This section contains technical information on using SMURPH in the UNIX (Linux) environment and under Windows (under Cygwinor Windows *bash*). In principle, the package can be installed on any system running a POSIX-compatible UNIX system, equipped with the GNU C++ compiler. The package used to run on many machines and UNIX systems with several versions of C/C++ compilers. These days, the popular C++ environments have become considerably more unified than they were in the early days of the package. Nonetheless, for the last fifteen (or so) years, Linux (Ubuntu), Windows + Cygwin, and Windows 10 + *bash*, running the GNU C/C++ set of development tools, have been the only systems where SMURPH was run, tested, and maintained. These are the environments recommended for running the package.

The options are in fact quite flexible, owing to the availability of virtual machines (e.g., VM VirtualBox from Oracle [6, 7], or VMWare [8, 9]) which make it easy to have full-fledged, efficient, and complete OS environments independent of the host system. From this point of view, the best solution for running SMURPH is probably a Linux (Ubuntu) virtual machine (assuming that a reasonably modern version of Windows is the host system of choice).[9] My personal, perhaps somewhat masochistic, setup is the Cygwin environment under Windows [10]. I have also tried the Windows *bash* environment (available under Windows 10) which (effectively enough for SMURPH) emulates a Linux system under Windows.

To take advantage of DSD (see Appendix C), a Java environment is needed, which can be provided on the same platform, or on another system with which SMURPH is able to communicate over the (local) network.

---

[9]I have never used Apple OS, but there is no reason why a Ubuntu virtual machine hosted by Apple OS would perform any worse than one hosted by Windows.

© Springer Nature Switzerland AG 2019
P. Gburzyński, *Modeling Communication Networks and Protocols*,
Lecture Notes in Networks and Systems 61,
https://doi.org/10.1007/978-3-030-15391-5

B.1   Package Structure

The package comes as a collection of files organized into the directory structure presented in Fig. B.1. The root directory of the package contains the following items:

MANUAL	This directory contains the documentation
SOURCES	This directory contains the source code of the package
Examples	This directory contains a collection of sample programs in SMURPH and the include library
README.txt	This file contains the log of modifications introduced to the package since version 0.9 until the end of August 2006
RTAGS_PG	This is the log of changes starting at March 3, 2006 and extending until present
INSTALL.txt	The installation instructions

There may be some other files and scripts of lesser relevance, e.g., related to integrating SMURPH with VUEE as part of the PICOS platform [3]. Such files include comments explaining their purpose. One of them is *deploy*, which is a Tcl script installing SMURPH alongside the remaining components of the PICOS platform. That script can also be used to install SMURPH (in its standalone variant) in the most typical way (see Sect. B.2).

Figure B.1 The structure of SMURPH directories

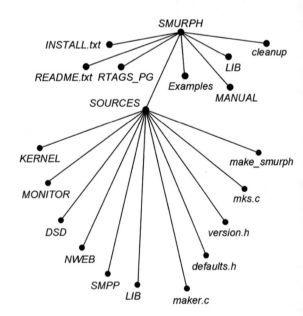

The essential parts of the package are contained in directory *SOURCES* which encompasses the following components:

KERNEL	This is a directory that contains the source code of the SMURPH kernel; those files are used to create SMURPH libraries that are configured with the user-supplied protocol files into executable SMURPH programs
MONITOR	This directory contains the source files of the monitor used to keep track of SMURPH programs in execution and to connect them to DSD (see Sect. C.2)
DSD	This directory contains the source and Java-machine code of the display applet DSD (Sect. C.1)
NWEB	This directory contains the source code of a tiny web server to facilitate the execution of the DSD applet from a browser (Sect. B.3)
SMPP	This directory contains the source code of the preprocessor (SMURPH compiler) which is run before the C++ compiler to turn SMURPH constructs into C++ code (see Fig. 1.4)
LIB	This directory contains the source code of the SMURPH I/O and XML library (Sects. 3.2.2 and 3.2.3)
maker.c	This is the source code of *maker*, i.e., the program to configure the package (see below)
defaults.h	This file contains definitions of default values to be used by the configuration program *maker*
version.h	This file contains the version number of the package and the code for recognizing the machine architecture (e.g., 32-bit vs. 64-bit) and the C++ compiler version
mks.c	This is the source code of *mks*, the driver for the SMURPH compiler

Another directory, named *LIB*, appears at the root level of the directory tree. It is initially empty, and its purpose is to contain binary libraries of the kernel created from the files in *KERNEL* (Sect. B.5).

Before the package is installed, the user must make sure that the target machine is equipped with the tools needed to compile the package. For a Linux platform, they amount to *gcc/g++* plus the standard (low-brow) development tools, like *make* and *gdb*. If anything is missing, which will become apparent during the installation, the missing packages should be installed and the installation tried again.

The version of DSD included with the package assumes JDK (Java Development Kit) which can be downloaded from http://www.java.com/. Note that SMURPH can be used without DSD, so the installation of JDK is not critical.

SMURPH can be obtained as a packaged (e.g., zip'ped and tar'red) directory or via a pointer to the GIT repository. The operation of unpacking depends on the mode of delivery, but in any case, the result of that operation will be a tree of files and directories as shown in Fig. B.1.

B.2 Installation

In this section, we assume that our operating system is Linux. I have tested the instructions from this section on 64-bit Ubuntu 18.04 running in a virtual machine (with Windows 10 as the host system). In Sect. B.4, we comment on running SMURPH under the Cygwin environment on Windows.

The installation consists in precompiling some components and configuring the SMURPH compiler, named mks, to reference them from where they are. The package doesn't in fact *install* itself anywhere to speak of, e.g., where it would need special permissions. The executable file with the *mks* program is put into a user-defined directory (typically named *bin* or *BIN*) which should be included in the standard *PATH* searched for executable commands. In the simplest, vanilla case, the procedure can be made extremely simple. By executing:

./deploy

at the topmost level of the package tree, the user will install the package in the most standard (and often quite reasonable) way. The *mks* executable is put into the subdirectory *bin* or *BIN* in the user's home directory, whichever of the two directories is found (*bin* is attempted first, in case both are present). Java and the DSD applet are not configured, which means that the DSD applet cannot be used with this type of installation.

The more elaborate installation procedure, allowing the user to select a few parameters of the installation, consists of the following steps:

1. Move to subdirectory *SOURCES* and compile the *maker* program. That program will carry out the installation asking the user for answers to a few questions. Execute:

   g++ −o maker maker.cc

2. Execute the maker by simply running the executable in the SOURCES directory:

   ./maker

   The program asks about the following things:

- The name of the Java compiler. This information is only relevant and needed if the user wishes to recompile the DSD applet. Normally, the applet need not be recompiled because the precompiled Java code is platform-independent. Note that the Java compiler is not needed to run the DSD applet. The default answer (an empty line) indicates no Java compiler (which only means that the Java applet will not be recompiled).
- The path to the SOURCES directory, i.e., to the directory containing the source files of the package. The default (which practically never should be changed) is the current directory.
- The path to the directory where binary libraries of the kernel modules will be kept. The default path is *../LIB*, i.e., the libraries will be kept in subdirectory *LIB* in the root directory of the package. This default practically never needs to be changed.

- The maximum number of versions of the binary library to be kept in the library directory (see Sect. B.5). The default number is 25, which is probably more than adequate.
- The name of the directory containing include files for protocol programs. It is possible to have multiple directories with include files, which will be searched in the order of their specification. By default, the standard library in *Examples/IncLib* is made available. That collection of include files is needed by the example programs in *Examples*, and it also contains many handy pieces that may be of general value, so it always makes sense to have it on the list. The default answer (an empty line) in response to this prompt, selects the standard include library as the only library automatically searched for includes. The user can also enter paths to other libraries, one path per line, and all those paths will append to the list headed by the standard library. To eliminate the standard library, the user can enter a line containing−(minus) as the only character. An empty line always terminates the list.
- The name of the directory containing the (optional) libraryfor XML data files included (see Sect. 3.2.3) from XML data sets. Multiple directories can be specified: they will be scanned in the order of their definition. An empty line terminates the sequence of data include directories.
- The Internet name of the host running the SMURPH monitor (Sect. C.2). The monitor is needed to provide an interface between programs in SMURPH and the DSD applet. Even if all three parties are to be run on the same machine, the host name of that machine must be specified here. By default, there is no monitor host, which means that the DSD applet will not be used. To run all three components on the same machine, the user should enter *localhost*.
- The number of the monitor socket port. This should be an unused port number for setting up an *AF_INET* socket available to non-privileged users. The default port number is *4442*.
- The name of the SMURPH compiler. This is the name under which the *mks* executable will be stored. The default name is *mks* and there is little point in changing it, unless the user dislikes the standard name for some reason.
- The directory where the *mks* executable is to be put. The default is ∼/*bin* where ∼ stands for the user's home directory. This directory should be included in *PATH*.
- Whether the monitor is to be built. If the answer is *yes* (which is the default), *maker* will create the executable version of the monitor and put it into directory *MONITOR*. It will also create the executable version of the tiny web server and put it in *NWEB*.
- The last question is about moving the DSD applet to a place other than its standard residence (the *DSD* directory). The reasonable thing to do is to leave the applet where it is.

When *maker* is done, it updates the contents of file *version.h* in directory *SOURCES* to reflect some of the user's selections. Also, it erases any files that were put into the SMURPH tree by any previous installation. This concerns the executables in directories *MONITOR*, *NWEB*, *SMPP*, as well as any old kernel libraries in *LIB*. Different versions of those libraries result from the different configurations of compilation options for a SMURPH program (causing recompilations of the files in *KERNEL*). Having a ready, precompiled library for a given configuration of options saves on the compilation time, because the kernel files need not be recompiled for that configuration. The number of possible library versions is equal to the number of combinations of compilation options, which is of order 10^6. Thus, *mks* bounds the number of library versions simultaneously stored (cached) in *LIB* to the user-definable limit (25 by default). When the limit is about to be exceeded, the least recently used library is removed from the cache. Note that a typical user, including myself, is extremely unlikely to ever reach the limit.

The monitor should be created on the host designated to run it. This means that when the host is different from the one running the simulator, the package should also be installed on the monitor host, at least to the extent of creating there the monitor's executable. The machine architectures and/or the operating systems of the hosts involved need not be identical.

The installation procedure doesn't start the monitor. Ideally, the monitor should be running (on the designated host) constantly, i.e., as a daemon. As our goal is a setup that is as little invasive as possible, the default execution mode of the monitor is as a regular program. For example, this command:

./monitor standard.t

executed in the *MONITOR* directory will run the monitor in the most straightforward way. File *standard.t*, present in the *MONITOR* directory, contains the set of predefined templates for DSD corresponding to the collection of standard exposure modes described in Sect. 12.5.

Needless to say, the monitor can be run in the background (by putting & at the end of the command line). It is possible to run it as a daemon by including the −*b* argument in its call line, e.g.,

./monitor −b standard.t

which, in addition to putting the program into the background, has the additional effect of detaching it from the terminal.

The standard monitor port is just some (random) number from the pool of TCP/IP port numbers available to non-privileged programs. There is a small chance that the monitor will exit immediately after startup with the following message:

port probably taken, rebuild with another socket port

which means that the monitor has failed to open its socket. This will happen, e.g., if the port number assigned by *maker* to the monitor socket is in use.[10] The user should then execute *maker* again selecting a different port number. Typically, legal values are between 2000 and 32000.

---

[10]Trying to start the monitor twice on the same host will also cause this problem.

B.3 Running DSD

The DSD applet (Sect. C.1) can be invoked from the Java applet viewer (aka appletviewer) or from a web browser. File *index.html* in directory DSD provides an HTML anchor to the applet. Most browsers will not allow the applet to connect to the monitor, unless the web server delivering the applet runs on the same host as the monitor. As an alternative to installing an actual, full-fledged web server (e.g., Apache), which most people would consider a gross overkill, SMURPH comes with a tiny, rudimentary web server, which I found on the network and slightly adapted to the task at hand.[11]

The web server is not needed when the applet viewer is used, because the applet viewer directly accepts the *index.html* file (in the DSD directory) as the encapsulation of the Java program to run. A web browser, on the other hand, needs an URL which can only be interpreted by a web server.

The first thing to do is to tweak Java permissions, which tend to be obsessively restrictive by default. This is needed, so the applet viewer (generally a Java applet) can communicate with other parties over a socket, even within the framework of a single host. For that, we need to locate a file named java.policy. On Linux systems (Ubuntu), with the most typical installation of Java (as one of the standard packages) the path to this file looks like */etc/.../lib/security/java.policy*, where the component represented by the three dots has some Java-related name. Root permissions are needed to edit the file. If everything is to be run on the same local system, then we should locate the "grant" section in the file, i.e., a chunk beginning with "grant {" and terminated with the matching closing bracket, and make sure that there is a line within that section that looks like this:

permission java.net.SocketPermission "127.0.0.1:", "connect,accept,resolve";*

The feat is usually accomplished by editing an existing line that appears close to the one above. A cognizant user can easily extrapolate the idea over more complicated communication scenarios, should a need for them arise.

The above should do for the case of the applet viewer. The *DSD* directory includes a *README.txt* file which provides additional hints. For access from a browser, we need to set up the tiny web server. For that, the user should move to the *NWEB* directory and execute:

./nweb24 -d ../DSD

The argument indicates to the server the root directory of the file system covered by the service. If DSD access is the only reason to run the server, then it makes sense to root the service in the DSD directory, as to 1) minimize the complexity of the URL, 2) avoid exposing too much of the (global) file system to the (apparent) security lapse caused by the presence of a poorly guarded service on the host. The

---

[11]As far as I could see, the server was not copyrighted in any way. Its author is Nigel Griffiths (from IBM, UK), and this is where the credits are due.

default port on which the service is made available is 8001. Thus, to access the DSD applet from a browser, we must point it to:

http://localhost:8001/index.html

Two of the examples that come with the SMURPH package, *Conveyor* and *Tanks* in *Examples/REALTIME*, use their own Java applets. Those examples are not described in this book, but they come with their own documentation. For access to those applets (together with the access to DSD), the root directory for the web server should be set to the root of the package tree. Here is a sample invocation of the server that also illustrates all the arguments that the program can possibly accept:

./nweb24 –d ../../ –p 8003 –l logfile.txt –b

So, the default port number can be changed, a log file can be set up (there is no log by default), and (with *–b*) the server can be run as a daemon. With the above specification of the root directory for service (recall that the command is executed in *SOURCES/NWEB*), the URL to the DSD applet will look like this:

http://localhost:8003/SOURCES/DSD/index.html

Note that the applets from SMURPH examples can also be run from the applet viewer, so the whole drill involved in setting up the web server is only needed when the applet(s) must be run from a browser.

One minor problem related to the DSD applet (as well as the other applets) is the different look of windows on the different Java platforms. This problem mostly concerns exposure windows and is not very serious, as the window contents can be scrolled manually if they appear misaligned. Ultimately, the user can edit file *DSD.java* in *SOURCES/DSD* and modify the constants *TOPMARGIN* and *BOTTOMMARGIN* declared at the beginning of class *DisplayWindow*, to adjust the initial location of the window contents permanently for the preferred platform.

B.4   Comments on Windows

SMURPH can also run on Windows under Cygwin, as well as under the *bash* environment available on Windows 10. Both these environments effectively emulate POSIX UNIX, with the latter providing a Ubuntu Linux framework, where most of the aspects relevant for SMURPH are indistinguishable from an actual Ubuntu installation on a real or virtual machine. Thus, once the requisite environment is properly set up, the basic installation procedures are exactly as described in Sect. B.2.

For one minor problem, I couldn't get the Java applet viewer to run under Windows. Thus, the only way to run the DSD applet in the Cygwin environment, where Java is necessarily Windows-native, is from a browser. The Internet Explorer works fine with the permission adjustments described in Sect. B.3 in *java.policy*. Under Windows, the file can be found in:

${user.home}\java.policy
${java.home}\lib\security\java.policy

with the first file applying to the current user and the second globally to all users of Java.

B.5 Creating Executable Programs in SMURPH

A protocol program in SMURPH must be compiled and linked with the kernel files to create a stand-alone executable file. The protocol program may consist of a number of C++ files, the name of each file ending with the suffix *.cc*. All those files should be kept in one directory. They may *#include* some user-created header files with names ending with *.h*. The SMURPH preprocessor (*smpp*, Sect. 1.3.2) will also automatically insert some standard include files (from *SOURCES/KERNEL*) in front of each protocol file.

A program using a component from the include library (or from multiple such libraries) configured with the package (Sect. B.2) must *#include* the respective header file in the header section of any source (*.cc*) file that references the types defined by the component. If the component comes with a source (a *.cc* file that must be compiled and linked with the program), then that *.cc* file must also be included by exactly one source file of the program. In the simplest case, the user can create a new source file whose only statement includes the respective source component (or components) from the include library. As this probably sounds a bit unorthodox, let us illustrate the idea with an example. Suppose that the program wants to use the "sampled" wireless channel model from Sect. A.3, represented by the files *wchansd.h* and *wchansd.cc* in the include library. Then, rather obviously, all the source and header files that refer to the types declared in *wchansd.h* must include that file. There is no need to provide any path components other than the file name (to indicate that the file comes from a different directory than the one containing the compiled program), as long as the include library was configured into the package (Sect. B.2). Then, to make sure that the *.cc* component of the model is compiled with the program, one of the *.cc* files in the program's directory must include the file *wchansd.cc*, like this:

#include "wchansd.cc"

Alternatively, it is also OK to add an extra *.cc* file to the program's directory, e.g., named *includes.cc*, and make it consist of the single line above. If the program uses multiple components from the include library, coming with other *.cc* files, all the include statements for those files can be put into *includes.cc*.

The reason why the source components of the libraries must be explicitly included by the program is that SMURPH does not precompile the library components at installation, because their object shape depends on the configuration of compilation options, as explained in Sect. B.2 (and also below). Note that both *wchansd.h* and *wchansd.c* include other components of the include library, describing the base model from which the sampled model is derived (Sects. A.1 and A.3). Those components need not be explicitly included by the protocol program, because the two files of the sampled model take care of that.

To create an executable program for a given source set, the user should move to the directory containing the protocol files and execute the SMURPH compiler *mks* (we assume that the default name of the compiler has not been changed at installation, Sect. B.2). The program accepts the following arguments (in an arbitrary order):

−a	All references to *assert* are turned into empty statements (see Sect. 3.2.7). This may speed up the execution a little (typically by a negligible amount) at the expense of some potential errors passing undetected. If this argument is not used, *asserts* are active
−b	This argument must be followed by a single digit indicating the selected precision of type *BIG* (see Sect. 3.1.3). The digit should be separated from −b by a space. The default precision of type *BIG* is 2 (yielding at least 62 bits for the representation of the requisite unsigned integer type)
−d	The argument selects the definition of type *DISTANCE* (see Sect. 3.1.6). By default, if the argument is absent, type *DISTANCE* is equivalent to *LONG*. If −d is followed by one of the letters *B*, *L*, or *l* (lower case L), separated from −d by space, then type *DISTANCE* is defined, respectively, as *BIG*, *LONG*, and *Long* (the second case corresponding to the default, i.e., no −d). If −d occurs alone (is not followed by any of the three letters), it selects *BIG* for the type (being the same as −dB)
−f	The C++ compiler will be called with the optimization option. The default is "no special optimization"
−g	This argument produces a debug-friendly version of the executable program. The C++ compiler will be called with the debugging option. Simulator-level tracing (Sect. 11.2) will be enabled. An alternative variant of this option is −G. It is almost identical to −g, but it additionally disables catching signals (like *SIGSEGV*) by the SMURPH kernel and makes sure to trigger a hard error (like *SIGSEGV*) when *excptn* (Sect. 3.2.7) is called. This makes it easier to catch and diagnose some errors (including violations of internal assertions) in *gdb* (the GNU debugger)
−i	This argument determines the definition of type *BITCOUNT* (Sect. 3.1.6), in the same way that −d determines the definition of type *DISTANCE* (see above)

(continued)

(continued)

−m	For a setting of *BIG_precision* (see −b above) that requires software emulation (typically for *BIG_precision* > 2), this argument removes the code for software error checks. It may slightly reduce the execution time of the model at the risk of making some potential errors more confusing
−n	The argument forces the clock tolerance (Sect. 5.3.3) to be 0, regardless of the settings of the protocol program, effectively removing all code for randomizing time delays, and makes all local clocks absolutely accurate
−o	This argument must be followed by a file name (separated from −o by a space). The specified file will contain the compilation output, i.e., the executable SMURPH program. The default file name, assumed in the absence of −o, is *side*[12]
−p	This option enables prioritized *wait* requests with a third argument (the so-called *order*, see Sects. 5.1.1 and 5.1.4)
−q	This option enables setting the message queue size limits described in Sect. 6.6.5. By default, the message queue size limits cannot be set
−r	This argument determines the definition of type *RATE* (Sect. 3.1.6), in the same way that −d determines the definition of type *DISTANCE* (see above)
−t	Forces recompilation of the protocol files, even if their corresponding object files are up to date. This option can be used after a library file included by the protocol program has been modified, to force the recompilation of the protocol files, even if they appear unchanged
−u	The standard *Client* is permanently disabled, i.e., no traffic will be generated automatically, regardless of how the traffic patterns are defined (see the *SCL_off* option in Sect. 6.6.2)
−v	Observers are disabled, i.e., they are never started and they will not monitor the protocol execution. This argument can be used to speed up the execution without having to physically remove the observers (see Sect. 11.3)
−z	This option enables faulty links described in Sect. 7.3. By default, faulty links are disabled
−C	The *Client* code is never compiled and −L is implied (see below). It is assumed that the program never uses the traffic generator, messages, and/or packets, as well as links and/or ports
−D	This option implies −n (perfect clocks) and removes nondeterminism from the program. Event scheduling (Sect. 5.1.4) becomes deterministic, i.e., no attempt is made to randomize the scheduling order of events occurring within the same *ITU*
−F	This option selects the no-floating-point version of the resulting executable. It only makes sense for the real mode of SMURPH and automatically forces −R, −S, as well as −D. One feature that does not work with −F is the *setEtu/setItu* pair (Sect. 3.1.2)
−I	This argument (it is capital i) must be followed by a path pointing to a directory which will be searched for include files. Multiple arguments of this form are permitted (up to 8), and the order of their occurrence determines the search order. The include directories specified this way are searched after those declared with *maker* when the package was installed (Sect. B.2)

(continued)

---

[12]On Windows, under Cygwin, the name is *side.exe*.

(continued)

–J	This option compiles in the code for journaling (Sect. 9.5) and is only effective together with –R or –W. The journal-related call arguments of SMURPH programs (Sects. 9.5.1 and B.6) are not available without this option
–L	The code dealing with links and link activities is never compiled. It is assumed that the program does not want to use links and/or ports
–R	This option selects the real-time version of the kernel. By default, the protocol program executes in virtual time, i.e., as a simulator
–S	The code dealing with random variables is never compiled and –C (including –L) is implied. It is assumed that the program does not use random variables, the traffic generator, messages, and/or packets, as well as links and/or ports. This option is often appropriate for control programs
–V	The program writes the package's version number to the standard output and exits
–W	This option compiles in the code for handling the visualization mode. It is needed, if the program ever wants to run in the visualization mode (i.e., call *setResync*, Sect. 5.3.4). Similar to –R, –W makes it possible to bind mailboxes, journal them, and drive them from journal files (Sect. 9.5). Note that –W and –R cannot be specified together
–X	The code dealing with radio channels and radio activities is never compiled. It is assumed that the program does not want to use radio channels and/or transceivers
–3	This option selects the 3d variant of the node deployment model for transceivers and radio channels (Sect. 4.5.2). It affects the headers of these *Transceiver* methods: *setup* (Sect. 4.5.1), *setLocation*, *getLocation* (Sect. 4.5.2), and the *getRange* method of *RFChannel* (Sect. 4.5.2). An attempt to combine –3 with –X results in an error
–8	This option replaces SMURPH's internal random number generator with a generator based on the UNIX *rand48* family. The *rand48*-based generator is slightly slower than the SMURPH generator, but it has a longer cycle

In consequence of running *mks*, all protocol files from the current directory are compiled and merged with the kernel files. The program operates similarly to the standard UNIX utility *make*, i.e., it only recompiles the files whose binary versions are not up to date. The resulting executable protocol program is written to the file named *side*,[13] unless the user has changed this default with the –o argument (see above).

Formally, the kernel files of SMURPH should be combined with the user-supplied protocol files, then compiled, and finally linked into the executable program. Note that it would be quite expensive to keep all the possible binary versions of the standard files: each different configuration of *mks* arguments (except –o and –t) needs a separate binary version of practically all files. Thus, only a few most recently used versions are kept, up to the maximum declared with *maker* (Sect. B.2). They are stored in directory *LIB*; each version in a separate subdirectory labeled with a character string obtained from a combination of the *mks* arguments. When *mks* has to create a SMURPH program with a combination of arguments that has no matching subdirectory in *LIB*, it recompiles the kernel files

---

[13]It is *side.exe* on Windows under Cygwin.

and creates a new *LIB* subdirectory. If the total number of subdirectories of *LIB* exceeds the limit, the least recently used subdirectory is removed before the new one is added.

It is safe to run concurrently multiple copies of *mks* (within the domain of a single installed copy of the package) as long as these copies run in different source directories. The program uses file locks to ensure the consistency of *LIB* subdirectories.

B.6   Running the Program

The binary file created by *mks* is a stand-alone, executable program for modeling the behavior of the network and protocol programmed by the user. Simple and typical ways to run such a program were shown in Sects. 2.1.6, 2.2.6, 2.3.3, and 2.4.7. In general, the program accepts the following arguments:

–b	This argument indicates that the simulator should be run in the background, as a daemon, detached from the parent process (shell) and from the terminal. If the argument is followed by 0 (separated from it by a space), the program will additionally make sure to direct all three standard streams (*stdin*, *stdout*, and *stderr*) to /dev/null. In that case, if the output file has not been specified (see below), which means that the program's results are being directed to *stdout* or *stderr*, the output will be effectively switched off and irretrievably discarded
–c	This argument affects the interpretation of the message number limit (see Sect. 10.5.1)
–d	With this option, the program will suspend its execution immediately after *Root* completes the initialization stage (Sect. 5.1.8), and wait for a connection from DSD before proceeding. This is useful for stepping through the protocol program (Sect. C.6.2) from the very beginning of its (proper) execution
–e	When this argument is present, it will be illegal to (implicitly) terminate a process by failing to issue at least one *wait* request from its current state before putting the process to sleep (Sect. 5.1.4). An occurrence of such a scenario will be treated as an error aborting the simulation[14]
–k	This argument must be followed by exactly two numbers: a nonnegative *double* number less than 1.0 and a small nonnegative integer number. These values will be used as the default clock tolerance parameters, i.e., *deviation* and *quality*, respectively (Sect. 5.3.3)

–o	If this argument is used, SMURPH will write to the output file the description of the network configuration and traffic. This is accomplished by calling:
	
	    System->printTop ();
	    Client->printDef ();
	(see Sects. 12.5.12 and 12.5.4) immediately after the initialization stage

–r	This argument should be followed by at least one and at most three nonnegative integer numbers defining the starting values of seeds for the random number generators. The

(continued)

---

[14]The option exists for PicOS/VUEE compatibility [3, 11]

(continued)

	three numbers correspond (in this order) to *SEED_traffic*, *SEED_delay*, and *SEED_toss* (Sect. 3.2.1). If less than three numbers are provided, then only the first seed is (or the first two seeds are) set. A seed that is not explicitly initialized is assigned a default value which is always the same
−s	This argument indicates that information about internal (system) events should be included in exposures (Sect. 12.5)
−t	This argument relates to user-level tracing (Sect. 11.1) and specifies the time interval during which the tracing should be on, and/or the set of station to which it should be restricted. The string following −t (after a space) should have one of the following forms:

start
start–stop
start/stations
start–stop/stations
/stations

where "start" specifies the first moment of the modeled time when the tracing should become active, and "stop" indicates the first moment when the tracing should stop. If those parameters look like integers (i.e., they only contain digits), they are interpreted in *ITUs*. If a decimal point appears within any of the two numbers, then both are assumed to be in *ETUs*. A single number defines the starting time, with the ending time being undefined, i.e., equal to infinity. No numbers (e.g., in the last variant in the above list) means no time restriction (start at zero, never stop)
The optional list of station identifiers (Sect. 4.1.2) separated by commas (not including blanks) appearing at the end of the argument and preceded by a slash restricts the tracing to the specified set of stations. This effectively means that a trace message will only be printed if *TheStation->getId ()* matches one of the indicated station *Ids*.
Here are a few examples:

−t 1.0–2.0
−t 3–3.5/2
−t 1000000
−t /3,7,123

Note that in the second case the first number will be interpreted in *ETUs*, even though it has no decimal point, because the second number has it
If the string following −t is terminated with the letter *s* (*q* works as well), it has the effect of setting the virtual time limit of the program (Sect. 10.5.2) to the end boundary of the tracing interval. For example, with this sequence:

−t 12.33–47.99/0,1s

the tracing is turned on from time 12.33 until 47.99 (expressed in *ETUs*) for stations 0 and 1. At the end of the tracing interval, the program will terminate. This feature is useful for quickly producing narrow traces surrounding the place of an error.
The interpretation of the argument string of −t is postponed until the network has been built, i.e., until the program leaves the first state of *Root* (Sect. 5.1.8). As we explain in Sect. 11.1, this is required because the interpretation of a time bound expressed in *ETUs* can only be meaningful after the *ETU* has been set by the program (Sect. 3.1.2), which usually happens in the first state of *Root*. For example, if the program issues a call to *setLimit* in the first state of *Root*, to define the virtual time limit for the run, that call will be overridden by the *s* flag of −t, but only if the end bound in the specification is finite. Otherwise, the *s* flag will be ignored. If the flag has been effective, i.e., once the time

(continued)

(continued)

	limit has been reset to the end bound of the tracing interval, any *setLimit* request from the program can only reduce the time limit (an attempt to extend it will be ignored) By default, if −*t* is absent, or if it is not followed by any specification string, the user-level tracing is on for all time and for all stations. This simply means that all trace commands appearing in the program are always effective. Note, however, that an empty −*t* specification is different from is total absence, as it nullifies any *settraceTime* or *settraceStation* invocations issued by the program in the first state of Root (Sect. 11.1)
−*u*	When this argument is used, the standard *Client* is disabled, and it will not generate any messages, regardless of how the traffic patterns are defined. Note that the *Client* can also be disabled permanently when the SMURPH program is created (Sect. B.5)

−*D*	This argument, which is only available in the real and visualization modes, and requires journaling to be compiled in (the −*J* option of *mks*, Sect. B.5), can be used to shift the real time of the program to the specified date/time. −*D* should be followed by a space followed in turn by the time specification. In its complete format, that specification looks like this:
	
	  yr/mo/dy,ho:mi:sc
	
	where the different components, respectively, stand for the year, month, day, hour, minute, and second. Each field should consist of exactly two digits. The following abbreviations are permitted:
	
	  mo/dy,ho:mi:sc
	  dy,ho:mi:sc
	  ho:mi:sc
	  mi:sc
	  sc
	  yr/mo/dy
	  mo/dy
	
	with the absent fields defaulting to their current values. One special date/time specification is *J* (the entire argument is −*D J*) which shifts the time to the earliest creation time of a journal driving the program's mailboxes (Sect. 9.5.2)

−*I*	This argument (capital i), which is only available in the real and visualization modes with journaling, must be followed by the specification of a mailbox. It declares a mailbox to be driven from the input section of a journal file (Sect. 9.5.2)
−*M*	This argument must be followed by a file name. The file is expected to contain extra window templates needed by the protocol program (see Sect. C.4), which apply to non-standard exposures defined by the user and/or, possibly, override some of the standard templates of the monitor (Sect. B.2)
−*O*	This argument, which is only available in the real and visualization modes with journaling, must be followed by the specification of a mailbox. It declares a mailbox to be driven from the output section of a journal file (Sect. 9.5.2)

−*T*	This argument is similar to −*t*, except that it refers to simulator-level tracing (Sect. 11.2). It is only available if the simulator has been compiled with the −*g* (or −*G*) option (Sect. B.5). Its format is exactly as for −*t*, with one extra option represented by +, which can appear at the end of the specification string (preceding or following the optional *s*). The time interval determines the range of simulator-level tracing, which consists in dumping to the output file detailed information about the states executed by the protocol processes. Here are a few examples:
	
	  −*T 667888388880*
	  −*T /12+*
	  −*T 47.5–49.5/8,6*
	  −*T 0–999999999/1,3,5,7s+*

(continued)

(continued)

	The + option (in the second and last cases) selects the so-called "full tracing" (Sect. 11.2), whereby the state information is accompanied by dumps of activities in all links and radio channels accessible via ports and transceivers of the current station. In contrast to –*t*, the lack of –*T* means no tracing. A single time value selects the starting time with the ending time being unbounded. The appearance of –*T* not followed by any specification string is equivalent to –*T* 0, i.e., all stations being traced all the time
– –	A double dash terminates the argument list interpreted by the simulator's kernel. Any remaining arguments, if present, can be accessed by the user program via these two global variables:

int PCArgc;
const char **PCArgv;

	interpreted as the standard arguments of *main* in a C program, i.e., *PCArgc* tells the number of arguments left, and *PCArgv* is an array of strings consisting of *PCArgc* elements containing those arguments

The first argument from the left that does not start with –, and is not part of a compound argument starting with –, is assumed to be the name of the input data file. The second argument with this property is interpreted as the output file name. If the input file name is . (period), or if no file name is specified at all, SMURPH assumes that the input data is to be read from the standard input. If no output file is specified or the output file name is . (period), the results will be written to the standard output. Using + as the output file name directs the simulator's output to the standard error.[15]

Below are three sample command lines invoking the simulator:

side . out
side –r 11 12 datafile –d outfile
side –k 0.000001 3 < data > out1234

In the first case, the program reads its input data from the standard input and writes the results to file *out*. In the second example, the simulator is called with *SEED_traffic* and *SEED_delay* initialized to 11 and 12, respectively. The simulation data is read from file *datafile*, the results are written to *outfile*. Before the protocol execution is started, the program will suspend itself awaiting connection from the display program. Note that although –*r* expects (up to) three numbers to follow it, the string following the second number does not start with a digit (it does not look like a number), so only two seeds are initialized. In the last example, the clock tolerance parameters are set to 0.000001 (deviation, amounting to 1 ppm) and 3 (quality). The input data is read from file *data* and the simulation results are written to file *out1234*.

---

[15]This exotic feature was used under Cygwin to make sure that a flush operation on the output stream had the desired effect of actually flushing all the pending output to the terminal window.

Appendix C
DSD: The Dynamic Status Display Program

DSD is a stand-alone program implemented as a Java applet which can be used to monitor the execution of a SMURPH program on-line.

C.1 Basic Principles

A SMURPH program communicates with DSD by receiving requests from the display program and responding with information in a transparent format. The SMURPH program and DSD need not execute on the same machine; therefore, the information sent between the two parties is also machine-independent.

A typical unit of information comprising a group of logically related data items is a window. At the SMURPH end, a window is represented by the following parameters:

- the standard name of the exposed object (Sect. 3.3.2)
- the display mode of the exposure (an integer number, Sect. 12.1)
- the *Id* of the station to which the information displayed in the window is related

The last attribute is generally optional and may not apply to some windows (see Sect. 12.5). If the station *Id* is absent, it is assumed that the window is global, i.e., unrelated to a specific station.

The above elements correspond to one mode of the "screen" exposure associated with the object, or rather with the object's type (Sect. 12.4). Additionally, this exposure mode can be made station-relative. Whenever the contents of a window are to be updated (refreshed) the corresponding exposure mode is invoked.

The graphical layout of a window is of no interest to the SMURPH program (and isn't its concern). The program only knows which data items must be sent to DSD to produce a window. Those data items are sent periodically, in response to an initial request from the display program specifying the window's parameters. The layout of the information, as displayed on the screen, is then arranged by DSD based on window templates (Sect. C.4).

© Springer Nature Switzerland AG 2019
P. Gburzyński, *Modeling Communication Networks and Protocols*,
Lecture Notes in Networks and Systems 61,
https://doi.org/10.1007/978-3-030-15391-5

A window can be put into a step mode where the SMURPH program will update its contents after processing every event that "has something to do with the window" and halt for an explicit command to proceed to the next step. For an *AI*-related window, the stepping event is "any event that awakes one of the processes waiting on the *AI*."

At any moment during its execution, a SMURPH program may be connected to DSD (we say that the display is active) or not. While the display is active, the SMURPH program maintains the list of active windows whose contents are periodically sent to DSD by invoking the exposures of the exposed objects. The SMURPH program only sends to the display program the raw data representing the information described by the object's exposure method.

DSD may wish to activate a new window, in which case the SMURPH program adds it to the active list, or to deactivate one of the active windows, in which case the window is removed from the pool. The SMURPH program never makes any decisions on its own regarding which windows are to be made active or inactive.

DSD is menu driven. The objects to be exposed and their modes (i.e., windows to be displayed) are selected from the current menu of objects available for exposure. That menu can be navigated through, based on the concept of the ownership hierarchy of objects maintained by the SMURPH program. To simplify things, it is assumed that each object belongs to at most one other object. This assumption is generally reasonable, although, e.g., a port naturally belongs to two objects: a station and a link. In this case, it is assumed that ports belong to stations, as this relationship is usually more relevant from the user's (and protocol program's) point of view.

Objects (the nodes of the object tree) are represented by their standard names supplemented by base names and nicknames (Sect. 3.3.2) wherever nicknames are defined. The display program uses this structure to create menus for locating individual, exposable objects. A typical operation performed by the user of DSD is a request to display a window associated with a specific object. Using the menu hierarchy of the program, the user navigates to the proper object and issues the request. This operation may involve descending or ascending through the ownership hierarchy of objects, which in turn may result in information exchange between DSD and the SMURPH program. When the user finally makes a selection, the request is turned by DSD into a window description understandable by SMURPH. That description is then sent to the SMURPH program which adds the requested window to its list of active windows. From now on, the information representing the window contents will be periodically sent to the display program (Sect. 12.4).

C.2   The Monitor

The SMURPH program and DSD need not run on the same machine to be able to communicate. DSD does not even have to know the network address of the host on which the SMURPH program is running. When the package is installed, the user should designate one host to run the so-called SMURPH monitor, a daemon providing identification service for all SMURPH programs to be run by the user. This host can be an arbitrary machine visible through the Internet, but, for a simple, natural, local setup, the monitor host is just the user's PC.

The user may want to run a web server on the same monitor host, to make it possible for DSD (called from Java-capable browsers) to set up a TCP/IP connection to the monitor. For security reasons, most browsers unconditionally restrict applet-initiated TCP/IP connections to the host running the web server that has delivered the applet page.[16] In Sect. B.3, we discuss these issues and explain the technicalities involved in running the DSD applet within the tight confines of the contemporary Java security rules.

The purpose of the monitor is to keep track of all SMURPH programs started by the user who owns the monitor.[17] Whenever a SMURPH program is invoked (on any machine visible to the monitor), it reports to the monitor; when the program terminates, the monitor also learns about that fact and removes the program's description from its directory. The following information is kept by the monitor for every active SMURPH program:

1. the Internet name of the host on which the program is running
2. the program call line, i.e., the program's command name and the call arguments
3. the date and time when the program was started
4. the (system) process Id of the program
5. a description of the channel (socket) connecting the monitor to the program.

The first four elements identify the run for the user; the last item makes it possible for the monitor to pass requests to the program.

---

[16]In the olden days, when DSD was first programmed, things were more permissive, and it was possible to easily and conscientiously override security restrictions for applets. These days things are a bit more complicated. Had I known about these problems in advance, DSD would have been programmed as a Tcl/Tk script.

[17]As an interspersed historical note, let me mention that the monitor's second role, which from my personal perspective proved more useful than that of providing a link to DSD, used to be interfacing multiple experiments in SMURPH, run concurrently on different workstations, to a central script responsible for their distribution and coordination. For that, SMURPH was equipped with checkpointing code that allowed it to swap out an experiment in progress upon demand and move it to another, compatible, workstation for continuation (load balancing). The setup, dubbed SERDEL (a Supervisor for Executing Remote Distributed Experiments on a LAN) was scrapped at some point, because 1) the checkpointing software became unreliable (and the maintenance cost of keeping it up to date with system upgrades became prohibitive), and 2) the (then popular) departmental networks of uniform, globally accessible workstations lost their appeal, yielding to the not-so-local networks of strictly private PCs.

C.3   Invoking DSD

DSD is a Java applet that can be run, e.g., from any Java-capable web browser (see Sect. B.3). Its startup panel consists of two buttons: one to start the applet, the other to unconditionally terminate it and remove all its windows from the screen.

When started, DSD opens its main control panel (the so-called root window),[18] which looks as shown in Fig. C.1. The window consists of a menu bar at the top and two display areas (panes). Initially, both panes are empty and the single item selectable from the menu bar is *Get List* from the *Navigate* menu.

The lower display area of the root window is the so-called alert area where the applet displays messages addressed to the user. Some of those messages originate in DSD itself (they may, but don't have to, refer to errors or problems encountered by the applet); some others may be notes (e.g., sent by *displayNote*, Sect. 12.4) or error diagnostics arriving from the SMURPH program.

The role of the upper display area (the selection area) is to present the list of objects that the user can select from at the time (the list is generally dynamic). The only sensible thing that can be done at the beginning, right after the root window has been brought up, is to select *Get List* from the *Navigate* menu. This will ask DSD to try to connect to the monitor and acquire information about the SMURPH programs currently running. This information will be displayed in a list (one program per line) in the selection area. If no SMURPH programs are running at the time, the message *Nothing there* will appear in the alert area and nothing more will happen.

If the list of programs displayed in the selection area is nonempty, the user can select one program from the list and choose an item from the Navigate menu. This menu now consists of the following two items:

Connect	By hitting this item, the user asks DSD to establish a display session with the indicated program
Status	By hitting this item, the user asks DSD to display short status information about the indicated program

In the latter case, the applet will poll the program (using the monitor to mediate this communication) for its status. In response, the program will send the following information that will appear in the alert area of the root window:

1. the number of messages received so far by the standard *Client* (Sect. 10.2.2)[19]
2. the amount of virtual time (in *ITUs*) elapsed from the time the program was started
3. the amount of CPU time (in seconds) used up by the program.

---

[18]This shouldn't be confused with the *Root* process discussed in Sect. 5.1.8.

[19]In time-consuming performance studies, this is often an important indicator of progress reflecting the amount of "work" accomplished by the model so far.

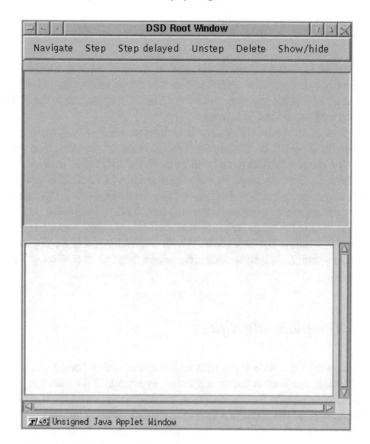

Figure C.1 The root window of DSD

If the user selects *Connect* from the *Navigate* menu, the applet will try to set up a display session with the selected SMURPH program.

During a display session, DSD presents in the selection area the list of objects whose dynamic exposures can be requested at the time. The list starts up with an initial selection of objects sent by the SMURPH program upon connection, and the *Navigate* menu provides a collection of operations for navigating through these objects and requesting their exposures. Depending on the current configuration of those exposures, other menus from the menu bar may also become useful.

Each object exposure requested by DSD appears in a separate window. As explained in Sect. 12.1, such an exposure corresponds to a triplet: (object, display mode, station). The raw exposure information sent by the SMURPH program is organized by DSD into a window based on the template associated with the exposure.

C.4   Window Templates

The material in this section may be helpful to the user who would like to create
non-standard windows and/or define non-standard exposable types (Sect. 12.2). It
explains how the raw information sent by the SMURPH program to the DSD applet
is formatted and turned into windows.

The layout of a window is determined by the *window template*, which is a
textual pattern supplied in a template file. A standard template file (providing
templates for all standard exposable objects of SMURPH) is provided with the
monitor (Sect. C.2) which reads that file upon startup. A SMURPH program itself
may use a private template file (Sect. B.6) to declare templates for its non-standard
exposable objects and/or override (some of) the standard templates.

A template file consists of template definitions, each template definition
describing one window layout associated with a combination of the object type, the
display mode, and a designator that determines whether the window is station-
relative or not.

C.4.1   Template Identifiers

The definition of a template starts with its identifier, which consists of up to three
parts. Whenever the user requests a specific exposure, DSD searches the list of
templates associated with the currently selectable objects, matching their identifiers
to the parameters of the requested exposure. These parameters specify the object,
the display mode, and (optionally) the station to which the exposure should be
related. Not surprisingly, the general format of a template identifier reflects the three
parameters:

name mode sflag

The first (mandatory) part of the template identifier should be a type name or a
nickname. The template will be used to display windows associated with objects of
the given type, or objects whose nickname (Sect. 3.3.2) matches the given name.
The second argument specifies the display mode of the template. If this argument is
missing, the default mode 0 is assumed. The last argument, if present, indicates
whether the window can be made station-relative. If that argument is absent, the
window is global, i.e., the template cannot be used to display a station-relative
version of the window. An asterisk declares a station-relative window template.
Such a template can only be used, if the exposure request identifies a station (any
station) to which the window is to be related. Two asterisks mean that the window
represented by the template can (but does not have to) be station-relative. We say
that such a template is *flexible*.

When the user requests a new exposure (identifying a selectable object, Sect. C.3), DSD presents a menu listing all templates that match the object to be exposed. That menu consists of the following templates (appearing in this order):

1. templates whose name matches the nickname of the object
2. templates whose name matches the type name of the object
3. templates whose name matches the base type name of the object (Sect. 3.3.2).

Within each class of templates, the ones defined privately by the SMURPH program take precedence over the standard ones (known by the monitor).

C.4.2 Template Structure

The best way to understand how templates are defined is to discuss a reasonably elaborate example. Figure C.2 shows the standard template describing the layout of the mode-0 *Timer* window (see Sect. 12.5.1). The first line contains the template identifier (Sect. C.4.1). The template is anchored to the *Timer* (this is a special object/type which occurs in a single copy, Sect. 5.3) and its covers the station-less variant of the mode-0 exposure. The second line must be a quoted string briefly describing the window's purpose. This description will identify the template on the menu shown by DSD, from where the user will be able to select the exposure mode. The description string must not contain newline characters. Any text following the closing quote of the description string and preceding the first line starting with B or b is assumed to be a comment and is ignored. Similarly, all empty lines, i.e., those containing nothing except possible blanks and/or tab characters are always skipped, as if they did not exist.

The next interpreted line of the template, the one starting with the letter B, begins the description of the window layout. The first and the last non-blank characters from this line are removed (in our case, this applies to the two letters B on both ends of the line)[20] and the length of whatever remains (the number of characters) determines the width of the window. Note that the width is expressed in characters. The tacit assumption is that the font used to display data in DSD windows is fixed. This may look ugly to the artistically inclined, but DSD has never claimed to be a masterpiece of graphical design.[21]

All characters in the remaining part of the width line other than + are ignored, except for being counted to the window width. A + character is also counted, but it has a special meaning: if present, it describes the initial width of the window (the column tagged with the + is included). At most one + may appear in a width line.

---

[20]Note that the second B could be any character at all. It would be ignored in precisely the same way as the first B.

[21]The very first version of the display interface, now defunct, absent, and forgotten, was for ASCII terminals. The somewhat primitive idea of templates was directly adapted from that interface. Simple and unappealing esthetically, as it is, it has proved effective for its practical purpose.

```
Timer 0
"Wait requests (absolute)"
// Displays Timer wait requests coming from all stations
B012345678901234567890123456789012345678901234567890123456789012345+B
|~~~~~~~~Time~~~St~~~~Process/Idn~~~~~TState~~~~~~~~AI/Idn~~~~~Event~~~~~State| r
|%%%%%%%%%%%%%& %%% %%%%%%%%%%/&&& %%%%%%%%% %%%%%%%%/&&& %%%%%%%%% %%%%%%%%%|
*  |
*  |
*8 |+
** |
E012345678901234567890123456789012345678901234567890123456789012345678901234567890123456E
```

Figure C.2 The mode-0 timer template

In our example, + occurs as the last (effective) character of the line, which means that the initial width of the window, as it should open in response to the exposure request, is full. This + tag is superfluous: by default, in the absence of a + character, the initial width of an exposure window is equal to the full effective length of the template's width line.

Some template elements represent fixed textual components, e.g., the header string following the width line. The opening (width) line (marked with *B*), however, does not belong to the window frame, it will not show up in the window, and it also does not count to the window's height. Similarly, the closing line of the template (the one starting with *E*) is not considered part of the window frame. Only the first character of that line is relevant, and the rest of it is ignored. The line starting with *E* terminates the description of the window layout.[22]

All nonempty lines, not beginning with *X* or *x* (see below), occurring between the starting and closing lines, describe the window contents. Each such a line should begin with | (a vertical bar) or * (an asterisk). A line beginning with | is a format line: it prescribes the layout of the corresponding window line. A layout line terminates with a matching vertical bar. The two bars do not count to the layout and act as delimiters. The terminating bar can be followed by additional information, which does not belong directly to the layout, but may associate attributes with the items defined within the line.

C.4.3 Special Characters

Within the layout portion of a template line, some characters have special meaning. All non-special characters stand for themselves, i.e., they will be displayed directly on the positions they occupy within the template. Below we list the special characters and their interpretation.

---

[22]Lower case *b* and *e* can be used (with the same effect) instead of the capital *B* and *E*.

~ *(tilde)*	This character stands for a virtual blank. It will be displayed as a blank in the window. Regular blanks are also displayed as blanks; however, two items separated by a sequence of regular blanks are considered separate, in the sense that each of them can have an individually definable set of attributes (Sect. C.4.8), while the blanks represented by a sequence of ~ characters glue the words separated by them into a single item treated as a unit from the viewpoint of attribute assignment
%	A sequence of % characters reserves space for one right-justified data item arriving from the SMURPH program to be displayed within the window exactly in place of the sequence
&	A sequence of & characters reserves space for one left-justified data item arriving from the SMURPH program to be displayed within the window in place of the sequence
@	This character is used to define a region, i.e., a semi-graphic subwindow (Sect. 12.3, C.4.7)

Non-blank fields separated by a sequence of ~ characters appear as different parts, but they are treated as a single item from the viewpoint of attributes. Moreover, the separating sequence of virtual blanks receives the same attributes as the fields it separates. For example, the header line of the *Timer* template contains multiple items separated by virtual blanks. The letter r occurring after the terminating bar of the header line says that the first item of this line is to be displayed in reverse video. From the viewpoint of this attribute, all the items of the header line amount to a single item. Thus, the entire header line will be displayed in reverse.[23]

A sequence of % characters reserves an area to contain a single data item sent to DSD by the SMURPH program as part of the information representing the object's exposure (Sect. 12.3). The type of that item is irrelevant from the viewpoint of the template layout and will be determined upon arrival.[24] The program will attempt to contain the item within its prescribed field as to maintain a regular layout of the window. The item will be truncated, if it does not fit into the field, and right-justified, if some space is left. The only difference between & and % is that an item displayed in a field described by a sequence of & characters is left-justified.

Data items arriving from the SMURPH program are assigned to their fields in the order in which they arrive. The order of fields is from left to right within a line and then, when the line becomes filled, DSD switches to the next line. Superfluous data items are ignored.

---

[23]The concept of "reverse video" is another relic of the ancient era of ASCII terminals.

[24]For some exposures, whose contents depend on the program call parameters or the actual (flexible) configuration of the dynamic objects (e.g., see Sect. 12.5.3), a dummy textual item (like "—") may be sent instead of a (normally expected) numeric item. The fact that the template determines the item type on its arrival helps DSD handle such cases. It is important that the number of items arriving from the simulator agrees with the expectations of the template, but whether they are textual or numeric makes little difference.

C.4.4 Exception Lines

A special character can be escaped, i.e., turned into a non-special one in a way that does not affect the visible length of the layout line. Namely, a layout line can be preceded by a line starting with the letter X or x, for exceptions. A position marked in the exceptions line by | (a vertical bar) indicates that the special character occurring on that position in the next layout line is to be treated verbatim as a regular character.

Sometimes two fields that are supposed to contain two different data items arriving from the protocol program must be adjacent (i.e., they touch with no separation by blanks). A column marked with a + in an exception line indicates the forced end of a field occurring in the next layout line. Any field crossing the position pointed to by the + will be split into two separate fields; the second field starting at the position immediately following the location of the + mark.

A single layout line can be preceded by multiple exception lines whose effect is cumulative.

C.4.5 Replication of Layout Lines

Often the same layout line has to be replicated a few times. In some cases, this number is arbitrary and variable. Such a situation occurs with the *Timer* window whose contents, except for the single header line, consist of a variable number of rows with identical format.

A line starting with an asterisk indicates a replication of the last regular layout line. The asterisk can be followed by a positive integer number that specifies the exact number of needed replications. If the number is missing, 1 is assumed by default. Two consecutive asterisks mean that the number of replications is undetermined: the program is free to assume any nonnegative number.

A replication line need not contain any characters other than the asterisk possibly followed by a number, or two asterisks. However, if such a line is to contain an initial height indicator (see below), it should be closed by the vertical bar, as the height indicator can only appear after the bar, as a formal attribute.

C.4.6 Window Height

The window height is determined by the number of the proper layout lines (excluding the B-line, the E-line, empty lines, and exception lines). A layout line whose content part terminates with a vertical bar can include +, the initial height indicator, which should immediately follow the terminating bar. If the height indicator does not occur, the initial number of rows to be displayed in the window is equal to the number of the proper layout lines in the template.

A height indicator occurring at a replication line with replication count bigger than 1 is associated with the last line of this count. A height indicator appearing at a ** line, or past that line, is ignored.

If the window height is undefined and no default height indicator occurs within the defined part, the initial height of the window is determined by the number of the proper layout lines preceding the first replication line with the undefined count.

The initial width and height of an exposure window (described by the window's template) are hints to DSD how the window should be (initially) opened. The user can resize the window with the mouse, e.g., to reveal the parts that have been cropped out in the initial trim of the window. The user can also scroll in the hidden parts of the window contents, e.g., if the contents are too large to fit on the screen all at the same time.

C.4.7 Regions

A region is a rectangular fragment of a window used to display graphic information. For illustration, let us have a look at the template shown in Fig. C.3 describing a window for displaying the contents of a random variable. The template defines one region delimited by the four @ characters that mark the region's corners. The rows and columns containing these characters belong to the region. Except for the delimiting characters, the rest of the rectangle representing the region is ignored, i.e., any characters appearing there are treated as a comment.

A single template may define multiple regions. Regions must not overlap and they must be perfectly rectangular. The ordering of regions among themselves and other fields (important for the correct interpretation of the data items arriving from SMURPH) is determined by the positions of their left top corners.

C.4.8 Field Attributes

A layout line terminated by | can specify attributes to be associated with the display items described in the line. The optional specification of these attributes should follow the initial height indicator (+), if one is associated with the line.

At most one attribute specification can be associated with a single item, including regions. Specifications for different items are separated by blanks. The correspondence between specifications and items is determined by the order of their occurrence from left to right. Note that regions are represented for this purpose by their left top corners. Superfluous specifications are ignored. Items without specifications are assigned default attributes.

The following attributes are understood by DSD:

Figure C.3 The mode-0
RVariable exposure

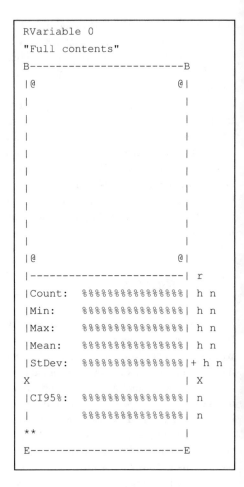

```
RVariable 0

"Full contents"

B----------------------B
| @                   @ |
|                       |
|                       |
|                       |
|                       |
|                       |
|                       |
|                       |
|                       |
|                       |
| @                   @ |
| ----------------------|  r
|Count:  %%%%%%%%%%%%%%%% | h n
|Min:    %%%%%%%%%%%%%%%% | h n
|Max:    %%%%%%%%%%%%%%%% | h n
|Mean:   %%%%%%%%%%%%%%%% | h n
|StDev:  %%%%%%%%%%%%%%%% |+ h n
X                       | X
|CI95%:  %%%%%%%%%%%%%%%% | n
|        %%%%%%%%%%%%%%%% | n
**                      |
E----------------------E
```

n	Normal display (the default)
r	Reverse (inverted) video
h	Highlighted display
b	Blinking

For illustration, Fig. C.4 shows a relatively complex template with two regions. The *h* (highlighted) attribute in the first layout line refers to the header string "Station " (extending over the blank following the word). The sequence of # characters in the first layout line stands just for itself. The *r* (reverse) attribute of the second layout line applies to the region (which is the first item formally defined in the line). The *b* (blink) attribute in the same line applies to the (right-justified) number displayed as the last (third) item. The role of *n* (which means "normal") preceding *b* is to skip one item (i.e., the header "Number of packets:"). The

```
Mytype 0 **
"Just an example"
B                                                                    B
!Station~%%%%~~~~~~~~~~~~~~~~############################| h
|@                          @  Number~of~packets:  %%%%%%%%|  r  n  b
|                              Number~of~bits:     %%%%%%%%|  n  b
|                              Acknowledgments:    %%%%%%%%|  n  h
|                              Retransmissions:    %%%%%%%%|  n  n
|                              Errors:             %%%%%%%%|  n  b
|                              Average~delay:      %%%%%%%%|  n  h
|                              ==========================|  r
|@                          @  @                   @  Status|  h  r
|=======Variance========                             A %%%%|  r  h  n
|Hits:       %%%%%%%%%%%%                             B %%%%|  h  n  h  n
|Misses:     %%%%%%%%%%%%                             C %%%%|  h  n  h  n
|Total:      %%%%%%%%%%%%                             D %%%%|  h  n  h  n
|Successes:  %%%%%%%%%%%%                             E %%%%|  h  n  h  n
|Failures:   %%%%%%%%%%%%                             F %%%%|  h  n  h  n
|Total:      %%%%%%%%%%%%  @                       @  G %%%%|  h  n  h  n
|Fairness:   %%%%%%%%%%%%  =======Mean========       H %%%%|  h  n  r  h  n
|~~~~~~~~~~~~~~~~~~~~~~~~~~~~~~~~~~~~~~~~~~~~~~~~~~~~~~~~~| r
E                                                                    E
```

Figure C.4 A sample template with two regions

h attribute in the eight layout line, the one in which the second region begins, applies to the second region. Even though a portion of the first region is still present in the line, that region doesn't count as an item defined within the eight line, and the second region is the first item defined there, with the "Status" header being the second.

C.5   Requesting Exposures

As soon as a display session has been established, the SMURPH program sends to DSD an initial list of exposable objects. These objects are elements of the tree of all exposable objects currently defined within the SMURPH program. The user can browse through that tree using the *Navigate* menu from the root window (Fig. C.1).

C.5.1   The Hierarchy of Exposable Objects

All exposable objects known to the SMURPH program are logically organized into a tree reflecting their ownership relation. The root of this tree is the *System* station (Sect. 4.1.2) which is assumed to own (directly or indirectly) all exposable objects. The ownership relation is illustrated in Fig. C.5. The ordering of subnodes reflects the order in which the objects represented by the subnodes appear in the selection area of the root window (Sect. C.3).

The first object owned by *System* is *Kernel*, the root of the process hierarchy (Sect. 5.1.8). Note that the parent-child relationship of processes need not coincide with the ownership tree structure. The *System* station also owns all regular stations created by the protocol program. These stations constitute a flat layer: they are all direct descendants of *System*. Similarly, all links, rfchannels, traffic patterns, the *Client*, and the *Timer* also belong to the *System* station.

A regular station owns its mailboxes, ports, and transceivers. It also owns all regular processes, *RVariables*, and *EObjects* that were created by *Root* in the context of the station at the initialization stage (Sect. 5.1.8). The rules for

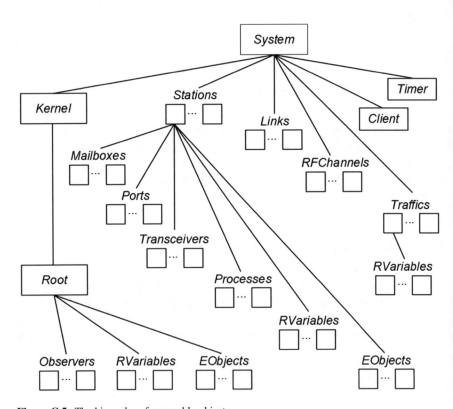

Figure C.5 The hierarchy of exposable objects

determining the ownership of such objects created while the protocol is running (past the initialization stage) are different (see below).

Each traffic pattern is the owner of its collection of random variables used to keep track of the standard performance measures (Sect. 10.2.1). The *Kernel* process owns all system processes as well as the user-defined *Root* process responsible for initialization (Sect. 5.1.8). The system processes are invisible to the user, unless the SMURPH program has been invoked with the –s option (see Sects. 12.5 and B.6).

The *Root* process is the owner of all observers (Sect. 11.3). It also owns all *RVariables* and *EObjects* that were created by *Root*, but outside the context of any regular station.

There are three categories of exposable objects that can be created (and possibly destroyed) dynamically by the protocol program after the initialization stage, namely: processes, random variables, and *EObjects*. Any such object is owned by the process directly responsible for creating it. Thus, the ownership structure of processes created by other protocol processes coincides with the parent-child relationship.

C.5.2 Navigating the Ownership Tree

At any time during a display session, the list of objects presented in the selection area consists of all the descendants of exactly one object in the ownership tree. The top item on that list (displayed in a different color) is the owner itself. Initially, immediately after the display session has been established, the SMURPH program sends to DSD the list of descendants of the *System* station, i.e., the first layer of the tree.

Each line in the selection area corresponds to one object and includes up to three names of the object (if they all exist and are different): the standard name, the base name, and the nickname (Sect. 3.3.2). An object can be selected by clicking on its line and unselected by clicking once again. The following three items from the *Navigate* menu refer to the objects listed in the selection area:

Descend	DSD asks the SMURPH program to send the list of descendants of the selected object; this way the user descends one level down in the ownership tree
Ascend	DSD asks the program to send the list of descendants of the object whose immediate descendant is the owner of the currently presented list of objects; this way the user ascends one level up in the ownership tree (of course, it is impossible to ascend above the *System* station)
Display	DSD opens a dialog to select the display mode and/or the station-relative status of an exposure for the selected object

Each list of objects arriving from the SMURPH program in response to a descend/ascend request is accompanied by a list of templates applicable to those

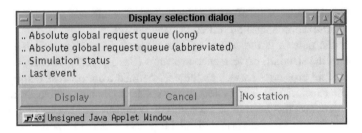

Figure C.6 A sample display dialog

objects. Those templates are selected by the SMURPH program itself (if it uses a private template file) and/or by the monitor (using the standard set of templates).

In response to *Display*, DSD shows a dialog listing all the templates applicable to the selected object. Figure C.6 presents the display dialog for the *Kernel* process (the first child item on the initial selection list). The scrollable list includes all templates (their description strings, Sect. C.4.2) potentially applicable to the selected object. Note that each template implicitly determines a display mode. Thus, the only additional element that must be specified to complete the exposure request is the station *Id*. This item can be typed in the text area (the rightmost area in the bottom panel).

The template description string shown in the display dialog begins with two characters that indicate the station-relative status of the template. Two periods mark an absolute template, i.e., one that cannot be related to a station. In such a case, the station field in the dialog is irrelevant and the template itself fully describes the exposure. The remaining two cases are .* (period + asterisk) meaning "station-relative," and ** (two asterisks) meaning "flexible" (Sect. C.4.1). The station Id field is only interpreted in one of these two cases. If it contains a legitimate station number (it must for an unconditionally station-relative template), the exposure is station-relative; otherwise, it is absolute.

Having selected the template and (possibly) the station Id, the user may click *Display* to request the exposure. In response to this action, DSD will create a new window (based on the template, Sect. C.4) and notify the SMURPH program that a new exposure should be added to its internal list of active exposures. DSD refuses to create a new exposure if an identical exposure is already active.

C.6 Exposure Windows

Typically, an exposure window contains several static components (extracted from the template, Sect. C.4.2) and some dynamic components that are filled dynamically by the SMURPH program. Recall that the SMURPH program only sends the raw data; the layout of the dynamic components within the window is determined by the template.

C.6.1 Basic Operations

As soon as it is created, a new exposure appears in four menus selectable from the root window: *Step*, *Step delayed*, *Delete*, and *Show/hide*. The exposure is identified in the menu by the name of the exposed object,[25] followed by the exposure mode in square brackets. If the exposure is station-relative, this string is further followed by the station Id in parentheses.

By selecting the exposure from the *Show/hide* menu, the user toggles its "visible" status, i.e., the window becomes invisible if it was visible and vice versa. A window that becomes visible is automatically forced to the top. An invisible exposure window remains active, and it continues receiving updates from the protocol program.

An exposure can be canceled by selecting it from the *Delete* menu. In response to this action, DSD removes the exposure's window from the screen and notifies the SMURPH program to remove the exposure from its internal list.

An exposure window can be naturally resized. If the entire contents of the exposure cannot fit into whatever frame is made available for the window,[26] the visible area can be scrolled. This is done, in a somewhat unorthodox way, by clicking and dragging in the visible area.

To reduce the amount of traffic between the SMURPH program and DSD, the former maintains two counters for each active exposure. One of them tells the number of the first visible item in the current version of the exposure's window, the other specifies the number of the first item that is not going to show up. Whenever a window is resized or scrolled, DSD sends to the protocol program new values of the two counters. As only those items that in fact appear in the window are sent in window updates, the scrolled-in area of a window may initially appear blank. It will become filled as soon as the SMURPH program learns about the new parameters of the window and starts sending the missing items.

C.6.2 Stepping

The SMURPH program can be put into the so-called *step mode* whereby it stops on some (or all) events. An explicit user action is then required to continue execution. The step mode is always requested for a specific exposure window (i.e., windows are the things being formally stepped). If a window is stepped, the SMURPH program halts whenever an event occurs that is somehow "related" to the exposure presented in the window. At that moment, the contents of the window reflect the program state immediately after the event has been processed.

---

[25]This is the nickname, if one is defined, or the standard name, otherwise.
[26]This is rather typical for windows with undefined height (Sect. C.4.6).

Multiple exposures can be stepped at the same time. The occurrence of any event related to any of the stepped windows halts the protocol program.

The following rules describe what we mean by an "event related to the exposure":

- If the exposed object is a station, the related event is any event waking any process owned by the station.
- If the exposed object is a process, the related event is any event waking the process. One exception is the *Kernel* process. It is assumed that all events are related to *Kernel*; thus, by stepping any Kernel exposure we effectively intercept all events.
- If the object is a random variable or an exposable object of a user-defined type, the related event is any event related to the object's owner (in the sense of Sect. C.5.1).
- If the object is an observer, the related event is any event that results in awakening the observer.
- If the object is an activity interpreter not mentioned above, the related event is any waking event triggered by the activity interpreter.

To put an exposure window immediately into the step mode the user should select it from the *Step* menu.[27] The frame of a stepped window changes its color from blue to red.

It is also possible to step an exposure at some later moment, by selecting it from the *Step delayed* menu. This may be useful for debugging: the user may decide to run the program at full speed until it reaches some interesting stage. In response to *Step delayed*, DSD presents a dialog shown in Fig. C.7. The dialog allows the user to select the virtual time (in *ITUs*)[28] when the stepping should commence, or the number of events to be processed before it happens. This number can be either absolute or relative, i.e., represent an offset with respect to the current time or event count. Note that the first stepped event must "have something to do" with the stepped exposure, so it may occur later (but not earlier) than the specified time.

The occurrence of a stepped event forces the protocol program to immediately (regardless of the value of *DisplayInterval*, Sect. 12.4) send the update information for all active exposures and halt. The program will not resume its execution until the user selects *ADVANCE* from the *Step* menu. Then the program will continue until the next stepped event and halt again. Each time that happens, the message *STEP* is displayed in the alert area of the root window (Sect. C.3).

The user can cancel the stepping for one exposure, by selecting it from the *Unstep* menu, or globally for all stepped exposures, by selecting *ALL* from that menu. If this is done when the program is halted, *ADVANCE* is still required to let it

---

[27]Key shortcuts are available and may be handy here (see Sect. C.7.3).
[28]Delayed event stepping, to make sense at all, calls for a precise specification of time. Therefore there is no *ETU* option.

Figure C.7 The step dialog

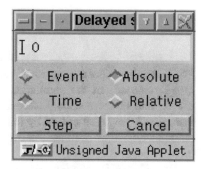

go. Alternatively, the user may select *ALL+GO* from the *Unstep* menu to cancel all stepping and resume normal execution of the program.

Needless to say, stepping affects the real-time character of the SMURPH program and is not recommended for monitoring the behavior of a control program driving real equipment.

When the display session is requested by the SMURPH program (Sect. 12.4), also when the program has been called with –*d* (Sect. B.6), the program appears halted at the very beginning of the display session, even though no exposure is formally being stepped. *ADVANCE*, or *ALL*, or *ALL+GO*, selected from the *Unstep* menu will resume the program's execution.

C.6.3   Segment Attributes

The attribute pattern of a segment (Sect. 12.3) is interpreted by DSD in the following way (bits are numbered from zero starting from the less significant end):

Bits 0–1	This field is interpreted as a number from 0 to 3 specifying the segment type. The following types are supported by DSD: 0–discrete points, 1–points connected with lines, 2–discrete stripes extending to the bottom of the region area
Bits 2–5	Thickness. This field is ignored for type–1 segments (i.e., points connected with lines). Otherwise, it represents the thickness of points or stripes expressed in pixels which is equal to the numerical value of this field $+2$
Bits 6–13	Color. The numerical value of this field is used as an index into the internal color table of DSD and determines the color of the segment

C.7   Other Commands

In this section, we discuss some other commands of DSD that are not directly related to exposure windows.

C.7.1 Display Interval

As mentioned in Sect. 12.4, a SMURPH program engaged in a display session with DSD sends exposure updates at intervals determined by the value of *DisplayInterval*. In the simulation (virtual) mode, this interval is expressed in events, and in the real (and also visualization) mode it refers to milliseconds. In Sect. C.6.2, we noted one exception from this rule: when the program is stepped, it sends an update as soon as it becomes halted, regardless of the prescribed interval between regular updates.

The value of *DisplayInterval* can be changed by selecting *Reset Refresh Interval* from the *Navigate* menu. In response to this selection, DSD presents a simple dialog shown in Fig. C.8. The number entered in the text area represents the new requested value of the display interval, which is either in events or in milliseconds, depending on whether the program runs as a simulator or executes in the real mode. The display interval cannot be reset in the visualization mode where it is coupled to the resync grain (Sect. 3.1.2).

C.7.2 Disconnection and Termination

By selecting *Disconnect* from the *Navigate* menu, the user terminates the display session. DSD notifies the SMURPH program that it should not send any more updates and it should not expect any more commands from DSD.

All the windows of DSD remain on the screen until either the root window is canceled (by pushing the *Stop* button on the applet's startup panel), or a new selection is made from the *Navigate* menu (which at this stage can only be *Get List*, Sect. C.3).

It is possible to terminate the SMURPH program from DSD, by selecting *Terminate* from the *Navigate* menu (during a display session). As this action is potentially dangerous, *Terminate* must be clicked twice in a row to be effective.

Figure C.8 The display interval dialog

C.7.3 Shortcuts

The following key strokes can be applied as shortcuts for some DSD actions (normally selectable from menus). They are only effective if pressed while within an exposure window.

s	immediate step for this exposure (*Step*)
S	delayed step for this exposure (Step delayed)
u	unstep this exposure (*Unstep*)
U	unstep this exposure and go
g	continue until next step (*ADVANCE*)

The last key (*g*) can be hit in any window, including the root window, and its interpretation is always the same: "continue until the next stepped event." The upper case *G* and the space bar can be used instead of *g* and their meaning is the same.

Bibliography

1. P. Gburzyński, *Protocol Design for Local and Metropolitan Area Networks* (Prentice-Hall, 1996)
2. N. Boers, I. Nikolaidis, P. Gburzyński, Sampling and classifying interference patterns in a wireless sensor network. Trans. Sensor Netw. **9** (2012)
3. N.M. Boers, P. Gburzyński, I. Nikolaidis, W. Olesiński, Developing wireless sensor network applications in a virtual environment. Telecommun. Syst. **45**, 165–176 (2010)
4. D.R. Boggs, J.C. Mogul, C.A. Kent, *Measured Capacity of an Ethernet: Myths and Reality* (Digital Equipment Corporation, Western Research Laboratory, Palo Alto, California, 1988)
5. F.P. Fontán, P.M. Espiñeira, *Modeling the Wireless Propagation Channel, a Simulation Approach with MATLAB* (Wiley, 2008)
6. P. Dash, *Getting Started with Oracle VM VirtualBox, PACKT* (2013)
7. R. Collins, *VirtualBox Guide for Beginners* (CreateSpace Independent Publishing Platform, 2017)
8. S. Van Vugt, *VMware Workstation: No Experience Necessary, PAKT* (2013)
9. A. Mauro, P. Valsecchi, K. Novak, *Mastering VMware vSphere 6.5, PACKT* (2018)
10. C. Team, *Cygwin User Guide* (Samurai Media Limited, 2016)
11. P. Gburzyński, W. Olesiński, On a practical approach to low-cost ad hoc wireless networking. J. Telecommun. Inf. Technol. **2008**, 29–42, 1 (2008)

© Springer Nature Switzerland AG 2019

479

P. Gburzyński, *Modeling Communication Networks and Protocols*,
Lecture Notes in Networks and Systems 61,
https://doi.org/10.1007/978-3-030-15391-5

Index

© Springer Nature Switzerland AG 2019
P. Gburzyński, *Modeling Communication Networks and Protocols*,
Lecture Notes in Networks and Systems 61,
https://doi.org/10.1007/978-3-030-15391-5

Printed in the United States
By Bookmasters